Introduction to Ecological Aesthetics

Fanren Zeng

Introduction to Ecological Aesthetics

商務印書館
创于1897
The Commercial Press

Springer

Fanren Zeng
Shandong University
Beijing, China

Translated by Wu Lihuan and Chad Austin Meyers

ISBN 978-981-13-8983-2 ISBN 978-981-13-8984-9 (eBook)
https://doi.org/10.1007/978-981-13-8984-9

Jointly published with The Commercial Press, Beijing, China
The print edition is not for sale in the Mainland of China. Customers from the Mainland of China please order the print book from: The Commercial Press, Ltd.

This Springer imprint is published by the registered company Springer Nature Singapore Pte Ltd.
The registered company address is: 152 Beach Road, #21-01/04 Gateway East, Singapore 189721, Singapore

Preface

A New Breakthrough of Ecological Aesthetics in Contemporary Aesthetics

Some scholars have asked in what areas ecological aesthetics can have new breakthroughs in the end? Our answer is that the emergence of ecological aesthetics is not only a timely and real development, but also provides a full range of breakthroughs in contemporary aesthetic disciplines with revolutionary significance. In general, we summarize this breakthrough in six areas.

The first is the breakthrough of the philosophical basis of aesthetics. This is the transition from traditional epistemology to the existentialism of materialism and the transition from anthropocentrism to ecological holism. As we all know, the relationship between human beings and nature is the most basic philosophical relationship. For a long time, for historical reasons, our aesthetic circles in this basic philosophy have been still in the pre-Marxist philosophical stage—that is, in the traditional epistemological stage. The advocating of practical aesthetics in our country emphasizes that "the basic problem of philosophy of aesthetic science is the epistemological problem," and that the beauty is "the result of the objectification of human essence" and so on. In fact, as early as 1844–1845, Marx had a breakthrough in traditional epistemology and anthropocentrism, focusing on understanding things from the perspective of human perceptual practice, and explaining the law of beauty from the point of view of unity of 'internal scale' and 'kind of scale'. It is shown that the Marxist concept of materialist practice not only transcends traditional epistemology, but also contains and transcends the modern ontology. It is a kind of new material practice theory to understand the essence of human nature from the perspective of 'practicing world' in the process of 'the reality that is carried out under certain conditions and can be observed through experience'. The binary opposition between the activity of the person in the 'world of practice' and the subject and object is two completely different forms of life in life, thus reflecting the relationship between two different people and nature. The latter is the opposition of human beings and nature, the former is the relationship between the human 'life of a personal existence' and the nature of a few moments away from it. Only in such a relationship, human beings, nature and beauty can be unified. Marx's concept of materialist practice and its connotation of existential philosophy is the philosophical foundation of contemporary ecological aesthetics, which is a simple epistemology and anthropocentrism very different from traditional, practical aesthetics.

Second, there has been an important breakthrough in the aesthetic research object. Due to the influence of Hegel's long-lasting idea that 'aesthetics is the philosophy of art', our aesthetic circles regarded the art as the only object of aesthetic research, or at least the most important research object, and thus neglected natural aesthetics. Practical aesthetics advocates that 'aesthetics should basically include the study of the beauty of objective reality, the general law of human aesthetics, and artistic beauty. Among them, art should be the main object and purpose of the study, because humans are to reflect and grasp beauty and make it transform the world of the great cause mainly through the art'. But, in fact, the aesthetic relationship between human beings and nature is the most basic and most original. Its importance is not under the aesthetic art. Especially in the circumstances that contemporary environmental pollution is becoming increasingly serious, the desire of human beings to wander in the arms of pure nature is more of an aesthetic ideal. So, as early as 1966, the American theorist Hepburn published the famous essay 'Contemporary Aesthetics and the Loss of Natural Beauty', which presented a strong attack on the serious neglect for the natural beauty in aesthetic world, which gave birth to contemporary Western environmental aesthetics developed later. The important breakthrough of ecological aesthetics on the object of aesthetics is the breakthrough of artistic centrism caused by anthropocentrism. It is clear that ecological aesthetics is a kind of aesthetics that contains an ecological dimension. It not only contains natural aesthetics, but also contains the natural dimension of art and life on the aesthetic.

The third is the breakthrough in the natural aesthetic. Traditional aesthetics always looks at natural aesthetics from the perspective of anthropocentrism, and summarizes natural aesthetics with 'human nature'. This is a question that has confused the aesthetic world for a long time. There are two reasons: First, is there natural, physical beauty? Second, is the nature of natural beauty 'Natural humanization'? In view of these theoretical perplexities, ecological aesthetics believes that the so-called aesthetic is a relationship between people and objects. It is a kind of activity or process. There is absolutely no existence of any kind of physical, 'natural beauty'. Third, ecological aesthetics believes that natural aesthetic is the result of the interaction between the aesthetic attributes of natural objects and the aesthetic ability of the people. The two are indispensable and it is absolutely not a simple 'natural humanization'.

Fourth, there has been an important breakthrough in aesthetic attributes. Influenced by Kantian's aesthetics, traditional aesthetics thinks that the basic attributes of aesthetics are a super utilitarian, without an interest in waiting and seeing. The practitioners of practical aesthetics believe that 'Aesthetic is the need and enjoyment of this super-organisms'. The unity of truth and goodness is 'Manifestation of the main body of the psychological feelings of freedom (seeing, hearing and imagination) which is aesthetics'. We can clearly see that the theory of perspective of aesthetic attributes, as well as the idea that aesthetic senses are audiovisuals are basically accepted. Ecological aesthetics does not oppose the aesthetic characteristics of artistic aesthetics, but emphasizes all the senses in the natural aesthetic. Contemporary Western environmental aesthetics put forward the famous concept of 'participation in aesthetics'.

Fifth, there has been the breakthrough of the aesthetic paradigm. The paradigm of traditional aesthetics emphasizes the beauty, symmetry, harmony, proportion of formal beauty, etc. Ecological aesthetics is a kind of contemporary life aesthetics, or existential aesthetics. Its aesthetic paradigm has broken through the traditional aesthetic form of the beautiful and harmonious, and into the human poetic dwelling and good living level. It is a unique aesthetic paradigm for its own aesthetic survival, poetic dwelling, quartet game, homeland consciousness, place consciousness, participation in aesthetics, lofty ecology, ecological criticism, ecological poetics, green reading, environmental imagination and ecological aesthetic education.

The sixth is the breakthrough of traditional Chinese aesthetic status. Affected by Euro-centrism, the West has traditionally held contemptuous attitude for oriental aesthetics and art, including China. Hegel and Bao Sang Kwai have published similar remarks. For example, Bao Sang Kwai believes regarding China and Japan and other oriental art and aesthetics that 'aesthetic consciousness has not yet reached the point of speculation theory'. This is really to evaluate the ancient Chinese non-tools, non-speculative aesthetic theory with the Western modern instrumentalism aesthetic theory. Along with the emergence of ecological aesthetics, a large number of valuable ecological and aesthetic wisdom in ancient Chinese aesthetics have been noticed and highly appraised, thus providing unprecedented wisdom resources for the construction of contemporary ecological aesthetics. Moreover, there are sufficient facts to prove that Western contemporary ecological aesthetics and environmental aesthetics and the development of ecological literature are greatly inspired by ancient Chinese ecological wisdom. For example, there is the Confucian thought of 'harmony between man and nature' The discussion in The Zhou Book of Changes about 'producing life is what we call change' (sheng sheng zi wei yi 生生之谓易), 'beginning, growing, expanding and rectifying', (yuan heng li zhen 元亨利贞) and 'a virtuous person can undertake great tasks' (kun hou zai wu 坤厚载物). In Taoism, there is the theory that 'is the Dao models itself on what is natural' (dao fa zi ran 道法自然) and 'everything as the same' (wan wu qi yi 万物齐一). The concept such as 'equality of all sentient beings' (zhong sheng ping deng 众生平等) belong to Buddhism. These ideas and doctrines reflect the philosophical and aesthetic wisdom of Chinese cultural traditions in the relationship between human beings and nature. It will come to the construction of contemporary ecological aesthetics of the ideological resources and wisdom enlightenment through our dialogue between China and the West. For example, the combination of the theory that 'human beings are an integral part of nature' (tian ren he yi 天人合一), the idea in aesthetical existence of ecological ontology, 'the beauty of striking harmony' (zhong he mei 中和美), 'Poetically Dwelling', the theory that there are four major domains, mankind is one of it, and

'The fourfold game', homesick poetry, peaceful auspicious image and sense of home, the freedom to live where we please, sense of place, Büchen, Beede, nature, artistic conception, other ancient poetic wisdom and ecological poetics, etc. The historical task of the development of contemporary ecological aesthetics is to build an ecological aesthetic system that contains Chinese ancient ecological wisdom, resources and discourse and conforms to China's national conditions and has a Chinese style.

Ecological aesthetics is a new discipline under construction, and there are many immature places that need to be further refined through criticism, discussion and research. The emergence of ecological aesthetics, although in the development of aesthetics is or strive to achieve a series of breakthroughs, but that does not mean that we do not pay tribute to the exploration and achievement of previous aesthetic research. Academic development is related to history and the time, today's evaluation can't deny the important value of its history, and the nourishment it has given us.

Beijing, China Fanren Zeng

Introduction: The Significance of Ecological Aesthetics, Research Status and Research Methods

Since the 1990s, ecological aesthetics has become a vital new form of theory and point of growth in the field of aesthetics in China, and has increasingly drawn widespread attention throughout the academic world. The facts prove that ecological aesthetics is one of the important achievements of the new era of Chinese research into aesthetics. It is important to note that the ecological aesthetics described in this book include the ecological aesthetics that emerged in the new era along with environmental aesthetics, ecological literature, ecological art, ecological criticism, etc., and although they go by different names, on the whole, they are all an aesthetic and artistic kind of research that contains some ecological dimension. Complementing one another and co existing with one another, they collectively compose the vivid landscape of a new era in the field of aesthetics and arts in China, the study of ecological aesthetics.

The Significance of Ecological Aesthetics

The Need of Reality

In 1894, Engels pointed out in his letter to Borgieius: 'Once society has the technical needs, this need will be more than ten universities to advance science'.[1] This shows that the real needs of society constitute the most fundamental impetus of scientific advancement. From the 1960s up to today, the negative factors in the modern industrial revolution—the destructive pollution of ecosystems and the natural environment brought about by technologism and anthropocentrism—have only gained in seriousness, and now directly threaten the survival and safety of hundreds of millions of people.

[1]Marx and Engels, *Collected Works,* vol. 4, People's Publishing House, 1972, p. 505.

Since the new era, through 20 years of opening reforms and economic development, China has obtained economic achievements that took the Western nations nearly 200 years to accomplish, but the single model of economic development also makes our country pay a heavy price in terms of environment and resources—the many years of environmental problems that came about over nearly 200 years of Western development have concentrated into 20 years of Chinese development. The seriousness of the problem has already reached a shocking magnitude. Under such circumstances, ecological and environmental problems have gradually attracted the attention of national leaders and academics. After the policy of 'sustainable development' that the leadership proposed in the 1990s, China recently brought up the goals of constructing 'environmentally-friendly society' and 'the building of ecological civilization', which have become important components of the innovative theory of Marxism with Chinese characteristics and the construction of harmonious society. In this scenario, under the guidance of 'the construction of ecological civilization', all of the ecological theories, including ecological aesthetics, necessarily become organic components of the construction of a progressive socialist civilization that is compatible with social development, and become the necessary result and trend of development of 'the building of environmentally friendly society' and 'the construction of ecological civilization'.

The Need to Construct Branches of Study

In our country, the branch of aesthetics continuously attracts the attention of the academic world and the broader masses of people, especially the younger generation, because of its unique content pertaining to the humanities. Over the past century, a large number of famous theorists on aesthetics with unique styles have emerged in China, such as Wang Guowei, Cai Yuanpei, Zhu Guangqian, Zong Baihua, Cai Yi, Jiang Kongyang and Li Zehou, whose many vital achievements are still nourishing a generation of scholars today. But the vast transformations that have taken place in accompaniment with the change of times have made some limitations of the original theories harder and harder to disregard. Even 'practical' aesthetics whose accumulation of theoretical content extends deeply and widely has also shown obvious theoretical drawbacks. What concretely expresses its philosophical foundation is largely limited to mechanical epistemology, and with only epistemology to guide it, what comes under the scope of its concern is more the cognitive function of aesthetics and not so much the function of aesthetics geared towards more profoundly revealing the meaning of human existence and human values. With respect to aesthetic theory itself, 'practical' aesthetics overly emphasizes that aesthetics is a 'humanization of the object', while ignoring the value of the object itself, especially the value of nature itself, and this is what exhibits the obvious tendency of anthropocentricism. On the question of natural beauty, under the influence of Hegel's idea of natural beauty, which largely dismisses it, along with Max Weber's theory of disenchantment, practical aesthetics largely disregards the value that nature uniquely possesses and the unique position that nature occupies in aesthetics. As a way of thought; 'practical' aesthetics on the whole never entirely succeeded in breaking free from the influence and constraint of the enlightenment model of thought based on the subject/object and mind/body dichotomy; in terms of the object of aesthetic research, due to long-term contamination with Hegel's thought, 'practical' aesthetics has continuously equated aesthetics with philosophy of art, and naturally, it has largely been excluded from the scope of aesthetics, and this effectively clashes with the increasing importance of reality for the contemporary aesthetics of nature.

In summary, 'practical' aesthetics is on the whole a form of aesthetics whose core concept is humanization and which disregards the ecological dimension with a thick scent of 'anthropocentrism'. In the current situation, we should say that it has largely fallen behind the development of the times. The reforming and overcoming of this form of aesthetics has already become a historical necessity. The proposing of 'ecological aesthetics' is precisely a reforming and overcoming of practical aesthetics and the need of the time for the field of aesthetics itself to develop.

The Need to Promote Traditional Chinese Culture in the Context of Globalization

At present, the world economy faces the new trend of globalization. In real life, the convenience of transportation and the development of internet communications have greatly narrowed the distance between countries and peoples. It is in this new circumstance that the self-development of ethnic culture has become the unique mark of whether or not a nation can stand out with the makeup that it uniquely possesses independently from all of the world's ethnicities. Spanning over 5,000 years, Chinese civilization has a long enduring history of development, which through the great river of history has given shape to the uniquely brilliant style of Chinese culture, and this has become the symbol and pride of the Chinese nation. Moreover, ecological wisdom that envelops the wisdom of ecological aesthetics is the unique wealth and important inheritance of Chinese culture. In Confucianism, 'Nature and humanity integrate into one' (tian ren he yi 天人合一) and 'the people are my companions and I share my life with all creatures' (min pao wu yu 民胞物与); in Daoism, 'Dao emulates what is naturally so' (dao fa zi ran 道法自然), 'all beings even out into one' (wan wu qi yi 万物齐一), and 'all beings are one body' (wan wu yi ti 万物一体); in Buddhism, 'all sentient beings cross over together in common' (pu du zong sheng 普度众生) and so on; all of such thoughts express the classical form of an extremely precious ecological wisdom, and have attracted the attention and appreciation of the international academic world. Contemporarily, the construction and development of all of the Chinese forms of ecological theory including ecological aesthetics will benefit the promotion of traditional Chinese culture and the transmission of the great spirit of the Chinese people, and will also make its due contribution to the great rejuvenation of the Chinese people in the twenty-first century.

The History of Development of Contemporary Chinese Ecological Aesthetics

In general, we divide the development of contemporary Chinese ecological aesthetics into three stages.

The Budding Period (1987–2000)

In the academic circles of China, the first introduction of 'literary ecology' related to ecological aesthetics is found in *the New Terminological Dictionary of Literature and Art* edited by Bao Chang in 1987. The dictionary defines 'literary ecology' from the perspective of the combination of literature and ecology. And the first article on 'ecological aesthetics' in China, *Foreign Ecological Aesthetics*, was written by the Russian Mankovskaya (written in 1992, translated from наука философии, The Science of Philosophy, No. 2). The author introduces and comments in exhaustive detail on the emerging ecological aesthetics (in fact, environmental aesthetics) in Europe and America, and introduces Seppama and Carlsson's view of ecological aesthetics, believing that foreign ecological aesthetics 'has been far beyond the scope of traditional research on the problem of nature in art' and that 'ecological aesthetics conceptually speaking has already been established'. Of course, he also points out some of the problems in the development of ecological aesthetics as well.

In 1994, Prof. Li Xinfu published the article 'On Ecological Aesthetics' in *Nanjing Social Sciences*, No. 12. This paper could be said to be the first academic dissertation on ecological aesthetics with theoretical depth in China. This paper discusses the background, basic principles and prospects of development of ecological aesthetics. In the discussion on the emergence of ecological aesthetics, this paper points out that ecological aesthetics 'is a new discipline developed and inspired by the global environmental protection movement and green movement accompanying the ecological crisis. It investigates the beauty of the Earth's ecological environment as its main task and object, which is the core component of environmental aesthetics, whose constitutive content includes the three big systematic levels of natural ecology, the ecology of social material production and the ecology of spiritual cultural production' and points out the three great aesthetic ideas that ecological aesthetics must establish, namely 'ecological balance is beauty of the highest value', 'the harmonious coordinated development of all natural beings' and 'the great enterprise of building a new ecological civilization' along with the method of the three great principles, namely 'Dao emulates what is naturally so' (dao fa zi ran 道法自然), 'return to simple and genuine being' (fan pu gui zhen 返朴归真) and 'moderating control'. When discussing the prospects of ecological aesthetics, this article points out that 'as a young emerging discipline, ecological aesthetics in terms of composition of cognitive theoretical content has its own unique system, standards and principles, and although it is not yet mature, the scientific and progressive character that it implies determines that it possesses robust vitality and prospects of development, and we should cheer for its birth and construction and contribute our power to it'. This article quotes large quantities of traditional ecological wisdom to theoretically treat the problems of ecological aesthetics in the early 1990s, and its value and status are extremely important. These two articles count as signs of the budding germination of ecological aesthetics in China.

In October 1999, the Hainan Provincial Association held an international symposium on *Ecology and Literature*, which was the first international symposium on ecological literature in China, reflecting the high degree of ecological consciousness in the academic and literary circles of our country. In 1999, Prof. Lu Shiyuan founded *The Communication of Spiritual Ecology*, which continues to this day. The journal has become a unique camp in the research of ecological literature in China, and its important contribution is obvious to all.

The Developing Period (2000–2007)

During this period, ecological aesthetics, as well as the ecological literature and environmental aesthetics closely related to it, show positive momentum of continuous development. In December 2000, Shanxi People's Education Publishing House published *Ecological Aesthetics* by Xu Hengchun, and *Ecological Literature and Art* by Lu Shiyuan. In the same year, the People's Literature Publishing House published *Green Thought of Literature and Art—An Introduction to the Ecological Study of Literature and Art* by Zeng Yongcheng. In January 2001, a seminar entitled 'Ecology and the Study of Literature and Art in the 21st Century' was held in Wuhan, and subsequently, 'Symposium on Globalization and Ecological Criticism' was held in Beijing. In November 2001, 'the First National Ecological Aesthetics Seminar' was jointly held in Xi'an by the Youth Conference of the Chinese Society of Aesthetics and Shaanxi Normal University College of Arts. Since then, a number of meetings were respectively held in Guizhou, Nanning and Wuhan. In June 2002, 'China's First Seminar on the Construction of the Field of Ecological Literature and Art' was hosted by the Institute of Literature of Suzhou University. In December of the same year, Jianghan University and Wuhan University co-hosted the 'Seminar on the Ecological Change of National Cultural and the Development of Literature and Art'. In June 2002, Zhang Hao published *Research on the Ecological Thought of Chinese Literature and Art*. The book systematically explains the wisdom of ecological aesthetics in Confucianism, Daoism, Buddhism and Chan Buddhism in Ancient China. In November 2002, Yuan Dingsheng published *Ecological Aesthetics*, which comprehensively works through a series of problems related to ecological aesthetics such as ‘the field of ecological aesthetics’, ‘ecological beauty’, ‘the aesthetic effect of ecology’ and ‘ecological aesthetics education’. In June 2002, Zeng Fanren published the article ‘Ecological Aesthetics: The New Perspective of Ecological Ontology in the Postmodern Context’ in *the Journal of Shaanxi Normal University,* No. 3, in which the author suggests that ecological aesthetics is the product of reflecting on and overcoming the modern industrial revolution in the postmodern context, and its basic theory consists of an aesthetic view of ecological ontology. The text was reprinted by *Xinhua Digest* in the same year. In August 2003, Wang Nuo’s *European and American Ecological literature* was published by Peking University Press. The book is the result of the author’s research during and after his visit to Harvard. In October 2003, Zeng Fanren’s *The Aesthetic Theory of Ecological Ontology* was published by Jilin People’s Publishing House; the book is divided into two volumes, ‘Ecological Aesthetics’ and ‘The Aesthetic Theory of Contemporary Ontology’ respectively, and records 14 articles on the Aesthetics of Contemporary Ontology and the Aesthetics of Ecological Ontology that the author penned since 2001. In May 2004, the International Symposium on ‘Beauty and the Contemporary Way of Life’ was held at Wuhan University. In November 2005, Peng Feng of Peking University published the book *Perfect Nature—The Philosophical Basis of Contemporary Environmental Aesthetics*. In March of the same year, Fudan University Press published Zhang Hairong’s *Ecological Ethics and Ecological Aesthetics,* which combines contemporary ecological ethics with ecological aesthetics, and examines the construction and significance of ecological aesthetics in the context of intense social reality. In August 2005, Shandong University’s Center for Literary and Artistic Aesthetics held the International Symposium on Aesthetics and Literature in the View of Contemporary Ecological Civilization in Qingdao. More than 170 scholars from home and abroad attended the conference; it launched more in-depth and open discussions about ‘the condition of contemporary Chinese research on ecological literature and ecological aesthetics’, ‘Western ecological criticism and environmental aesthetics’, ‘the ecological wisdom and ecological culture of China’, ‘ecological ethics and ecological aesthetics’ and other important problems revolving around the relationship between ‘ecology, the humanities and aesthetics’. After the symposium collected articles were published: *Humanity and Nature: Aesthetics and Literature in Contemporary Ecological Civilization*. In March 2006, Hunan Science and Technology Publishing House published the series of translations, Environmental Aesthetics, co-edited by the famous American environmental aesthetician, Arnold Berleant, and Prof. Chen Wangheng of Wuhan University; the series includes Arnold Berleant *Environmental Aesthetics* and John Seppama’s *The Beauty of the Environment*. In June of the same year, the series entitled *Aesthetics-Design-Art Education* edited by Prof. Teng Shouyao of the Institute of Philosophy of the Chinese Academy of Social Sciences was published by Sichuan People’s Press, and this series includes

Carlson's *Environmental Aesthetics*. The major writings of the top three famous international contemporary environmental aestheticians were translated and introduced in China. In July 2006, Chinese Social Sciences Press published Dr. Hu Zhihong's doctoral thesis, the book *Western Ecological Criticism*, in Sichuan University's Chinese Journal of Social Sciences. From the standpoint of ecocentrism, the book makes an in-depth elucidation of western ecological criticism from different perspectives such as ground of thought, construction of theory, literary research, cross-cultural research and comparative literature; it has a unique standpoint and strong academic value. In July 2006, East China Normal University Press published Prof. Lu Shiyuan's *The Space of Ecological Criticism*, which from the perspectives of the age of ecology, ecology of spirit and the horizon of ecology explains a series of the author's new insights into ecological literature and ecological criticism. In April 2007, Liu Yuedi and others at the Institute of Philosophy of the Chinese Academy of Social Sciences translated and published Prof. Arnold Bridel's *Environment and Art: A Multidimensional Perspective of Environmental Aesthetics,* which is the collection of the latest achievements of 12 influential aestheticians in the field of contemporary international environmental aesthetics. The book's greatest value lies in its rich and innovative content, and it especially has theoretical value. In July 2007, Wuhan University Press published Prof. Chen Wangheng's *Environmental Aesthetics,* which is China's first book that goes by the name of environmental aesthetics. In December 2007, the Commercial Press published Zeng Fanren's nearly 10 years of essays on the aesthetics: *Chinese Aesthetics in the Transitional Period,* the third part of the book is 'Ecological Aesthetics-from the Anthropocentric to the Ecological Whole', a total of 17 articles on ecological aesthetics were collected. Prof. Ruxin, President of the Chinese Aesthetics Society, points out in the book's 'Preface': 'so I think the proposing of ecological aesthetics is our academic community's first innovation, and it perfectly fills in a blank space in ecological research, which is of real significance both in theory and in practice'. Prof. Ye Long, Vice President of the Chinese Aesthetics Society and Professor of Philosophy at Peking University, gave a speech on 15 November 2007 on 'Building a Spiritual Homeland for the Chinese Nation' in *Guangming Daily* 'associated with this consciousness of ecological philosophy and ecological ethics, traditional Chinese culture also has a consciousness of ecological aesthetics'.

The New Period of Construction

In October 2007, President Hu Jintao and China's top levels of decision-making bodies clearly put forward the important thesis of the construction of 'ecological civilization' after the material civilization, spiritual civilization and political civilization of socialism, and in making it the important goal of building harmonious socialist society, the content of 'ecological civilization is explained in-depth. This thesis is of great significance for China's construction of ecological theory, which includes ecological aesthetics, and it marks the entrance of China's ecological theory from the margins into the mainstream, thereby initiating the new period of construction.

What we are calling the 'new construction' includes two aspects, namely conclusion of reflection and construction. With respect to conclusion of reflection, we should recall the 20 years of twists and turns that ecological aesthetics and ecological literature went through since 1987. First of all, it is the translation, introduction and combing through the achievements and materials from ecological aesthetics and ecological literature in China and abroad. Abroad, nearly, all of the important achievements of the contemporary theorists of ecological aesthetics and environmental aesthetics have already been translated and introduced; in China, we have mainly been excavating and combing through the wisdom of ecological aesthetics in ancient Chinese Confucianism, Daoism and Buddhism, and have made preliminary investigations into the wisdom and resources of ecological aesthetics found in ancient, modern and contemporary literature; second, we have held about ten academic seminars on ecological aesthetics, ecological art and ecological literature, have published more than a dozen works on ecological aesthetics and ecological literature, have proposed a series of important academic viewpoints, and theoretical research is progressing in-depth; with respect to the influence of ecological aesthetics and ecological literature, these fields have moved gradually from a state of not being understood to one of the appropriate recognitions. With respect to the national social science project, there have already been five young scholars whose projects related to ecological literature and ecological criticism have been approved. There are nearly 20 MBA and Ph.D. dissertations with ecological literature, ecological criticism and ecological aesthetics as their titles, and finally, with respect to team building, the number of scholars who are currently engaging in ecological aesthetics, ecological art and ecological literature continues to expand. In August 2005, the international academic seminar on ecological aesthetics held in Qingdao is currently China's largest academic seminar on ecological aesthetics and ecological literature.

Representative Works

Xu Hengchun: Ecological Aesthetics (Shaanxi People's Education Press, 2000 Edition)

This book is the first complete, self-contained work on ecological aesthetics in China, and the researcher Xu Hengchun visited Germany twice, gaining inspiration from the excellent ecological environment of Germany and from the book *Ecological Psychology*, and he wrote a book after that inspiration ripened over a long period of time. The contributions the book offers are many. First of all, the book explicitly proposes the viewpoint of 'the arrival of ecological civilization' which reflects on and overcomes industrial civilization. The author points out: 'The dawn of a new ecological civilization has already emerged, which is the harmonious symbiosis of humanity and nature in the era of ecological civilization'.[2] Second, the book proposes the two theoretical prerequisites for establishing ecological aesthetics: one is an ecological world view, one is the ancient Chinese consciousness of life. The author believes that the ecological worldview is opposed to the mechanistic worldview, and contains three major principles of thought, namely the organic whole, ordered whole and natural evolution.[3] With respect to consciousness of life, the author points out: 'The examination of human ecosystems that ecological aesthetics makes presupposes the being of human life and starts with the interrelationship and movement of all biological systems. Therefore, a conception of human life becomes the theoretical basis of this examination'.[4] Third, the book proposes a very important category of 'ecological beauty'. The author points out that 'what we call ecological beauty is not natural beauty because natural beauty is only the aesthetic value of nature itself, while ecological beauty is the product of the harmony of ecological relations between humanity and nature, and its object of aesthetic regard is the ecological processes of humanity and ecological systems. Ecological beauty first and foremost embodies the participatory nature of the subject and the interdependent relationship between the subject and the natural environment, it is a common feeling of life and rejoice inspired by the biological interconnections between humanity and nature. It is the biological chord between humanity and nature, and not the soliloquy of nature'.[5] Here, the author insists that the object of ecological aesthetics is human ecosystems, its distinctive features are participation and interdependence, and its symbol is the common feeling of life and biological chord between humanity and nature. All of this shows that 'ecological beauty' is the beauty of a human being's aesthetic existence. Fourth, the book further explains the meaning of aesthetic beauty, which includes promoting the reformation of modern aesthetic theory and establishing the values of healthy life; it benefits overcoming the technological alienation from ecology, and it helps reform unreasonable means of production and irrational ways of life.[6] Fifth, the author presents constructive and developing opinions with practical value on humanity's living environment, urban landscape and way of life from the perspective of the theory of 'ecological beauty'.

[2] Xu Hengchun, *Ecological Aesthetics*, Shaanxi People's Education Press, 2000, p. 7.

[3] Ibid, p. 44.

[4] Xu Hengchun, *Ecological Aesthetics*, Shaanxi People's Education Press, 2000, p. 14.

[5] Ibid, p. 119.

[6] Ibid, p. 150.

Lu Shuyuan: Ecological Literature and Art *(Shaanxi People's Education Press, 2000 Edition)*

Professor Lu Shuyuan's book is the first work of 'ecological literature and art' in China. This monograph is written in Prof. Lu's excellent and unique prosaic writing style, possesses a readability that theoretical work rarely has, and of course inevitably also generated the reserved character of some theoretical viewpoints. Part one of the book is conclusive in nature, part two is the application of the theory, and in total has 14 chapters. The main points go as follows: first, the main aim of the book is to 'try to explore the relationship between literary art and the entirety of the earth's ecosystems, and furthermore use the perspective of modern ecology to examine literary art'.[7] Second, it discusses the new era of 'ecological civilization' on which ecological literature depends. As the author says: 'we agree with the idea that the coming era is an era of human ecology',[8] and formulate it as the '"post-modern" era is the era of ecology'.[9] And then from the aesthetic viewpoint, the book proposes moving towards the era of aesthetic ecology; third, the book gives a brief summary of the object and content of research on ecological literature and art: 'A complete branch of ecological literature and art should face the entirety of humanity's literary artistic activities, and moreover, explain them. And literary artistic activity as one of the most important spiritual activities of human being must through and through have an intimate relationship with the living conditions of all human beings, and because this is even more so for excellent works of literature and art, we should bring them under the scope of ecological literature and art to examine and investigate them'.[10] Fourth, the book proposes that the earth is the life of the entire ecosystem itself; it is a life cycle that contains a series of organic links. In addition, it also focuses on explaining the natural identity between the partial charm of nature, feminine nature and art. Nostalgia is attachment to habitat and the series of important propositions of ecological poetics. Fifth, the book proposes the famous tripartite division of ecology: natural ecology, social ecology and spiritual ecology;[11] Sixth, it proposes the artist as the problem of the growth of the natural organism in the natural, familial, social and literary soil.[12] Seventh, the book proposes the meaning of ecological literary criticism: nature is an organic whole; humankind takes more responsibility for maintaining the holistic perfection of nature; the ecological disasters that humankind is currently about to face are manmade; the resolution of the ecological crisis requires human beings to fundamentally adjust their values and way of life; the harmonious consistency between the spirit of humanity and the spirit of nature is ecological and aesthetic; poetry and

[7] Lu shuyuan, *Ecological literature and art,* Shaanxi People's Education Press, 2000, p. 2.

[8] Lu shuyuan, Ecological literature and art, p. 24.

[9] Ibid, p. 169.

[10] Ibid, p. 28.

[11] Ibid, p. 132.

[12] Ibid, p. 203.

art are plants that are rooted in the soil of nature and that blossom in the heavens of spirit; the true spirit of art amounts to the spirit of ecology; ecological art criticism regards the philosophy of art as the spirit of utopia; ecological literary criticism is a literary criticism that gives more respect to the content of literature and art; ecological literary criticism does not exclude other types of criticism, including formal criticism, and its basic principle is 'pluralistic coexistence'; eighth, the new understanding of natural beauty: natural beauty is not 'humanized nature'; it is both naturally generated and objectively existing, and is also inherent in the mechanism of human aesthetics, and sometimes it is strengthened by the person's consciousness, experience, and state of mind, 'humans and nature coexist in an "organic organization"'.[13]

Chen Wangheng: Environmental Aesthetics *(Wuhan University Press, 2007 Edition)*

Professor Chen Wangheng's *Environmental Aesthetics* is the first monograph on 'environmental aesthetics' in China; it sums up the results of relevant domestic and foreign research, and puts forward his own view with completeness and a strong ability to generalize; it is an important achievement in the study of environmental aesthetics (ecological aesthetics) in China. We summarize the main points as follows: First, the basic content of the book; the author makes a generalization at the end of the preface, that is, he proposes the concept of 'a landscape'. If the original substance of artistic beauty is a mood, then the original substance of environmental aesthetics is a landscape. They are the concrete manifestations of the general substance of beauty. The creation of a landscape is the effect of the subject and the object. Natural, rural and urban environments are the three major spheres of aesthetic research, and the contradiction and unity between the ecological and the humanistic, between the natural and the manmade are the basic problems of research of environmental aesthetics. The environment is our home. So, the most fundamental quality of a beautiful environment is the feeling of home. In the horizon of environmental aesthetics, 'habitable' and even more so 'enjoying habitation' is the primary function of environmental aesthetics, while 'enjoying a tour' can only be its second function. The function of the environment is of course not just 'habitation' and 'touring'; the environment is the root of human existence, and the source of human life; it involves every sphere of human life, and any enterprise of humanity changing the environment will involve an aesthetic component; changing the environment into a landscape is also creating the world in accordance with the laws of beauty. In the construction of our living environment, the concept of a garden has important aesthetic value, and constructing garden-style cities and garden-style rural areas is the highest state of our habitation. We should

[13]Ibid, p. 90.

say that the author's self-summarization is quite comprehensive. Second, the distinction between environmental aesthetics and traditional aesthetics: The author points out that 'the focus of aesthetic research shifts from art to nature, and its philosophical foundation extends from traditional humanism and scientism to humanism, scientism and ecologicalism; aesthetics is moving toward everyday life and is moreover applied to practice. It is not difficult to foresee that environmental aesthetics will inject fresh blood into aesthetic research, and it is bound point out for human practice a path toward beautiful harmony between humanity and the environment: "The emergence of environmental aesthetics is an extension of the traditional field of aesthetic research, and implies the birth of a new environment-centered aesthetic theory"'[14]. Third, what is the environment? The author points out: 'The environment can only be humanized nature. With respect to the ontological sense, humans and the environment coexist simultaneously, and without an environment that is suitable for human survival, human beings cannot exist; and an environment without human existence, cannot be called an environment at all'.[15] Fourth is the relationship between the environment and art. The author does not entirely agree with the viewpoint of the Western theorists of environmental aesthetics who strictly distinguish the two, and thinks that it should be a combination of the two. He states: 'therefore, the task of environmental aesthetics is not to distinguish between art and the environment, but is rather precisely to combine art and the environment, and take the path of environmental art or artistic environment'.[16] Fifth, the author thinks that 'the feeling of home is the basis of environmental aesthetics'; 'it focuses on the dimension of sensibility, and includes the fusion of sensual enjoyment and passion'.[17] Sixth is the horizon of environmental aesthetics. The author thinks that explaining the interrelationship between the three types of environment—natural, rural and urban—will be a completely new horizon that we can access in environmental aesthetics.[18] Seventh is the relationship between environmental aesthetics and ecological aesthetics. The authors argue that 'these two kinds of aesthetics research the environment, but they are not the same, the problem of ecology does not only occur in the environment, the study of ecology is a dimension, a theoretical system, or rather, a perspective from which to observe the world, whose integration with aesthetics opens up a new scene of aesthetics indeed, but it can not belong to environmental aesthetics … both of these two kinds of aesthetics have the value of being; they cooperate with each other, and jointly promote the development of aesthetics'.[19] Eighth is the philosophical keynote of environmental aesthetics. The author explains from four different levels—the ecological, cultural, ethical and philosophical. Ninth, the original substance of environmental beauty is 'a landscape'. What we call 'a landscape', the

[14]Chen Wangheng. Environmental Aesthetics, Wuhan University Press, 2007, p. 10.

[15]Ibid, p. 17.

[16]Chen Wangheng. *Environmental Aesthetics*, Wuhan University Press, 2007, p. 22.

[17]Ibid, p. 24–25.

[18]Ibid, p. 26.

[19]Chen Wangheng. Environmental Aesthetics, p. 44–45.

author believes is mainly composed of two factors: the first is 'scenery', which refers to all of the perceptible material factors that objectively exist. The second is 'a view', which refers to all of the subjective psychological factors at work when the aesthetic subject experiences a landscape. These psychological factors and all of the material factors as the object agree with each other, and thereby makes scenic things that are originally material become a landscape the unifies the subjective mind and objective influences. The beauty of the environment is in a landscape, a landscape is the way of being of environmental beauty, and is also the original substance of environmental beauty.[20] Ninth is the enjoyment of environmental beauty. The author thinks that the appreciation of the environment is the perceptual synthesis of many perspectives, and simultaneously a holistic appreciation; not only do the sense organs have to participate in it, some kind of utilitarian value judgment is involved with it as well; tenth, the author forcefully criticizes 'the theory of natural beauty'. thinking that this is an anti-humanist viewpoint that dichotomizes subject and object. So what is natural beauty? The author thinks that 'in the final analysis it is human activity that makes natural beauty emerge, but humanity does not reduce all of nature to aesthetic objects, and it is not the case that all of nature can be transformed into an aesthetic object, and nature as an aesthetic object must be that part of nature which affirms human existence, human life and human emotion'.[21] Eleventh, he advances the theory that 'nature is the perfection of beauty'. Nature is the source of human life, the source of the laws of beauty and the source of aesthetic creation. The theory that 'nature is the perfection of beauty' must bring great changes in the philosophical foundation of aesthetics, the focus of aesthetic research and the form of aesthetic theory. Twelfth, the author also discusses the beauty of the agricultural environment, the beauty of gardens and the beauty of the urban environment, which is an important attempt to move aesthetic theory towards application in practice.

Wang Nuo: European and American Ecological Literature *(Peking University Press, 2003 Edition)*

This book is the first academic work of research on European and American ecological literature in China, and moreover, it elucidates the author's own viewpoint on the basic problems concerning ecological literature. Concerning 'ecological literature', the author thinks that, 'ecological holism is the foundation of thought of ecological literature, which is the literary study that takes the interests of the whole of the ecosystem as the most valuable examination, that expresses humanity's relationship to nature and that explores the social source of ecological crises. Ecological responsibility, critique of civilization, ecological ideals and ecological

[20]Ibid, p. 136.

[21]Chen Wangheng. *Environmental Aesthetics*, p. 222.

warnings are its salient features'.[22] Second, it sketches out the basic course of development of European and American ecological literature since 1974. The first chapter mainly discusses the development of ecological philosophy and ecological ethics in Europe and America as the foundation of ecological literature. From ancient times all the way to the twentieth century, from Schweitzer's 'reverence for life' to Leopold's 'holistic view of ecology', from Westering's 'ecological humanism' to Singh's animal liberationism, and from Ecological socialism to 'ecological justice', etc., it nearly touches on and comments on all of the important ecological thought in the West since modern times. The second chapter discusses the development of ecological literature, and predominantly investigates the development of the variety of periods of ecological literature in Europe and their representative authors' works. The third chapter trenchantly analyses the thought content of ecological literature in Europe and America, including 'the critique of conquering and dominating nature', 'the critique of industry and technology', 'ecological holism', 'the return to harmony with nature' and so on. This book is an important achievement of Chinese scholarly research on European and American ecological literature.

Peng Feng: Perfect Nature—The Philosophical Basis of Contemporary Environmental Aesthetics *(Peking University Press, 2005 Edition)*

This book is the reflections of a scholar with a philosophical background on the problems of natural beauty from the perspective of philosophical ontology, and moreover, theoretically tackles the problem of the perfect beauty of nature from a completely new perspective. The author first brings up the theoretical deficiencies of the contemporary theorists of ecology on the issue of 'the perfect beauty of nature' from the perspective of philosophical ontology, believing that this deficiency will bring about a contradiction between the connectivity of ecological science and the uniqueness and isolated nature of aesthetics. He states: 'We have a way of understanding natural beauty that differs entirely from that of environmental scientists and ecologists, and for the time being I'll call it the aesthetic way. In accordance with this aesthetic way of understanding, natural beings are beautiful, because natural beings are beings which are completely identical to themselves, they are not repeatable, cannot be compared, do not obey any understanding based on pre-existing concepts. This aesthetic way is precisely the opposite of the ecologist's manner of making universal connections, so it can be appropriately called a completely isolated way'.[23] Second, the author questions this formulation of

[22]Wang nuo, *European and American Ecological Literature,* Peking University Press, 2003, 11.

[23] Peng feng, *Perfect Nature—The Philosophical Basis of Contemporary Environmental Aesthetics,* Peking University Press, 2005, preface, p. 4.

'ecological aesthetics'. He states: 'While ecological science can better use the viewpoint of universal connections to explain the value of natural beings, it cannot easily explain the beauty of natural beings. It is precisely in this sense that we can very appropriately say that there is a new ecological ethics, but it is very hard to say that there is a new ecological aesthetics'.[24] At the same time, the author gives his own formulation of the basic viewpoint of his reflection on nature from the perspective of philosophical ontology. He states, 'the aesthetic experience is the original experience of life in the world, and the aesthetic object is the original mode of things; the relationship between human being and object in aesthetic appreciation is a pre-real bodily relation', and 'aesthetics exists in this tension between deconstruction and construction'.[25] So what is natural beauty? The author points out that, 'the reason why natural things are beautiful, is not because all natural things fit the same form of beauty, but because all natural things are similarly not of the same beauty. From the perspective that natural things are completely identical to themselves, their beauty is incomparable, unclassifiable, and completely equal'.[26] The author jointly cites relevant thoughts from Zhuangzi, Chan Buddhism and Confucianism along with the relevant theories of Kant, Adorno and Dufrenne to demonstrate. The author's discourse is unique, but there are some contradictions.

Zeng Yongcheng: Green Ideas of Literature and Art—Introduction to Literary and Artistic Ecology *(People's Literature Publishing House 2000 Edition)*

This book is the first Chinese work of literary ecology to unfold an exposition with the Marxian conception of ecology as the guide and main thread. As the author states: 'The green thought of art should absorb the wisdom from Marxian ecological ideas, and find the most solid and most viable theoretical basis for itself'. The author summarizes the ecological view of Marxism, especially that of Marx in *The Economic and Philosophic Manuscripts of 1844*, as the humanist conception of ecology, practical materialist anthropology and view of life, the becoming ecological of aesthetics, and the ecological thought of literary thinking, etc., and then discusses them from different levels—the ecological nature of literary aesthetic activity, the ideas of literary ecological thought, the ecological function of literary aesthetic activity, literary activity and ecological problems as well as the socialist market economy and literary ecology. The author investigates literary problems on the grounds of treating ecology as a new thought and method, and brings the idea of ecology, which originally emerged in the natural sciences, into literary research.

[24]Ibid, Preface, p. 2.
[25]Ibid, Preface, p. 3.
[26]Ibid, p. 4.

The Construction and Development of Ecological Aesthetics in the Future

A Serious Reconsideration and Summarization of History

Summarizing the development of ecological aesthetics in the past 20 years, we see that despite great achievements many problems still remain. At the theoretical level, we are still at the initial stage in both breadth and depth. We still have to fill in considerable gaps whether in grasping foreign materials on ecological aesthetics or in domestic research on ecological aesthetics. So far, we have still not seen the emergence of theoretical work with profound interpretive force, and greater depth is especially lacking in the excavation, sorting out and investigation of the rich wisdom of ecological aesthetics in traditional Chinese culture. So, Chinese academic circles have still not found their own unique voice in the international academic world of ecological aesthetics, environmental aesthetics and ecological literature.

A theoretical formation comes into possession of vitality not by dint of the novelty of its name and way of posing things, but most importantly by virtue of the theoretical level of its achievements. Whether ecological aesthetics is truly going to stand up on its own depends on whether it can make higher levels of achievement with theoretical interpretive force. What is still lacking in this respect is the integration of our domestic research on ecological aesthetics with our actuality in China, which is still far from adequate. First, in close connection with the actual conditions of China, we must expand and deepen the enterprise of the great revitalization of the Chinese people, of which so many generations of our Chinese people have dreamed; then, it is a matter of how to organically blend Chinese ingredients and Chinese resources into the construction of ecological aesthetics. Of course, the instituting of a theoretical formation happens through establishing theoretical categories that are uniquely its own. The vitality of ecological aesthetics ought to manifest itself in establishing a system of categories distinct from that of traditional aesthetics and artistic aesthetics. Although the academic world has already undertaken many effective explorations of this, it still has not attained any results with an explanatory theoretical force that is widely received. With respect to scholarly teamwork, although teams of researchers have continued to expand in the past 20 years and especially the last 5 years, the number of scholars who specialize in the study of ecological aesthetics still has not expanded adequately. With respect to the state of foreign exchanges, international dialogues about aspects of ecological aesthetics have developed greatly since the new century. Not only were the second and third international academic conferences held, but specialists also went abroad to do visiting research, but on the whole, the depth and breadth of exchange are still far from adequate.

Ecological aesthetics, ecological literary studies and ecological literature still encounter much questioning in the Chinese academic world at present. The existence of such doubts just shows that there are still considerable gaps in our investigative work, which we must deepen much further. Fortunately, we now enjoy a positive international atmosphere along with major policy initiatives under the participation of the Chinese government like 'the construction of ecological civilization' and 'the goal of building an environmentally friendly society'. On this favourable terrain, our efforts must move ahead in the direction of greater depth, greater comprehensiveness and increasing sinosization.

Striving for New Breakthroughs

Based on our summarization of the past, the striving for new breakthroughs should begin mainly with respect to the following three aspects: The first aspect is further integration based on previous research. We mainly have to synthesize and deepen as far as possible the existing research achievements in China and the West with respect to such aspects as the economic society, cultural philosophy and literary context generated by contemporary ecological aesthetics, the guidance of Marxist theory that we must uphold, the Eastern and Western resources and basic categories of ecological aesthetics that the construction of ecological aesthetics depends on, and the interpretation of Chinese and Western works on ecological aesthetics. The second aspect is trying to establish aesthetic categories that are uniquely the possession of ecological aesthetics, and most importantly establishing a basic aesthetics of ecological ontology in order to show the distinction between the aesthetics of ecological ontology and traditional aesthetics and explain the necessity of its independent existence. The third aspect is striving to blend the wisdom of ecological aesthetics from China and the West, and reconsidering the building of categories of ecological aesthetics. For example, integrating the aesthetics of ecological ontology with the wisdom of ecological aesthetics in the ancient Chinese saying 'production and reproduction is change' (sheng sheng wei yi 生生为易); integrating the unique object of ecological aesthetics with the ancient Chinese saying 'the heavens and the earth have a great beauty that is without words'; integrate research on phenomenology of ecology with the ancient Chinese concepts of 'purifying the affective mind' (xin zhai 心斋) and 'sitting in forgetfulness' (zuo wang 坐忘); integrating the theory of human nature in ecological aesthetics with the ancient Chinese philosophical aesthetics of life. Combining 'Nature and humanity integrating into one whole' with the 'mirror-game of the fourfold; combining the ancient Chinese concept of 'suitable dwelling' with the Heideggerian one of 'poetic dwelling'; combining the ancient Chinese 'homecoming motif' with the sense of home and consciousness of place; combining the theory of 'the spirit of art' with that of 'state of mind' and 'participatory aesthetics'; combining the Chinese theory of "metaphor and incitement" with ecological imagination; combining the ancient Chinese theory of 'monthly cycles' with ecological aesthetic practice. The list goes on. We must try to establish a newer theoretical system of ecological aesthetics with Chinese characteristics through such efforts as the ones just mentioned.

Research Methods

Follow the Guidance of the Ecology of Historical Materialism

Marxist theory provides the theoretical guidance of science and the most important position, viewpoint and method for us to accurately understand nature, society and humanity's spiritual life. Similarly, we must also self-consciously adhere to the guidance of Marxism in the contemporary investigation of ecological aesthetics. We first of all must adhere to the theoretical guidance of Marxian historical materialism by exploring and understanding the most basic economic and social causes of contemporary ecological problems and examine the necessary connection between the emergence and resolution of ecological problems and a certain social order from such important theoretical perspectives as social existence determining social consciousness and the connection between economic base and superstructure. From this, we must further distinguish the socio-economic ground that ecological aesthetics establishes. At the same time, we must diligently study the whole series of important statements Marxism makes about humanity's relationship to nature, and adopt them as important theoretical guidance for the construction of ecological aesthetics. Although Marx and Engels lived through the middle of the nineteenth century, which was the most prosperous period of the human industrial revolution, when anthropocentrism rose to overwhelming and absolute dominance, the scientific worldview of Marx and Engels determined that they must observe humanity's relationship with nature through the profound thought of dialectical materialism, and critique the capitalist system and bourgeois class for endlessly plundering nature, and resulting in behaviour that alienates humanity from nature. This dialectical materialist conception of nature becomes the most important theoretical guidance and cognitive resource for our investigation of ecological aesthetics today.

Adhere to Contemporary Ecological Holism and the Ecological Philosophy of Ecological Ontology

The most important feature of ecological aesthetics is that it is a new view of aesthetics that envelops the ecological dimension and follows the guidance of contemporary philosophy of ecology. The contents of 'contemporary philosophy of ecology' are extremely complex, including the two different factions of philosophy of ecology, namely 'anthropocentrism' and 'ecocentrism'. The ecological view of anthroporcentrism insists on humanity's absolute control over nature, and merely suggests some moderation under this premise, while 'ecocentrism' insists on the absolute value of all being and the absolute equality between nature and humanity, which necessarily results in its opposition to modern social development. The "ecological holism" that we subscribe to is a much more comprehensive and inclusive philosophy of ecology, which aims at the symbiosis between humanity and nature and the unity of the relative value of humanity and the relative value of nature. Such a philosophy of ecology is actually a philosophy of 'ecological ontology'. The construction of ecological aesthetics can only take the path of scientific development under the guidance of ecological philosophy.

Adhere to the Method of Reflecting on and Transcending the "Post-modern Age"

Ecological aesthetics is a new idea of aesthetics generated in the context of the postmodern age. We say the context of 'the postmodern age' to say that ecological aesthetics is the product of the age of post-industrial society, namely 'ecological civilization'. Because of this, we should adhere to the basic characteristics of postmodern theory in reflecting on and transcending modernity. Today, 'postmodern' has both the connotations of 'deconstruction' and 'construction', the basic characteristics of which are reflecting on and transcending 'modernity', but deconstructive postmodernism emphasizes critique and destruction, while constructive postmodern emphasizes overcoming and constructing. Ecological aesthetics is a form of aesthetics basically characterized by overcoming and constructing.

Contents

Part I
The Emergence of Ecological Aesthetics

Chapter 1
The Socio-economic Context that Ecological Aesthetics Produces

The necessity of the economy and society produced by ecological aesthetics must explain whether the form of human civilization should achieve a new transition from industrial civilization to ecological civilization. This chapter first introduces two deeply and widely influential books, *Silent Spring* and *The Limits of Growth*, then it explores the problems of "disenchantment" and "Reenchantment," and finally it gives concrete accounts of the arrival of the age of ecological civilization and the contemporary ecological condition of China.

1.1 Humanity is Already at the Crossroads—Humanity is Transitioning from Industrial Civilization to the Initial Awakening of Ecological Civilization

(1) Rachel Carson and *Silent Spring*

Rachel Carson (1907–1964) was a famous American marine biologist of her time, and author of *Under the Sea Breeze*, *Surrounded by Our Ocean* and *Ocean Edge*. Her most famous and most representative work is *Silent Spring*, published in 1962, which has already become the "milestone canonical work" in contemporary ecological theory, and is also a canonical work of literary ecological critique. The writing of this book took a good many years in the process of which Carson lived through two great misfortunes: the first is her mother's passing; Carson never married and depended on her mother for her whole life, so her mother's passing was a heavy blow to her; the other time was Carson herself suffering from cancer. Two years following the publication of *Silent Spring*, in 1964, she passed away. This book not only manifests the passion for literature and depth of thought that she had, but also exhibits her rigor and positivistic spirit as a scientist. She painstakingly compiled large amounts of evidence, and after putting it through large amounts of survey studies, the book fiercely attacks the damage wreaked on the earth and atmosphere by pesticides and points the sharp point of her pen directly at the production of DDT. The publication

F. Zeng, *Introduction to Ecological Aesthetics*,
https://doi.org/10.1007/978-981-13-8984-9_1

of this book gravely offended the interests of the agricultural capitalist and the groups of scientists related to them, who criticized, ganged up against her and even launched fatal blows. *The New York Times* used a lively and directive title when discussing her case—"The Silent Spring is Now Noisy Summer." Carson obstinately braved the pressure, made sure the publication of the book went smoothly, and it became one of the highest grossing books in America and the whole world at the time. It is precisely because of this book that America passed an environmental protection law. The publication of the book also genuinely became a turning point for ecological problems. "She deployed explicit, richly poetic and easy to understand language to concretely describe how pesticides damage the American resources of air, soil and water and how the damage incurred from the abuse of pesticides greatly outweighed the benefits they were bringing about", "she wakened people to ecological problems with the touch of her own pen", and many Americans saw Carson as a brave hero.[1]

(2) An Allegorical Revelation—Humanity Endangers Herself

The first chapter of Rachel Carson's *The Silent Spring* imagines the condition of a town in the central America suffering from the destructive effects of pesticides. What accompanied the modernization of American agriculture was initially the widespread use of such modern agricultural technologies as fertilizers and pesticides, which means putting tons of pressure on the ecological environment. This town that Carson envisions would hear a cacophony of birds singing the glory of blooming diversity, a cacophony that dramatically dims down to a dead silence, the reason being that the overuse of such insecticides as DDT wreak havoc on the fields and soil and harm insects and all species of birds, which is effectively people harming themselves. This town is a reflection of countless U.S. towns during the 60's; it is an extremely illustrative allegory of industrial civilization leading humanity to harm itself. Humanity not only destroys nature, but harms humanity itself, "this means that malignant disease will strike two out of three families."[2]

(3) The Merciless Denunciation of an Era Dominated by Industries Trying to Make Money at Whatever Cost

In Capital, Marx argues that the essence of capital is the pursuit of the unlimited proliferation of capital. In *Silent Spring*, Carson gives a vivid interpretation of this conclusion from her own perspective. In the book, she exposes the reason causing the silent spring to be the overuse of modern agrochemicals and specifically the pesticide DDT, but the ultimate reason behind this phenomenon is the domination of capitalist industries, which seek the maximization of profit. She states, "it is an age dominated by industry, in which the right to make a dollar at whatever cost is seldom challenged."[3] She traces this situation further back to the capitalist pursuit of money developing since the industrial revolution, which is precisely the essence of capital Marx spoke of as the pursuit of the proliferation of capital, but she elevates

[1]Peichao [1], p. 115.

[2]Carson [2], p. 192.

[3]Ibid., p. 11.

this problem to the level of public health. She states, "As the tide of chemicals born of the industrial age has arisen to engulf our environment, a drastic change has come about in the nature of the most serious public health problems."[4]

(4) **Revealing the Law of the Balance of Nature**

Rachel Carson was a marine biologist, who skillfully applied basic biological theories in her book, specifically the theory of the balance of nature, and advocated the thought of the food chain. The famous English historian Arnold Joseph Toynbee also advocated the theory of the food chain, stating that humanity's destruction of the natural environment commits the crime of "matricide." Carson states, "In some quarters today it is fashionable to dismiss the balance of nature", "but it is still there: a complex, precise, and highly integrated system of relationships between living things which cannot safely be ignored any more than the law of gravity can be defied with impunity by a man perched on the edge of a cliff. The balance of nature is not a status quo; it is fluid, ever shifting, in a constant state of adjustment. Man, too, is part of this balance. Sometimes the balance is in his favor; sometimes—and all too often through his own activities—it is shifted to his disadvantage".[5] Carson also states about the chains of life that, "from the small-as-dust green cells of the drifting plant plankton, through the minute water fleas to the fishes that strain plankton from the water and are in turn eaten by other fishes or by birds, mink, raccoons—in an endless cyclic transfer of materials from life to life."[6] The theory of chains is extremely important in contemporary biological thought. Every entity and biological being on earth, including man, is a link in this chain of life, and enjoys the rights that it deserves to enjoy in this chain, and should also take on the responsibility that the chain gives it whether consciously or unconsciously—which is to maintain the stability and balance of the chain of life. Today there is a saying that the chain of life may still be restored after being destroyed, but actually the chain of life is irrecoverable after it is destroyed. After an ecological milieu is destroyed and then repaired, the ecological chain in it has actually already transformed and is no longer the same ecological chain that existed prior to the destruction. After primeval forests have been felled, the ecological chain is destroyed, after which planting trees may bring about another new chain, but the original chain is no longer recoverable; After the Huang He river is polluted, we can treat it, but the chain of life in it is already altered, and the Huang He river changes into another Huang He river after treatment. The facts prove that ecological conditions are singular and unrepeatable. So, we have to protect and try to maintain balance in the chain of life.

(5) **The Critique of Humanity's Arrogant Fantasy of "Controlling Nature"**

The saying of "controlling nature" is all too common. In the last century we said "to hell with Nature, the earth can produce as much as man wills," which exudes from great confidence in controlling nature, but a confidence which is an ignorance, a

---

[4]Carson [2], p. 162.

[5]Carson [2], p. 215.

[6]Ibid., p. 39.

self-aggrandizing fantasy. The human species in fact cannot possibly have complete control of nature, and on the contrary can only respect nature and follow nature. Marx never respected controlling nature, "but respected shaping nature in accordance with the laws of beauty", which include two aspects: one is the needs intrinsic to species, another is the needs intrinsic to human being. In satisfying the needs of human beings, we also must respect the needs of species, which is to speak of respecting nature, not controlling it. When Carson speaks of man controlling nature, she states, "The control of nature is a phrase conceived in arrogance, born of the Neanderthal age of biology and philosophy, when it was supposed that nature exists for the convenience of man. The concepts and practices of applied entomology for the most part date from that Stone Age of science. It is our alarming misfortune that so primitive a science has armed itself with the most modern and terrible weapons, and that in turning them against the insects it has also turned them against the earth."[7] Using the pesticide DDT to control the growth of so-called pests leads to the "silent spring," which is but humanity reaping what it has sown. Similar to this, China once got caught up in the movement of "eliminating the four pests," when sparrows were thought to be one of the so-called "four pests," and the Chinese people wiped out the sparrows through many methods from spraying chemicals to trapping them with nets, practices which were likewise extremely infantile, resulting in the destruction of ecological chains and the balance of nature.

(6) **Warning that Humanity is Standing Where Two Roads Diverge**

In *Silent Spring*, Rachel Carson points out the seriousness of ecological problems and issues a solemn warning to humanity. She states, "we stand now where two roads diverge. But unlike the roads in Robert Frost's familiar poem, they are not equally fair. The road we have long been traveling is deceptively easy, a smooth superhighway on which we progress with great speed, but at its end lies disaster. The other fork of the road—the one 'less traveled by'—offers our last, our only chance to reach a destination that assures the preservation of our earth."[8] What she calls "the road we have long been travelling," is actually that of humanity destroying itself by polluting the ecological environment, while the other seldom travelled fork in the road is that of humanity and nature balancing, harmonizing and unifying.

1.2 *The Limits to Growth*—The Other Model of Development that Humanity Should Choose

(1) Thinking on the Model of Human Development—Peccei and the Club of Rome

The Club of Rome is an international, non-governmental, non-ideological, cross-cultural civilian group of international academic researchers, who do not serve the

[7] Carson [2], p. 263.

[8] Ibid., 244.

interests of any state or government. This club has a group of world renown physicists, biologists, mathematicians, economists, sociologists, philosophers and prospectors. Its founder is the Italian Aurelio Peccei (1908–1984). Peccei was born of a progressive cultural family from Torino, a doctorate in economics, a famous Italian industrialist, social activist and researcher of global problems. He once worked at Fiat Automobiles, dedicated himself to the anti-fascist leftist movement during WWII, and became a member of the Italian resistance movement. In 1944 he was thrown in prison for a year. After the war he continued to work at Fiat Automobiles, and resided in China for eight years. After 1957, with the support of Italian investors, government officials and industrialists, he founded an international non-profit engineering and economic consultation company that he presided over as CEO, and it later developed into Europe's largest and most vital economic consultation company. In the middle of the enterprise, Peccei felt accomplished enough to step back from the company, and put his focus on the human predicament, hoping to find a method to alter humanity's course leading to disaster. In April of 1968, under his initiative, more than 30 specialists, including scientists, educationalists, economists and entrepreneurs from ten countries, including Italy, Sweden, Japan, the Federal Republic of Germany, and Great Britain, attended a meeting at the Accademia dei Lincei in Rome, where they discussed present and future predicaments. This was the first international meeting where the problem of the human predicament was discussed. The Club of Rome was born on the ground of this meeting. So far, the Club of Rome's nearly 100 representatives coming from 40 countries engage in cross-disciplinary innovative research on such global problems as the population, resources, food, energy and environment in current societies, and wrote a series of synthetic investigative reports: *The Chasm Ahead, The Limits to Growth, Mankind at the Turning Point, Road Maps to the Future, Dialogue on Wealth and Welfare,* etc.

The Limits to Growth is a scientific report on the development of human social economy written by four Club of Rome entrusted young scientists from MIT. The book was published in 1972, which is precisely the year in which *The Declaration of the United Nations Conference on the Human Environment* was adopted, marking the point at which humanity stepped into the era of ecological civilization, when ecological problems became global problems. *The Limits to Growth* explains the limits of growth with scientifically vetted data and reasoning, and reveals to humankind for the first time the severe consequences of unlimited growth on a limited earth, sending shockwaves throughout the globe. The revised edition was published in 1992, and the 3rd edition in 2004. The 3rd edition once again sounds the alarm on the issue of growth with 30 years of additional data to back it up. Just as the author says, this book is not simply a book making predictions about humanity's future, but is rather a book that makes simulations of all kinds of scenarios and consequences on the basis of known data and hypothetical situations, a book that provides humanity with the correct path to choose, and also a book of warning to the world. The author repeatedly argues "that humanity has already overshot the carrying capacity of the global environment, but still has enough time to reflect on and choose active measures necessary to make corrections even at the global level".[9] The significance of this book for humanity cannot be underestimated, but at first, it met with great resistance, because it aimed at limiting growth, and advocated an "enough is enough" model of life, which for those craving growth and overconsumption was unbearable.

(2) **The Three Great Transformations of the Model of Development in Human History**

The Limits to Growth brings up the three great transformations of the model of development of human history. First is the transformation of the model of economic social development. The author remarks, "Human history has witnessed several structural transformations. The agricultural and industrial revolutions were the most profound examples".[10] These two transformations both arose out of "shortages" and "deficits," and were forced as the result of having no other alternative. The agricultural revolution was humanity's successful coping with a shortage of wildlife species.[11] Due to population growth, the decrease of wild animals in the hunting economy could no longer satisfy people's needs, so humanity settled down and became sedentary, choosing crop growing and the agricultural economy instead. Some nomadic regions of Chinese Inner Mongolia no longer adopt the nomadic model of production, but developed into a enclosure-raising model instead. I once went to a huge pasture in Inner Mongolia, but scenes of "wind blowing the grass and the sight of cows and goats" are already hard to find, the grass grows very short, and although the locals say that grass is very nutritious, it ultimately reflects the trend of insufficiency of soil fertility and an originally nomadic people being forced to turn toward agriculture under the pressure of the growth of population and demands.

In accompaniment with historical development, "greater population growth produced new shortages, specifically shortages of land and resources. Thus, another revolution became inevitable".[12] At this time, "machines and not land became the core means of production. Roads, railways, factories and chimneys littered the horizon, and the city also continuously inflated in size. Such changes are also a complicated enjoyment".[13] This is shows that because of rapid population growth, traditional agriculture could no longer sustain the life of such populations, thereby bringing about new shortages, which is the main reason behind the birth of the industrial revolution. But, "the success of the industrial revolution just like the hunting and agricultural revolutions before it ultimately brought about its own shortages as well,"[14] and such shortages include the aspects of "source" and "sink." "Source" refers to resources; "sink" refers to the earth's carrying capacity for pollution. This leads to a new revolution of "sustainable development." *The Limits to Growth* points out, "Like the other great revolutions, the coming sustainability revolution will also change the face of

[9]Meadows [3], Forward page.

[10]Ibid., p. 223.

[11]Ibid., p. 248.

[12]Ibid., p. 249.

[13]Ibid.

[14]Meadows etc. [3], p. 250.

the land and the foundations of human identities, institutions, and cultures. Like the previous revolutions, it will take centuries to unfold fully–though it is already under way."[15] The "sustainable" development revolution that is already underway is also the ecological revolution, which has the following characteristics:

1. This is a revolution in consumer society that makes people maintain a "suitable" consumption of material life;
2. This is a revolution in making "sustainable, efficient, plentiful, equal, better and collective" the highest values for society as a whole;
3. This is a revolution in respecting the following principles:

   ① The economy is a means and not an end, it serves the welfare of the environment and not the reverse;
   ② Build efficient and renewable energy systems;
   ③ Build efficient and closed-cycle material systems. That is build material systems that can recycle waste or render waste harmless. In June of 2008 China gradually began halting the manufacture and usage of non-decomposable plastic bags, which is an exceptionally good policy measure.
   ④ Technological designs reduce waste emissions and waste products to the absolute minimum, and the whole of society decides not to produce waste emissions and waste products that technology and nature cannot treat. In developed nations, the sorting of trash is very strict, differently colored plastic bags hold different trash, organic materials are separated from inorganic materials, and so is the decomposable from the non-decomposable, and trashed that is mixed together must be taken back;
   ⑤ Maintaining the diversity of ecosystems. The ancient adage goes: "what is harmonious grows with one another, what is homogeneous does not transfer." Any species on earth is a member of the ecological chain, no matter how extremely unimportant humanity might view them, every species has its right to survive, and plays its own unique function in the chain of life;
   ⑥ The reason for living and the reason for wanting to make your own living conditions improve no longer involve the increase of material enjoyment.[16] Of course, we want to guarantee necessary material satisfaction. We want to build a comfortable society, but cannot endlessly seek material enjoyment and possession.

(3) **Several Key Terms**

For the sake of explaining the relationship between development, resources and environment, *The Limits to Growth* creates a unique vocabulary, which reflects the author's basic thought and constitutes the key terms of the whole book.

[15] Ibid.

[16] Meadows etc. [3], p. 254.

"Sources" are the resources that maintain the development of human life. They can be divided into two kinds: renewable resources and non-renewable resources. Renewable resources include soil, water, forest, fish, etc; non-renewable resources include fossil fuels, rare minerals deposits, underground water. *The Limits to Growth* brings up two principles for "sources".

1. For renewable resources, demand the sustainable rate of use to be no higher than the rate of renewal. For instance, the amount of fish caught cannot be higher than the growth rate of the population of remaining schools of fish.
2. For non-renewable resources, demand the sustainable utilization rate to be no higher than the sustainable rate of utilization of the renewable resources used to replace them. For example, the utilization rate of oil cannot be higher than the production rate of such new resources that replace oil like wind, electricity, and photo-electricity, otherwise when oil is all used up, the resources used to replace it cannot satisfy demand.

"Sink" is nature's capacity of absorbing and purifying human waste. For "sink" there is also a principle of sustainable development—"the rate of sustainable emission cannot be higher than the speed at which the waste is recycled, absorbed and harmlessly decomposed in the sink."[17]

The economist Herman Dali describes these three previously mentioned principles "the three simple principles of the sustainable limits to the capacity of matter and energy to absorb and emit".[18]

The idea of an "ecological footprint" is a new concept arising from ecological philosophy, specifically from the book *Limits to Growth* (third edition) from the 1990s. The term becomes an important keyword for the book. The term "ecological footprint" refers to "the relationship between the human needs of the planet and the capacity that the earth can provide",[19] specifically "providing resources for the international community (food, feed, trees, fish and urban land) and the area of land needed to absorb emissions (carbon dioxide)".[20] 1.3 billion people survive on 9.6 million square kilometers of land in China. To contrast, only less than 40 million people live on more than 10 million square kilometers of land in Canada. China's ecological footprint is much smaller. In the past, Chinese textbooks often talk about China's "vast territory and abundant resources." However, from the perspective of ecological footprint theory, China's pressure on environmental resources is very large: China has neither a vast territory, nor abundant resources. Mathis Wackenagel predicted the following by using ecological footprint theory: in the 1980s (1980) human demand and the Earth's carrying capacity are roughly equal. By 1999, human demand had exceeded the Earth's carrying capacity by 20%. WWF also noted in a report that the world's average ecological footprint is 2.2 hectares per person, much more than the 1.8 hectares that the Earth can offer. Since the 1960s, China's

---

[17] Meadows etc. [3], p. 52.

[18] Ibid., 50.

[19] Meadows etc. [3], p. 3.

[20] Ibid., Forward page.

average ecological footprint has doubled. The current demand is more than double the amount of sustainable supply in the country,[21] which demonstrates the seriousness of the problem.

"Exponential growth" refers to a way in which populations, capital and economies grow. ""That is, the process of doubling, doubling, and doubling, is very surprising, and it is so fast to produce such a huge number."[22] This is to say that when a quantity increases with its existing ratio, it grows exponentially. For example, when somebody deposits $100 into the bank at 7% interest, it will to $107 in the first year. In the second year, the 7% rate will lead to a growth of $7.49. At this rate, the initial deposit will be able to grow to $196.72 by the tenth year, almost doubing in value. To contrast, if you put $ 100 in a jar and add $7 per year, you will only have $149 after ten years, which is linear growth. By the end of the 50th year, the money in the bank account would be 6.5 times greater than that of the jar at almost $2500. According to the author of *Limits to Growth*, the population growth in less developed regions has increased exponentially over the past half century, and the population has doubled in 19 years, while the population in more industrialized areas has grown linearly and more gently. Regarding the exponential growth of the population, there is a real problem in front of us: regarding our ecological footprint, is there a growth limit population demands and ratios of people to land? And if so, where is the limit?

The original meaning of "overshot" is excessive, overdone, over, overloaded, crossing the border."Overshooting, is unintentionally and unconsciously overshooting limits"[23] as the book says. In 1992, 1,600 scientists from over 70 countries, including 102 Nobel prize winners, signed the World Scientists' Warning to Humanity: humanity and the natural world are "overshooting", and human activity is bringing grave and irreparable damage to the environment and important resources. If it is not stopped, much of our actions at present will severely threaten the human society, earth and animal kingdom that we hope for, and will alter the world of human life to the point of being unable to continue life as we know it. If we are to avoid the conflicts our present path is bringing about, we urgently need some fundamental changes.

(4) **The World has a Choice**

The Limits to Growth does not predict the future of the world, but rather provides all kinds of models for humanity to decide through a scientific attitude and scientific method. It states that "what the world faces is not a predestined future, but rather a choice".[24] Three models are provided to humanity to choose from in this book:

① The world has no limits and can grow infinitely, the result of which is squeezing the earth dry to the point of collapse.

② The limit is closing in, and humanity cannot possibly respond appropriately, the result of which is also collapse.

---

[21]Cited from Reference News June 11, 2008.

[22]Meadows etc. [3], p. 19.

[23]Meadows etc. [3], p. 2.

[24]Ibid., p. 262.

③ The limit genuinely exists, and in some respect we have overshot it already, but with enough time we may possibly reduce humanity's ecological footprint, and continue to evolve toward a much better world for the absolute majority of human beings on earth.

The author of *Limits to Growth* tells us that for the third model, "from the evidence we have mastered, from the world data to the computer model, it shows that it is credible and achievable".[25] This is the conclusion of *Limits to Growth.*

1.3 The New Relationship between Human beings and Nature—From "Disenchantment" to "Reenchantment"

(1) On the "Disenchantment" of the world and some "Reenchantment"

The so-called "enchantment" view of the world holds that ghosts, demons and other superstitious entities exist. In ancient agricultural societies, science and technology was not yet developed, and people did not understand natural phenomena. They therefore believed that many mysterious phenomena arose due to the gods or ghosts, thereby "enchanting" nature. In some parts of China's rural areas in the past, snakes, foxes, and other animals are often sacred because of their color, and cannot be killed, as doing so would bring about disaster. In industrial society, science and technology have developed, and people's ability to know the nature has been greatly enhanced. As a result, nature's mystery began to fade away. During the Enlightenment, Francis Bacon proposed that "knowledge is power" and believed that knowledge can change everything, which is a heroic declaration of human respect for knowledge and the self. Rene Descartes put forward the idea that "I think so I am", and believed that reason is above all else, and therefore that the rational subject is above all else. They estimated too highly of experimental science as the representative of scientific forces, and believed that the power of relying on technology can do everything. They therefore put forward the idea of the "Disenchantment of the World". According to research, it is the famous German sociologist Max Weber who first proposed the idea of the "Disenchantment of the World." In fact, the "disenchantment" had already begun during the Enlightenment. Immanuel Kant's famous saying: "Man made legislation for nature ", clearly demonstrates this view position. Weber put forward this statement when introducing one of Christianity's reformed sects, "Calvinism" in in *Protestant Ethics and the Spirit of Capitalism* in 1904 to 1905. He said, "This great historical process of religious development - the magic is eliminated of the World, where it reached its logical end."[26] The Elimination of the World is also translated as "the loss of the world's magic." Weber specifically interpreted it as, "rejecting the magic of

---

[25] Ibid., p. 263.

[26] Weber [4], p. 79.

the Eucharist as a path to salvation."[27] "The magic of the Eucharist" means that Jesus will give his disciples wine and bread that will symbolize his own blood and flesh during the last supper so that they can be rescued after his martyrdom. Refusing to take part in this religious practice is also "disenchantment." The translator interprets it as "a very important aspect of Weber's more extensive rationalization process, where he sums up his philosophy of history."[28] In other words, "disenchantment" is an important aspect of Weber's rational process, not only confined to religion.

D.R. Griffin, the contemporary American philosopher directly criticized Weber's view of the "disenchantment of the world" in the article "Peace and Postmodern Paradigm", advocated some "Reenchantment". He writes, "Max Weber once pointed out that this kind of world's Disenchantment is an important feature of the present era, which naturally is seen as a dead thing, it is composed of lifeless objects, there is no divine nature of life in it. This kind of 'natural death' leads to a variety of catastrophic consequences."[29] On the contrary, he proposed a Reenchantment of the World, claiming that, "This requires the realization of the Reenchantment of the World, postmodern paradigm will help this ideal".[30] He will have "the Reenchantment of the World" as a theoretical achievement of postmodern paradigm. What does "the Reenchantment of the World" mean? Is it possible to return to the "everything has spirit" in the agricultural society?—of course not.

Rather, we understand it as part of the restoration of the magic, sacredness, and potential aesthetics of nature.

(2) **Nature is Magical, Sacred, and Worthy of Fear**

This is a matter of intense debate. Someone once raised the idea of "fearing nature" based on the relationship between human beings and nature some time ago. In response, it was strongly opposed by some well-known scientists and humanists, who instead believed that this is to have the relationship between human beings and nature turned upside down. It is only human beings who are great and worthy of fear. The author believes that we should maintain a modest fear of nature, because until now, nature still has some kind of mystery and magic for our human beings, and this phenomenon will continue forever.

When discussing how the universe was generated, Engels used two instances of "do not know" consecutively in the *Dialectics of Nature.* He said, " This much is certain: there was a time when the matter of our island universe had *transformed* a quantity of motion—of what kind we do not yet know—into heat, such that there could be developed from it the solar systems appertaining to (according to Mädler) at least twenty million stars, the gradual extinction of which is likewise certain. How did this transformation take place? he also said: We know just as little as Father Secchi knows whether the future *caput mortuum* of our solar system will once again

[27] Ibid., p. 185, note (19).

[28] Ibid., pp. 185–186, note (19).

[29] Ray Griffin [5], p. 218.

[30] Griffin [5], p. 222.

be converted into the raw material of a new solar system.[31] In fact, how the universe is generated in the end, at present we do not know. Currently the theory about the universe are just hypotheticals about "Nebulae," "particles," "the big bang," and so on. Regarding humankind's domination of nature, Engels had a very famous saying. He wrote, "We should not be too fascinated by our human victory in nature; for every such victory, nature has retaliated against us."[32]

The development of natural science is also a challenge to the "disenchanting" of nature. The first to be overridden is the scientific determinism represented by the natural mechanics of Isaac Newton. One substantialt representative is the French scientist Pierre Simon, marquis de Laplace. He claims in his *Introduction to Probability in Philosophy*: "If a wise man knows all the forces in the universe, knows the locations of all objects, then through 'the simple calculation', the future and the past will show in his eyes." However, this approach to scientific knowledge has been overthrown by modern natural science. In the first place, thermodynamics denies it. Thermodynamics indicates that a liter of gas contained in a sealed container contains tens of billions of atoms. However, it is impossible to know the position of each atom. One can only describe the state by probability theory and statistics. Werner Heisenberg's uncertainty principle also points out that we can't know exactly where a particle is and how fast it is moving. Then the meteorologist Hendrik Antoon Lorentz put forward "chaos" theory in 1961, which claims that a very simple system will eventually become unpredictable and contain very complex movements and consequences. This is, the famous "butterfly effect." A butterfly flapping its wings on the Amazon River can potentially set off a storm in the Mississippi River Basin. "Chaos" theory also pointed out that people can't know where the solar system or planet will be in one million years' time.

As for the sanctity of nature, this should also be self-evident. First of all, nature is the mother of humankind. Human beings came from nature, and will finally return to nature. At the same time, nature is also the home of humankind. Humans rely on the natural provisions of sunshine, air, water, and food to survive, and can't be separated from nature for a moment. Humankind is great, but is small in the face of nature. It is perfectly desirable to maintain a moderate fear of nature.

1.4 We Have Been in the Era of Post-industrial Civilization—The Era of Ecological Civilization

(1) Constructive Postmodernism and Ecological Holism

"Postmodernism" is a concept or paradigm that has been offered up since the 1960s. It has been popular since the 1980s and is still far reaching. Postmodernism's connotations are extremely rich and complex, but in general, they include "the post-

[31] Engels [6], (Vol. 9), p. 424.

[32] Ibid., pp. 559–560.

modernism of deconstruction" and "constructive" postmodernism. The main representative figure of the postmodernism of deconstruction is Derrida, though he does not admit it himself. His theory of deconstruction and has destruction as its focus. However, it does not completely deny everything. Under this premise, we retain certain "scratches." The opposite of postmodernist deconstructionism is constructive postmodern, one of whose representatives is D.R. Griffin of the United States. This constructive postmodernity is a reflection upon and transcendence of modernity, and includes reflection on the opposition between human beings and nature in modernity. From the reflection of human plundering and controlling nature develops an account of ecological holism. As Griffin says, "the postmodern view produces a spirit that combines the particular concern of human well-being with the consideration of ecology."[33] He added that "postmodern thought is completely ecological, it provides a philosophical and ideological basis for the lasting insights advocated by the ecological movement."[34] The basis of philosophy and ideology is a rejection of the binary opposition between human beings and nature, and advocacy of an organic, complete philosophy. Thus, constructive postmodernism must contain the advocacy of ecological civilization.

(2) **The Arrival of "Anthropocene" and the Era of Ecological Civilization**

The first is the advent of the "Anthropocene," wherein the impact of human activities on the environment has intensified. From a geological point of view, the evolution of the earth is historical. Geologists use the Jurassic, Cretaceous, and other concepts to express stages in the development of life on earth. Many scientists believe that the current Earth has entered a new historical period they call the "Anthropocene." According to the report of *China Environmental News* on August 31, 2004:

> It has been thought that the geological period in which we live should be "Holocene", which is about ten thousand years after the end of a recent glacial period. However, more and more scientists have begun to gradually accept the theory that the earth has entered another period of development – the "Anthropocene," during which the human impact on the environment is no less than that of nature itself.

Paul Grutzen, the Nobel Prize winner pointed out at the recent European Science Forum in Stockholm that humans are rapidly changing the physical, chemical, and biological characteristics of the planet on which they live, and the most significant "achievement" of theirs is causing climate change.

At the same time, the leader of the International Plan for Crust and Biosphere Research, Will Steven, believes that the"Anthropocene" and the development of human society in the early quiet environment has a huge difference. The future we will face will be in huge environmental turmoil.

Through the computerized "earth system" simulation experiment, scientists have revealed to humans the importance of protecting our planet from catastrophic changes. According to the computer simulation experiments with the global warming trends further intensified. The Amazon rain forest will disappear, while the Sahara

---

[33]Griffin [5], p. 23.

[34]Ibid., p. 227.

will become more moist and green. (This latter change will exacerbate the Amazon disaster.) In other words, in the foreseeable future, Amazon and Sahara may swap roles with one another.

Additionally, scientists are also closely watching the North Atlantic circulation, the western Antarctic glaciers, the Asian monsoon, and so on. These changes are due to changes in the Earth's environment prompted by human beings. The marine research professor, Catherine Richardson from Denmark pointed out that the ocean now contains 50% more carbonic acid than that in the air. Ocean acidification will also lead to a lack of marine plants, a lack of animal groups, and even mass extinction, which will accelerate global warming.

The "Anthropocene" era means that human beings should reflect and change their behavior, or they will be destroyed. At the same time, after the 1960s, humans began to enter the post-industrial era. In summary this era can be referred to as the information age, the knowledge economy era, etc. Recently it was also summarized as an era of ecological civilization. *Guangming Daily* published a paper "On Ecological Civilization," on April 30, 2004 which pointed out, "At present, human civilization is in the transition from industrial civilization to an ecological civilization stage," and that, "ecological civilization is a new stage of human civilization development, a form of human civilization that comes after the industrial revolution." In 1972, the promulgation of the United Nations *Declaration on the Human Environment* had marked the entry of humankind into the era of ecological civilization. China has also recently put forward the construction of "ecological civilization" to add to the three other types of civilization, including material civilization, spiritual civilization, and political civilization. In fact, China's current scientific development includes harmonious society goals, a concept of being "people-oriented," the sustainable development of the economy, "building an environment-friendly society," and other principles. In a sense, these are also a reflection of industrial civilization and transcendence, marking the establishment of a new ecological civilization.

(3) **Towards Ecological Modernization**

History tells us that the development of contemporary ecological theory has gone through three stages:

The first stage, occurred during the 20th century, and lasted for about 60–70 years. This stage refers to the sudden environmental problems and the awakening of human beings that it led to. With *Silent Spring, Limits to Growth* and other works published as a symbol of this stage, questions began to arise and were reflected upon. However, there have been cases of going too far, including such erroneous trends as "anti-modernization," "anti-industrialization," "anti-productive forces," and "anti-technology." Some even propose going back to the Middle Ages, which is not only wrong, but also unrealistic and does not work. Some people exaggerate that the happiness of our present life is not necessarily better than the medieval peasants in the United Kingdom, which is an attitude far removed real life. In fact, most of human rural life has been very hard, and has not improved until recent years. Additionally, the very low productivity of the Middle Ages cannot be ignored. Therefore, we should remain cool and calm regarding the ecological environment.

The second stage came after the 1980s with the idea of development and environmental synchronization being put forward. The idea of economic growth and environmental pressure is inversely proportional, this is the "ecological modernization" theory put forward by the German scholar Huber. Ecological modernization refers to a mutually beneficial coupling of modernization and the natural environment, and can be understood as the ecological transformation of the world's modernization. Europe, the United States and other developed countries have generally done this. They basically realized that economic growth rate exceeds the pressure of environmental growth.

The third stage refers to developing countries represented by China, who are on the road of ecological modernization. China is in a situation of rapid economic growth, and in the process of modernization. However, China is also taking the road of ecological modernization. In other words, besides the four areas of modernization of "industrial, agricultural, national defense, science and technology," there is a fifth, "ecological modernization." This goal is planned to be achieved in 2050 or so, so that China's economic development and environmental degradation can be completely decoupled, thereby ensuring that the habitat and environment fully meet the standard of the major developed countries.

To guarantee ecological modernization, one must construct an advanced ecological culture. Culture is the leader, supporter, and guarantor of ecological modernization, which is in need of an advanced ecological culture to pilot, support, and guarantee itself. *Limits to Growth* tells us that "the world is not a destined future, but a choice".[35] Choice is a kind of attitude, a kind of values, a culture. We should advocate an "aesthetic attitude towards nature," and a cultural attitude of "enough is enough" instead of the wrong attitude that treats nature as an enemy,"—"the more the better." Therefore, we must not only adjust the development model, but also construct ecological culture, including ecological philosophy, ecological ethics, ecological economics, ecological sociology, and ecological aesthetics. Ecological culture is an advanced culture in line with the direction of social development, and the development of ecological culture is where our responsibility lies.

1.5 The Realistic Urgency of Ecological Civilization Construction in China

(1) China as a resource-constrained country is facing increasing environmental and resource pressures

In the past, there have been discussions of China's "vast territory and abundant resources." In fact, China's environmental resources have sustained unprecedented pressure which becomes the "bottleneck" of China's modernization. China's 1.3 billion population accounts for 1/5 of the world's population, with a forest coverage

[35] Meadows, etc [3], p. 262.

rate of only 20%. This does not reach the level of the world's per capita average. China's national per capita of fresh water is 1/4 of the world's per capita. The North only has 1/18, while Shandong is 1/10. China's desertification land is equivalent to 14 Guangdong provinces, and also showing a growing trend. China's ecological footprint is limited.

China's basic contradiction, according to the summary of some authoritative statesmen, apart from the contradiction between the growing material and cultural needs of the people and the backwardness of productive forces, is one between the growing material and cultural needs of the people and the environment, ecology, and resources.

(2) **Recently, China's environmental problems have become increasingly serious, directly threatening the success of modernization and the survival of the people**

Over the past 30 years, China has implemented a large-scale modernization cause. The country is prosperous. The people are rich, and have made great progress. But China has also paid a huge environmental cost, and its environmental problems have become more and more serious. On December 21, 2007 *Guangming Daily* reported:

> According to authoritative statistics, China's annual environmental pollution caused by the economy accounted for the proportion of GDP is very alarming. Authoritative people point out more profoundly that the Environmental Problems in the developed capitalist Countries in the past hundred years occurred in China in 20 years suddenly. Accidents are frequent, and the problem is serious. Due to the impact of the development model, China has not properly accepted the transfer of environmental pollution in developed countries. To a certain extent, there should be more emphasis on the seriousness of environmental problems.

According to the *New York Times* report, China Handan Iron and Steel Plant introduced Germany's Ruhr iron and steel making equipment from the late 1990s, improving steel production by leaps and bounds. Now, China's steel production has more than that of Germany, Japan, and the United States combined steel production. However, with the transfer of this equipment, environmental pollution has also been transferred. Ruhr was a high pollution area. If you wear a white shirt in the morning, it will turn gray at night. But now the Ruhr people in Germany are in exchange for clear water and blue sky. In contrast, Handan in China has a serious problem of environmental pollution. The residents living in the west side of Handan City have been living in dusty smoke, containing carcinogenic substances in the gas.[36] A village in central Shandong is also the so-called "cancer village," as the incidence of esophageal cancer is extremely high because the village is close to the Dawen River polluted by the nearby city of industrial and mining enterprises. Now, young adults and children have been transferred out of the village. Only the elderly stay there and helplessly wait for fate to come. At present, there are some "cancer villages" similar to this in China. This situation shows that serious environmental problems not only directly threaten the success of the country's modernization, but also is a direct threat to people's lives and health.

[36]Quoted from the *Reference Message* December 22, 2007 "World Factory is also the world chimney."

(3) **The problems of resources and environment directly contradict the goal of the great rejuvenation of the Chinese nation and the development policy of a "people-oriented" in China**

In recent years, China's average annual growth is about 10%, but if you get rid of the economic losses caused by environmental pollution, the growth rate is very small. Practice has proved that the path of "post-pollution control" is impractical because it will pay a much higher price. It will not only pay an economic price, but also will pay the price of people's health, good life, and even lives. This is contradictory to the "people-oriented" approach in China's scientific development. The United Nations stipulates that environmental rights are also a human right, and everyone has the right to live with dignity in a beautiful environment.

Imagine if there is no clean drinking water: we would all go crazy to grab a small amount of mineral water or rush around for a bucket of water. Would people in this way live with dignity? When we are closed upstairs by "SARS" virus, only to get a little life supplies in the window through the basket, will we be living with dignity? In this case, can we take the kind of arrogant, conqueror's attitude to nature, or can we be moderately fearful of nature? That is to say, when a quantity increases in proportion to its existing one, it grows exponentially. For example, depositing $100 in a bank will increase to $107 in the first year at a 7% interest rate. The growth rate in the second year is 7% of $107, or $7.49. By the tenth year, it will grow to $196.72, almost doubled. If you put $100 in a jar, and invest $7 per year, then after 10 years, it will be $149. This is linear growth. By the end of the fifth decade, the money in the bank account will be 6.5 times more than in the jar, almost $2,500. According to the authors of *Limits to Growth,* the population growth of underdeveloped cities has grown exponentially over the past half century, and the population has doubled in 19 years. The population of industrialized areas has grown linearly and is relatively flat. For the exponential growth of the population, there is a real problem before us. That is to say, under a certain ecological footprint, is there limit to the growth of human needs and land ratio, and where is this limit?

References

1. Peichao, L. (2004). *The overturning of ethical expansionism: Research on the western trend of environmental ethics*. Hunan Normal University Press.
2. Carson, R. (1997). *Silent Spring*, Lv Rui Lan, Li Changsheng translation. Jilin People 's Publishing House.
3. [U.S.] Meadows, D. etc. ()2013. *The limits to growth,* Li Tao, Wang Zhiyong translation. Machinery Industry Press.
4. [Germany] Weber, M. (1987). Protestant ethics and the spirit of capitalism, Yu Xiao, Chen Weigang translation. Shanghai Sanlian Bookstore.
5. U.S.] Ray Griffin, D. (1998). *Postmodern spirituality and society*, translated by Wang Chengbing. Central Compiler Publishing House.
6. [English] Engels. (2009). *The dialectics of nature, The Collection of Marx and Engels*. People's Publishing House.

Chapter 2
Philosophical and Cultural Background of Ecological Aesthetics

In the previous chapter, we discussed the socioeconomic background of ecological aesthetics. We discussed the gradual transition from industrial civilization to ecological civilization from the middle of the 20th century. This chapter will discuss the philosophical and cultural background of ecological aesthetics. That is, from the middle of the 20th century, with the arrival of the era of ecological civilization, the transformation from anthropocentrism to the whole of ecology, and from traditional epistemology to modern ontology, gradually arose. Even in the field of Christianity, which has had a deep impact on Western ideology and culture, ecological changes have occurred accompanied by ecological theology.

2.1 The Transformation from Anthropocentrism to the Whole of Ecology

(1) Anthropocentrism as a historical form and its end

In the study of contemporary ecological theory, one sensitive issue concerns "anthropocentrism." Due to the abandonment of "anthropocentrism" in contemporary ecological theory, including ecological aesthetics, one often encounters the following concerns: if ecological theory is not human-centric, then what should be at its center? Is the contemporary ecological theory actually contrary to the idea of being "human-oriented"? To answer these questions, it is necessary to study "anthropocentrism" as a theoretical form.

Any theoretical category will generate and develop under certain historical conditions. Changes in social and historical conditions will inevitably lead to corresponding changes in the theoretical categories which grow within them. "Anthropocentrism" was formed in history, and is a historical, theoretical form. As we all know, the relationship between nature and human beings is the most basic theory of human beings. It contains various theoretical points of view regarding the pros and cons of humanity and nature. But in people's minds, the dominant form of theory

F. Zeng, *Introduction to Ecological Aesthetics*,
https://doi.org/10.1007/978-981-13-8984-9_2

is the concept of "anthropocentrism" from the 18th century industrial revolution. Before the industrial revolution, throughout the long history of humankind, the level of productivity was extremely low. People were at a disadvantage in their relationship with nature. Therefore, the concepts of ignorance and worship of nature dominated because of nature's overwhelming advantage. After the industrial revolution, science and technology progressed in leaps and bounds. The rapid development of productive forces enabled economic development to gradually transition from the agricultural era, the industrial era, the electronic age, and to today's information age. Human beings' power and control over nature is also growing rapidly. In this case, the theory of "anthropocentrism" is represented only as a scientific and technological doctrine. The starting point can be located in 1769 when Watt invented the steam engine. From that time to the 1960s, over 200 years had passed. During this period, there had been a strong sense of rationalism accompanying the trend towards "anthropocentrism." Its representatives include the famous British philosopher, Francis Bacon (Francis Bacoon, 1561–1626). Bacon is the ancestor of modern experimental science. He greatly respected the power of science, and put forward the idea that "knowledge is power." ("Knowledge itself is power.")[1] He believes that science can get rid of superstition, can affect morality, and, of course, can also understand nature, transform nature, and rule nature. The French philosopher Rene Descartes (1596–1650) was also an advocate of science and reason. In the *Discourse on Method*, Descartes claims that by virtue of practical philosophy, one can "make himself a master and possessor of nature." Descartes also put forward the very famous "I think. Therefore, I am" as the first proposition of its metaphysics. "I think" refers to his capacity as a skeptical and thoughtful subject to everything, and further demonstrates the decisive role of human reason. Descartes even said that animals are unconscious, irrational machines. Another enlightened thinker, Immanuel Kant (1724–1804), a pioneer of classical German philosophy, put forward the famous proposition of "knowledge as natural law" in his book *The Critique of Pure Reason*. He reaffirmed the idea of human beings' rational abilities, which go beyond nature.

These theoretical conceptions of "anthropocentrism" at the time undoubtedly played a great historical role. They acted not only as ways to promote the power of science and technology, but also as ways to greatly promote the development of science and technology. Because of their scientific spirit, and the spirit of the Enlightenment, these notions greatly expanded the wisdom of the people. In politics, because of the opposition of reason and tyranny, these notions also played an important role in rejecting feudalism and embracing bourgeois democratization. But any social form and theoretical form is historical. Once they complete their historical mission and exhaust their energy, they will come to their own negation, and thus to an end. The same is true of capitalist modernization and its concept of "anthropocentrism."

---

[1]In the foreign literary world and foreign philosophical circles, the commonly recognized source is Bacon's Religious Meditation Record written in Latin, which is also translated as Contemplative Record. In that book, Bacon's Latin original text is "nam et ipsa scientia potestases" and the English translation is "Knowledge itself is power". See Zhou Donglin's Bacon's Famous Saying "Knowledge is power" Three Solutions, Journal of Fudan University, No. 5, 2007.

From the late 19th century, and especially in the early twentieth century, the unavoidable expansion of capital not only greatly violated the interests of the majority of the working class, but also greatly violated nature, causing serious environmental pollution and local ecological disaster. Pittsburgh of the United States, Ruhr of Germany, and Tokyo of Japan were heavily polluted areas of the ecological environment. As the American eco-feminist philosopher Carolyn Merchant wrote in the Preface to *The Death of Nature* in 1990, "Today, a global ecological crisis that goes beyond the environmental crisis of the 1970s threatens the health of the entire planet. Ozone depletion, carbion dioxide buildup, chloroflurocarbon emissions, and acid rain upset the respiration and clog the pores and lungs of the ancient Earth Mother, rechristened "Gaia," by atmospheric chemist James Lovelock. Toxic wastes, pesticides, and herbicides seep into ground water, marshes, bays, and oceans, polluting Gaia's circulatory system. Tropical rainforests and northern old-growth forests disappear at alarming rates as lumberers shear Gaia of her tresses. Entire species of plants and animals become extinct each day. A new partnership between humans and the earth is urgently needed."[2] Merchant here describe humans under the guidance of the concept of "anthropocentrism" using affectionate language and imagery. "Industrialization" brought about serious disaster on the Earth, in nature and in life. The pace of this "industrialization" had reached a point where it must be curbed, and the accompanying concept of "anthropocentrism" also was to change at the time.

The French philosopher Michel Foucault (1926–1984), proclaimed the end of the "anthropocentric" philosophical era dominated by instrumental rationality in the book *The Order of Things: An Archaeology of the Human Sciences*, and declared in 1966 that humankind would usher in a new era of philosophy. Foucault points out that, "today we and Nietzsche still show the turning point from afar, which has been asserted, is not God's absence or death, but the end of man".[3] Here the so-called, "end of man" is the end of "anthropocentrism." He further elaborated: "We are easy to think that since people have found themselves not in the center of creation, not in the middle of space, and perhaps even not the top and the final stages of life, people have been liberated from themselves; The man is no longer the master of the kingdom of the world, and the man is no longer ruled in the center of existence."[4] Another French philosopher Jacques Derrida (1930–2004) arrive at conclusion of the "de-center" a "deconstruction" by using the structuralist approach in the book *Writing and Difference*. He reveals in the text a center which can be in the structure as well as outside the structure and therefore does not exist in the paradox, and thus concluded that: "In this way, people no doubt have to start thinking about the following Question: the center does not exist, the center can not be considered in the form of the presence of the center, there is no natural place, the center is not a fixed location but a function, a non-place, and in this non-place symbol Replace endless play with each other."[5] In Derrida's view, contemporary society is in an era

[2]Merchant [1], "Foreword", p. 1.

[3]Foucault [2], p. 503.

[4]Foucault [2], p. 454.

[5]Derrida [3], p. 505.

of pluralism. Human beings and all things should go to equality and symbiosis, any "center" should not be in. The concept of "anthropocentrism" collapses naturally in his "deconstruction" of the philosophical paradigm.

The end of "anthropocentrism" is of great significance, marking the end of an era. As the famous "green peace philosophy" claimed, "mankind is not the center of this planet. Ecology tells us that the whole earth is also part of our body, and we must respect it as our own".[6] This "Green peace philosophy" also described the collapse of "anthropocentrism" as a "Copernican revolution," which shows its significance.

The end of "anthropocentrism" refers to its end as a philosophical thought that dominated the history of mankind. This does not mean that it has never played a positive role in history, nor does it mean that this theory has no merit. History has shown that anthropocentrism, subjectivity, and capitalist industrialization have played major roles in human history. It can be said that the publicity of human reason, the progress of science and technology, and the development of industrialization changed the form of human civilization, human lifestyle and human ways of life, and greatly promoted the progress of history. In his famous Communist Manifesto, Marx also fully affirmed the bourgeoisie and its industrialization. He said:

> The productive forces created by the bourgeoisie in its less than a century of class rule are more than all the productive forces created by all generations. The conquest of nature, the adoption of machinery, the application of chemistry in industry and agriculture, the movement of ships, the passage of railways, the use of telegraphy, the reclamation of the whole continent, the navigation of the river, and large populations called out from the ground as if with spells—which of these over the past century can you think of such a productivity lurking in social work?[7]

Here, Marx fully affirms capitalist industrialization and the power of science and technology. But this also does not mean that the capitalist system and its associated concept of capitalist industrialization, "anthropocentrism," has permanent validity. Marx's understanding of the tendency of historical development in the *Communist Manifesto* is that the demise of capitalism is unavoidable. Likewise, the end of the "anthropocentrism" associated with the industrialization of capitalism is unavoidable. Of course, at the moment, although the concept of "anthropocentrism" as a dominant advantage tends towards collapse, the scientific spirit and humanistic spirit contained therein is still inherited, and will be carried forward into new situations and theoretical contexts.

(2) **Western modern ecological theory and its controversy**

In the first place, Darwin put forward the "theory of evolution." He argued that people are evolved from animals with a large number of scientific facts. He argued that people evolved from primates, as our physical constitution is very similar to these animals. Even the human spirit was not impressive to Darwin, it gave him a different impression. The book, *The Origin of Species*, published in 1859 proclaimed this view. He also said, "arrogant man believes that he is a great work, worthy of God

---

[6]Quoted from Huxiang [4], p. 532.

[7]Marx and Engels [5], (Vol. 1), p. 256.

to give care. I believe that it is more humble and true to see people as being from the evolution of animals".[8]

In the second place, contemporary environmental theorist Albert Schweitzer (1875–1965) came forward as the founder of the biocentric ethics, and in 1915 he proposed the ethical view of "reverence for life". The word "reverence" indicates the awe or humble consciousness that arises in the face of a great and mysterious force. But he clearly pointed out that having "reverence for life" is not only mean being in awe of human life. He has said, "The greatest flaw of all the ethics so far is that they believe that they only need to deal with the relationship between people." In his view, "a man, only when he sees the life of plants and animals as sacred as human life, he is morally".[9]

Third, there is the biological chain. In 1927, Charles Elton pioneered the term "food chain", revealing the dependence of organisms on nutrients, thus constituting an interdependent biological chain. The dependency of this chain begins with a reliance on the sun, moves through the plants to the herbivores, passes to the carnivores, and finally arrives at people. Elton used a pyramid as a metaphor: the simplest of the most simple food chain with the largest number of organisms is the foundation of the pyramid structure, and is also the most important. If you eliminate those at the top level of the food pyramid—for example, eagles or human beings—then the ecosystem will generally not be disrupted. However, if the base of the food pyramid is removed—for example, plants or soil bacteria—then the food pyramid will collapse. It is entirely conceivable that parasites on inhuman life will be gone without the earth, the air, and the water, and the inhabitants of their parasites will also be gone, leading humankind to die. We can completely imagine a world without human beings, but we can not imagine a world with only human beings without other beings. Similarly, we can not imagine a world with life only and without a life support system. The lower level exists as a support system for a higher level. Humans are at the top of the pyramid tower, which illustrates the fragility of mankind.

Fourth, there is earth ethics. Aldo Leopold (1887–1948), the pioneer of contemporary ecological ethics, presented earth ethics in his book the *Sand County Almanac*. *Sand County Almanac* is known as "A new Bible for the Modern Environmentalist Movement." Leopold presented the fruits of ecological science, believing that "the complexity of the earth's organism" is "an outstanding scientific discovery of the twentieth century." He argues that the idea of distinguishing species into good species and bad species is a product of anthropocentrism and utilitarian bias. Influenced by the Russian thinker Peter Osbinski, he claimed that nothing in nature is dead or mechanical, and that life and feeling exist in all things. The basic belief of holism is that the whole is greater than the sum of parts. Aldo Leopold believed that any inseparable existence was a living creature. The earth was alive, an organism of a certain type and degree of life. With this belief, the relationship between human beings and the earth can be understood as an ethical relationship. He insists that "the inseparability of" the earth—its soil, mountains, rivers, forests, climate, plants

[8]Quoted from Nash [6], p. 50.

[9]Quoted from Nash [6], p. 73.

and animals "is a sufficient reason for her to be respected. The integrity, stability, and beauty of the biological community are regarded by the earthly ethics as the highest. Humankind is not present on or above the community, but is only one of the members of the community". He has claimed that, "... a land ethic changes the role of *Homo sapiens* from conqueror of the land-community to plain member and citizen of it. It implies respect for his fellow-members, and also respect for the community as such."[10] He advocated "thinking like mountains", that is, objectively, to think in a holistic way, rather than just thinking from the human point of view. He also proposed the value of earth ethics: "A thing is right when it tends to preserve the integrity, stability, and beauty of the biotic community. It is wrong when it tends otherwise."[11]

Fifth is the animal liberation movement. Australian philosopher and activist Peter Singer (1946-) presented it in 1973. In the article "The Liberation of Animals," he argues that because people and animals are homogeneous both physically and mentally, have the ability to judge the value of happiness and pain, and have ability to feel equally, then respecting animal's "right to life", "protecting their freedom", "forbidding torture", advocating intergenerational equality, and opposing all "species discrimination", should be the methodological criterion for "communication" between human beings and animals. Similarly, based on the theory of the homogeneity of human and animal intelligence, by giving human beings natural rights unconditionally to animals, then animals should not be killed. The famous British novelist Thomas Hardy even advocated that the "golden rule" in the "Bible"—"how you are willing others to treat you, how you treat others"—should be applied to other species, especially to animals. Because "animals can not express their demands, they can not be organized, can not protest, can not march, can not impose political pressure, they can not improve our level of conscience—all of these facts can not weaken us." The sense of responsibility to defend their interests, on the contrary, their lonely helplessness makes us more responsible".[12] The participants in the "Animal Liberation Movement" are radical ecological practitioners who advocate non-violence or even use violence to resist ecological destruction.

Sixth, there is deep ecology. In 1973, the Norwegian philosopher Arne Naess proposed the concept of "deep ecology" in an article called "Shallow and Deep, Long-term Ecological Movement: a Summary". "Deep ecology" is to replace "shallow ecology". "Shallow ecology" refers to ecology that is confined to ecology and biology, which was proposed by the German biologist Heckel in 1866 to explore the relationship between biological communities. On the basis of that, Naess put forward the deep ecology, believing that shallow ecology is "anthropocentric", and only concerned about the interests of mankind. To contrast, deep ecology is non-anthropocentric and holistic, being concerned about the whole of nature. Shallow ecology focuses on environmental degradation, such as pollution, depletion of resources, etc. Deep ecology questions the root causes of environmental crises, including social, cultural,

---

[10]Leopold [7], p. 194.

[11]Ibid., 213.

[12]Quoted from Nuo's [8], p. 110.

human, and other factors. In practice, shallow ecology advocates the improvement of existing values and social systems, while deep ecology advocates the reconstruction of the order of human civilization, construing it as an organic part of the whole of nature. The core concept of "deep ecology" is "self-actualization", where the "self", according to him, " is based on the distinction between the more extensive capitalization of the 'self' and the self-identity of the narrow sense of the self, and in some oriental" self "(atman) tradition has been recognized. This 'great self' contains all the forms of life on earth with their own body. If we use the five words to express this supreme criterion, I will use 'maximized (long-term, universal) self-realization'! Another more popular statement is 'live and let live'. If I had to give up the term because of the unavoidable misunderstanding, I would use the term 'universal symbiosis' to replace".[13] In other words, this "self" is the expansion of the earth on all forms of life of the self, that is, the whole of ecology. This is one of the goals pursued by deep ecology—the goal that human beings and all life forms on earth can be achieved.

Seventh, there is Wilderness Philosophy, proposed by the contemporary ecological ethicist Holmes Rolston, a professor at the University of Colorado and the founder of the International Society for Environmental Ethics and the Journal of *Environmental Ethics*. In 1995, he presented the concept of "Wild Turning Philosophy" in his book *Philosophy Gone Wild*. From the 1960s, the environmental crisis intensified, and the value and meaning of the "wilderness" were referred to as an important agenda. "The Wilderness" is a synonym for natural ecosystems in the writings of Rolston, referring to the original ecological environment, including native nature, the wilderness, and so on. That is, "the wilderness" is not nature by human practice, but is rather "the least or undeveloped areas and ecosystems that are interfered with by humans".[14] His philosophy tended towards the wilderness, expressing "serious reflection of the philosophical community to the human and the Earth's ecological system". This shift is more unexpected than that of any other philosophical change. In Ralston's view, the wilderness is self-organized, and has the same objective, intrinsic value as human beings. He also regards natural self-creation as the source of the intrinsic value of natural matter. He makes it clear that "the creativity of the natural system is the mother of the value; all the creatures of nature are valuable to their realization of the natural creation. Where there is spontaneous creation, there is the value."[15] He regards human beings as the subject of value as a natural self-evolution, and as self-organized animals by the use of their natural self. This is organization theory. "Nature not only creates a variety of values, but also creates people who have the ability to evaluate. Nature is evolving in the direction of value; it is not that we give

[13]Quoted from Yi [9], pp. 47–48.

[14]Yi [9], p. 107.

[15]Rolston [10], pp. 269–270, 271.

the value to nature, but nature gives us the value".[16] The self-organization of the wilderness illustrates that "the nature of the ecosystem is not a bad 'wilderness', nor is it 'degenerate', nor is it worthless. On the contrary, she is a beautiful, complete and stable life community."[17]

2.2 Controversy over the Field of Modern Ecological Theory in the West

At present, the field of western ecological theory is facing a very sharp and fierce debate. The two sides adhere to opposed cultural philosophies regarding the position of the ecological center and anthropocentrism. They disagree mainly on the following three issues:

First, they disagree over the ecological value of the problem. Ecological value is one of the important concepts in contemporary ecological philosophy and ecological ethics. It mainly discusses whether natural objects have the same "intrinsic value" as human beings. Natural things obviously have an instrumental external value. But whether it has "merit and evil", "good or bad," or any such intrinsic value is being debated. It is generally believed that natural things are not subjects. There is no conscious purpose for them, and therefore they have no merit or badness, no good or bad intrinsic value. But contemporary ecological theorists, especially ecocentric theorists, believe that nature not only has indispensability in the ecosystem, but also has inherent rationality, implying that natural things have intrinsic value. This is a sharp, complex, and controversial issue. Then, the following question is, "Who is the valued subject of intrinsic value among natural things?" or, "Who is the bearer of value?" An additional question is whether there is an ethical relationship between human beings and nature, or if nature can be an ethical subject?

Whether it is anthropocentrism or ecocentrism, theorists first recognize the value of nature to human beings. Nature has a practical, aesthetic, cognitive and other value attributes, which affirms the status of human beings as the main value. But anthropocentrism does not involve the "intrinsic value" of nature, and denies the "intrinsic value" of nature. In other words, nature is only the object of human transformation, while ecocentrism believes that nature has its "intrinsic value," and that human beings should respect nature. But nature is unconscious and can not be the bearer of its own value. That is, it is not the subject of value, as a subject of value can only be people. In this way, they can not help but return to the position of anthropocentrism. Then, the legitimacy of the basic theory of "ecological value theory" and " ecocentrism" also have problems.

"The separation of value and worthlessness is always a trial or test, also a struggle of opinion or emotion, to use the power of faith or tradition in the struggle."[18] In fact,

[16]Rolston [10], p. 516.

[17]Rolston [10] p. 10.

[18]Alexander [11], p. 68.

human beings cannot judge the inherent value of nature. The natural value that human beings can judge can only be the value of nature to human beings, as embodied in human and natural relations. However, the existence of natural "intrinsic value" has a guiding significance for the transformation of the human world view, of the relationship between man and nature on the views of human behavior. Some scholars have suggested that "in real life, we can regard anthropocentrism as a universal social and ethical standard that requires all people to abide by, and the animal liberation/rights theory, bioequivism and ecological holism are understood as the individual moral ideal with the ultimate love, to encourage people to actively pursue."[19] This view of "ecocentrism" as a moral ideal has its own value, but it is not advisable to use "anthropocentrism" as a criterion that all human beings must follow.

It is clear that the debate between anthropocentrism and ecocentrism is of great importance, but the extremes of both views are obvious. Our position is that human beings and nature coexist towards ecological holism.

Second, on the Issue of Ecological Rights, the idea of "rights" is a human-specific concept. Since western modern times, the emphasis has been on "natural human rights", but do natural things have the right to determine whether or not they can accept people's moral care?

A view from the kind of mind concerned about the rights of natural things—that is, their living conditions—the benevolent movement contributed to the emergence of the first animal protection laws. In 1822, the British Parliament passed the Prohibition of Child Abuse Act (Martin Act), which was a milestone in the history of animal protection. The second is the utilization of feelings of the nerves and the feeling of pain as the standard for expanding the rights to animals. Singh believes that all species should be equal, and that the ethical principles of human equality should be applied to animals. He also points out that the perceived ability (ie, the ability to feel pain or experience pleasure) is the only reliable standard for caring for animal interests. But animal rights theory is only concerned with the individual welfare of animals, denying that the plants and the ecosystem have the right to life, which is its major flaw. The third view is the gradual extension of natural human rights, from the Jews, blacks, women, and the mentally challenged first to animals, and then to the inorganic. The fourth is to discuss rights from the perspective that the inorganic becomes the human and animal habitat, including bacteria that are harmful to humans.

From the development track of anthropocentrism and "natural human rights" to the theory of ecocentrism, contemporary western ecological ethics holds that the scope of the subject of rights must be extended from all the objects to the whole ecosystem. In this way, the relationship between man and nature can be regarded as a relationship of adjustment and restriction by ethical principles, and that nature has rights. We argue that the question of whether or not nature has the right, whether it can become a subject of rights, should be approached from the perspective of ecological existence theory and ecological holism.

---

[19] Yang Tongjin: *The Basic Concept of Environmental Ethics, Ethics and Civilization,* No. 1 of 2000.

Third, there is the issue of ecological equality, of whether natural things and human beings, and even relationships between natural things, can enjoy equal rights. Contemporary ecological philosophers advocate absolute equality, which is the idea that people and all things are absolutely equal, and that people can not touch all things. This, in fact, denies to human beings a life of eating and drinking, and means that human beings can not survive and develop. For example, former U.S. Vice President Al Gore said in *The Unbalanced Earth*: "There is a group called 'deep ecologists' who is now famous. It uses a disaster to describe the relationship between man and nature. According to this statement, the role of human beings just like pathogens, is a kind of bacteria to make the earth rash fever, threatening the basic life of the earth. They made a contrary mistake, that is to define the relationship between man and the earth almost entirely from the material sense - as if we are only some humanoid bladder, who is destined to dry instinctive bad things, do not have wisdom or free will to understand and change their own way of life."[20] Here, Gore showed some misunderstanding and prejudice. Ecocentrism's "absolute equality" does lead to what Gore says, but "ecological holism" does not lead to this result. First of all, the concept of ecological equality in ecological holism does not define the relationship between human beings and the earth in a material sense, but rather extends beyond current, material interests. Defining the relationship between the two, between human beingsand the earth continues to maintain a high degree of focus. What is more important is that the ecological equality view of ecological holism abandons anthropocentrism but is not anti-human. The "ecological equality" that many contemporary ecological theorists, including ecological aestheticians, try to assert is not the absolute equality of all things, but is rather the relative equality of people and all things. If there is absolute equality, then of course, it will limit the survival and development of human beings, and thus embark on the road of anti-human thought. The relative equality refers to the "biological chain of equality," according to which human beings will enjoy the same things with the universal, biological chain. However, they should not undermine the balance possessed by the biological chain. As Arne Naess said, the ecological equality of ecological holism is "the principle of biosphere egalitarianism, all things in the biosphere have the survival and prosperity of the equal rights".[21]

In contemporary times, although the debate between anthropocentrism and eco-centrism is very sharp, the trend of historical development is a combination of both, referring to the overall view of ecology. This kind of ecological, holistic view was first implied in the theoretical system of some theorists who hold the concept of ecocentrism. For example, everyone is familiar with Aldo Leopold who put forward the important concept of the "ecological community" in his famous ecological and philosophical work the *Sand County Almanac*. He believes that the whole ecosystem is a "biological pyramid" or "land pyramid," which is a "community" of collaboration and competition between different parts, including "soil, water, plants and animals, or summed them up: land". He said: " land ethic changes the role of Homo

[20]Quoted from Yi [9], pp. 136–137.

[21]Ibid., pp. 51, 50.

sapiens from conqueror of the land-community to plain member and citizen of it. It implies respect for his fellow-members, and also respect for the community as such"[22] His thinking of "ecological community" is based on the idea of a "biological chain". Rachel Carson specifically discusses the biological chain in *Silent Spring*, writing, "—from the small-as-dust green cells of the drifting plant plankton, through the minute water fleas to the fishes that strain plankton from the water and are in turn eaten by other fishes or by birds, mink, raccoons—in an endless cyclic transfer of materials from life to life."[23] It is because of the biological chain that there is the existence of ecological community. Nature can maintain balance only through the biological chain. Therefore, the contemporary ecological theorist David Ray Griffin pointed out that the human "must walk gently through the world, just use what we must use for our neighbors and future generations to maintain ecological balance"[24] Staring from the relative "ecological equality" in this "biological chain", the principle of "ecological whole" advocates the principle of "universal symbiosis" and "benevolence", and advocates that humankind and nature are in solidarity. Human beings should "regard nature" as their own brothers and compatriots, as both constitute an "interdependent" relationship. This ecological world view is not only not anti-human, but rather takes a more beautiful "survival" as its starting point. In fact, is a more extensive humanitarianism based in the universal spirit of love. Therefore, the "relative equality" view of this "ecological holism" principle in the biological chain has rationality in both theory and practice.

Thus, organic unity, the ecological whole, the biological chain, the great self, and symbiosis have all become important key words of modern ecological theory. This is an important trend in the development of its theory. Towards the unity of ecocentrism and anthropocentrism, and towards the whole of ecology is the historical track and prospective development of modern ecological theory.

This trend in ecological theory towards an "ecological whole" is also reflected in the recent rise in the western and in the international context of the gradual rise of ecological humanism. Originally, as early as the 1970s, biology professor David Ellenfield of Latour University had thought that the root causes of environmental problems can be attributed to "humanitarianism blasphemy".[25] In his view, the traditional humanitarianism is anthropocentrism, which is only concerned about the interests of people, and is not concerned with the interests of nature that are closely related to the interests of people.[2] Old humanism is inevitably disintegrating. The collapse of "anthropocentrism" calls for the emergence of a new humanism.

As we all know, since the Enlightenment, the usual humanism is "anthropocentrism." In this context, people and natural ecology are fated to oppose one another. It is impossible to unify an ecological outlook, to unify the humanistic and the aesthetic view. The new ecological humanism based on the ecological and aesthetic nature of human beings may be able to unify the above three. As Philip Regal, an

[22]Leopold [7], pp. 193, 194.

[23]Carson [12], p. 48.

[24]Ray Griffin [13], p. 227.

[25]See Nash [6], p. 63.

ecologist and professor of evolution and behavior at the University of Minnesota in the United States, says, "If knowledge about the human condition is humanistic, then understanding the larger system in which mankind exists is very important to the humanist. 'Ecological humanism' implies insight into the patterns of association between individuals, individuals and social institutions, individuals and non-human environments." This is what Ragal wrote in *the Ecological Humanism* published in the United States in 2002.

This so-called "ecological humanism" is actually a kind of integration and reconciliation of anthropocentrism and ecocentrism, and is a new development and extension of humanism in contemporary thinking. As we all know, nowadays, in the increasingly serious ecological crisis, there have been radical ecologists who advocate for the absolute value of and the excessive emphasis on nature, entering into intense debate with traditional anthropocentrism. In this case, "ecological humanism" is able to overcome these two theoretical tendencies, and can contribute to the unification of the two. "Ecological humanism" is based on the establishment of human beings' ecological and aesthetic nature. This is what we said above. People are born with a natural ecological affinity and love, and thus gain a better desire for it to survive. This kind of "ecological humanism" is the philosophical and theoretical basis of the construction of a new ecological aesthetic. It provides a new, humanistic spirit in the era of ecological civilization. It is a new humanistic spirit which contains the more thorough, more comprehensive, and more contemporary spirit of the "ecological dimension". It can also be called the spirit of ecological humanism.

In this context, the new ecological aesthetics includes two such levels of connotation. From the cultural point of view, it includes mainly the unity of the relative value of human beings and the relative value of nature. This is also a reconciliation through our "ecological humanism" of absolute ecology and anthropocentrism. Because absolute ecologicalism advocates the absolute value of natural ecology, which will inevitably lead to a complete denial of human needs and values, it thereby leads to the negation of human beings. This is a road that does not work. Likewise, "anthropocentrism" will unlimitedly expand human needs and values, resulting in serious damage to the natural ecology. History has shown that this will endanger the interests of mankind itself, and so is also an unreasonable path. There is only one correct path, and that is the one under the principle of ecological humanism, which recognizes the relativity of the value of both sides, and seeks to unify them. This is a "symbiotic," viable way. Under the guidance of the principle of "symbiosis", we can carry out a "win-win" policy of social and economic development, prioritize environmental protection in social development, and take the road of building an environmentally-friendly society. In intergenerational relations, we must carry out the principle of intergenerational equality, taking into account the interests of contemporary and future generations, and work towards truly sustainable development.

2.3 The Proposition of Heidegger's Ontology: The Transformation from Traditional Epistemology to Contemporary Ontology

Hegel passed away in 1831, marking the end of classical Western philosophy. The traditional epistemological philosophy characterized by the subject/object dichotomy also gradually came to an end, and was ultimately replaced by contemporary existentialism. Martin Heidegger (1889–1976), the famous German philosopher of our times, was also the founder of contemporary existentialist philosophy. The publication of his *Sein und Zeit* marked the maturation of contemporary existentialist philosophy, which began to replace traditional epistemology. Heidegger's philosophical thought bifurcated roughly in 1936, the point in time separating his early period from his late period. The early period is largely considered by the scholarly world to show the tendency of anthropocentrism due to emphasizing the priority of Dasein, while the later period truly shows a movement in the direction of ecological humanism. Heidegger's existentialist philosophy occupies an important position in Western ecological theory. Because humanity and Nature, the humanist viewpoint and the ecological viewpoint are only capable of reaching genuine unity on the theoretical basis of existentialist philosophy, contemporary ecological theorists generally call Heidegger an important forerunner of ecological theory and the metaphysical theorist of ecology. Michael Zimmerman thinks people see Heidegger as the deeper forerunner of ecology, because Heidegger proposed an even higher "humanism," which transcends the anthropocentric and dualistic humanism connected to controlling Nature and makes possible true human "dwelling," a genuine dwelling that is intimately tied to letting beings be what they are.[26]

The positing of Heidegger's existentialist philosophy contributes to and inspires ecological aesthetics in the following four ways:

(1) **Theoretical Ground: Existentialist Ontology**

Heidgger's discussion about the relationship between Being-there and Being reveals the ecological holism of humanity's harmonious symbiosis and being-with Nature; it lays the philosophical groundwork for the overcoming of anthropocentrism, and provides the theoretical foundation for the turn to aesthetic ontology.

Heidegger put forward the ontological model of being-in-the-world with "Dasein and the world," and uses it to replace the traditional epistemological model of being-in-the-world characterized by "the subject and the object," thereby laying the philosophical ground work for the unification of human being and Nature.

[26]See Yi [9], p. 69.

As is widely known, epistemology is a relationship of being-in-the-world characterized by the subject/object dichotomy between human being and the world, in which human being and Nature are fundamentally opposed and cannot possibly attain harmoniously coordinated unity. But, contemporary existentialist philosophy is a relationship of being-in-the-world characterized by Dasein and the world; being-in-the-world as the basic existential structure of Dasein is the transcendental determination of Dasein, and the relationship between Dasein and the world that it indicates is not a spatial relationship, but is rather a relationship of Dasein integrally fusing with the world, which is much more primordial than a spatial relationship. It is this relationship of being-in-the-world that finally provides the possibility and precondition of humanity's integral harmony with Nature. He states: "[s]ubject and object do not coincide with Dasein and the world."[27] This relationship of "being-in-the-world" between "Dasein and the world" can provide the precondition of unity between humanity and Nature, precisely because "Dasein" is but the relational existential state composed by human being's present moment and surrounding things, and Dasein exists and unfolds in this relational state. Here there is only "relation" and "involvement", but no "division" and "opposition." Just as Heidegger states, "The concept of 'facticity' [of the Being of Dasein] implies that an entity 'within-the-world' has Being-in-the-world in such a way that it can understand itself as bound up in its 'destiny' with the Being of those entities which it encounters within its own world."[28] He then goes on to explain that those entities which Dasein encounters and is "bound up" with in the world are "things ready-to-hand." People are faced with innumerable things in life, but only those things that people genuinely use and are concerned with are "ready-to-hand," while other things are "present-at-hand," that is even though such things are near at hand, people do not use and find concern with them, and thus no one establishes genuine relationships with them. He speaks of the "ready-to-hand" as an "involvement." He states, "[t]he character of Being which belongs to the ready-to-hand is just such an *involvement*. If something has an involvement, this implies letting it be involved in something."[29] This is to say, human being is involved with Nature through human being's factual existence; Nature is an indispensable component of human being's factual existence; Nature is implied in "Dasein," and is not outside of Dasein. This is precisely the philosophical ground of harmonious unity between humanity and Nature that contemporary existentialism puts forward, which marks the transition from "the subject/object dichotomy" to "Being there and the world" and from epistemology to contemporary ontology. Just as the contemporary ecological critique Harold Fromm states, "'the problem of the environment,' which many people persist in viewing as a peripheral arabesque drawn around the 'important' concerns of human life, must ultimately be seen as a central philosophic and ontological question about the self-definition of contemporary man."[30]

[27] Heidegger [14], p. 70.

[28] Ibid., pp. 65–66.

[29] Ibid., p. 98.

[30] Fromm [15], p. 38.

2. **Poetic Dwelling: The Aesthetic Ideal of Ecological Aesthetics**

The harmonious relationship of symbiotic coexistence of humanity and Nature is the theoretical starting point of the construction of ecological aesthetics. Heidegger calls it "poetic dwelling," which is a renowned proposition of contemporary ontological aesthetics that marks the timeliness and theoretical profundity unique to the latter. Heidegger wrote "Remembrance" for the 100 year anniversary memorializing the passing of Hölderlin. Engaging in a "remembrance" of Hölderlin's poem, Heidegger cites him, "[f]ull of merit, yet poetically, man dwells on the earth…" and elucidates it as follows, "All accomplishing and working, exploitation and cultivation, remains 'culture.' But culture is never more than the consequence of a mode of dwelling. Dwelling, however, is *poetic*."[31] Actually, "poetically dwelling" is the necessary intension of the philosophical aesthetics of Heidegger's ontology; it implies the meaning of residing and inhabiting. He states, "Nor does the term 'Being-in' mean a spatial 'in-one-another-ness' of things present-at-hand, any more than the word 'in' primordially signifies a spatial relationship of this kind. 'In' is derived from '*innan*'—'to reside,' '*habitare*,' 'to dwell' [sich auf halten]. 'An' signifies 'I am accustomed,' 'I am familiar with,' 'I look after something…' The entity to which Being-in in this signification belongs is one which we have characterized as that entity which in each case I myself am [bin]. The expression '*bin*' is connected with '*bei,*' and so '*ich bin*' ['I am'] means in its turn 'I reside' or 'dwell alongside' the world, as that which is familiar to me in such and such a way."[32] It is clear from this that "Being-in-the-world" is human being inhabiting and residing, which is to say dwelling in the world.

"To dwell" has the meaning of "to remain" and "to stay," but at the same time it has etymological roots in "to be at peace," "to be brought to peace," "to remain in peace" and "the free"; by extension it has the connotation of "to preserve from damage and danger" and "to safeguard." "To dwell, to be set at peace, means to remain at peace within the free, the preserve, the free sphere that safeguards each thing in its essence. The fundamental character of dwelling is this sparing."[33] Clearly, Heidegger's understanding of "dwelling" begins with the existential structure of "Being-in," draws freedom and safeguarding from the essence of "dwelling" and derives the ontological thought of "setting something free" and "sparing something's essence" from the existential structure of Dasein (namely the essence of human being), from which we can obviously spot the tracks of Heidegger's thought. Human being's dwelling is a safeguarding, "[i]n saving the earth, in receiving the sky, in awaiting the divinities, in initiating mortals, dwelling propriates as the fourfold preservation of the fourfold. To spare and preserve means: to take under our care, to look after the fourfold in its essence."[34] In the existential constitution, Dasein dwells alongside the world and as Dasein becomes the true being-at-home of human being (returning to the essence of human being), it has the responsibility of safeguarding and looking after, a respon-

[31] Heidegger and Hoeller [16], p. 113.

[32] Martin Heidegger, *Being and Time*, pp. 79–80.

[33] Heidegger [17], p. 246.

[34] Martin Heidegger, *Basic Writings*, p. 247

sibility which is found in the essence of the Being of Dasein, which is to say human being's existence. This is certainly the most extremely insightful part of Heidegger's thought. Due to safeguarding the world, that is safeguarding Dasein's existence itself, "That which maintains us in our essence only maintains us, however, so long as we ourselves, of our own accord, retain that which maintains us." (Vortrage und Aufsatze, p. 123) So, Heidegger can also necessarily draw the conclusion that "man is the guardian of being." All of this is already signaled in the existential analytic of Dasein, where there are already inklings of the thought openly exposed. This is the essential aim of contemporary ecological aesthetics.

3. **"The Fourfold Game of Sky, Earth, Man and Gods": The Only Path of Aestheticizing Existence**

"The Fourfold Game of Sky, Earth, Man and Gods" is the important category of ecological aesthetics that Heidegger puts forward, and the only path that Being as "Dasein" may take to unfold and succeed in the aestheticizing of existence through the "the structure of the fourfold world of sky, earth, man and gods."

Heidegger proposed "The Fourfold Game of Sky, Earth, Man and Gods" in a lecture for the Hölderlin Society held in a theatre in Munich. He states, "The appropriative mirror-game of the simplicity of earth and sky, divine and mortal, we call the world. World holds sway by worlding." (*The Heidegger Reader*, p. 265) In a lecture entitled "The Thing" he explains the fourfold game with the example of an average jug. He believes, the essence of the jug is not exhibited in the earth that is used during the making of it, but rather in the jug's void and in the gift that its void gives and holds. This holding is a collecting, which manifests the standing forth of the thing in the relationship of being: "The spring stays on in the water of the gift. In the spring the rock dwells, and in the rock dwells the dark slumber of the earth, which receives the rain and dew of the sky. In the water of the spring dwells the marriage of sky and earth. It stays in the wine given by the fruit of the vine, the fruit in which the earth's nourishment and the sky's sun are betrothed to one another."[35] He also states, "In the gift of the outpouring earth and sky, divinities and mortals dwell *together all at once*. These four, at one because of what they themselves are, belong together. Preceding everything that is present, they are enfolded into a single fourfold."[36] So, the jug and human being bond with the immortals and show the rich character of Being as the Being of beings. Only in this way can we understand Heidegger's words: "The jug is not a vessel because it was made; rather, the jug had to be made because it is this holding vessel."[37] The gift of spring water or wine pouring out of the jug and the fourfold content it implies is intimately related to the unfolding of Dasein; it is Dasein's mode of existence in the world; it is a relationship of "intimacy" like a wedding between human being and Nature, and beauty as coexisting with Truth sets in of its own accord through this relationship of intimacy in the movement toward the aestheticized Being of humanity.

---

[35] Martin Heidegger and Albert Hofstader (trans.), *Poetry, Language, Thought*, 1971, p. 170.

[36] Ibid., p. 171.

[37] Ibid., p. 166.

On the relationship of human being to the world and the thing, Heidegger states, "Men alone, as mortals, by dwelling attain to the world as world. Only what conjoins itself out of world becomes a thing."[38] Thus, "the earth as earth" similar to "the world as world" and "man as man" could all be seen as the existential constitution of Dasein—and may only thoroughly shake off the misfortunes brought about by representations in the existential ontology of Dasein. Only in this sense can we finally understand what Heidegger means by (men as mortals) let the earth become earth through dwelling. As human being returns home and truly dwells, the essence of this fourfold saving of earth, receiving of sky, awaiting on gods and initiating of mortals may finally be safeguarded. Or to borrow Heidegger's words, "[i]n saving the earth, in receiving the sky, in awaiting the divinities, in initiating mortals, dwelling occurs as the fourfold preservation of the fourfold." "There are four voices which ring out: heaven, earth, man and god. Destiny gathers the whole infinite relation in these four voices. Yet none of the four stays and goes one-sidedly by itself. In this sense, none is finite. None is without the others. *In*-finite, they hold themselves to each other, they are what they are from the *in*-finite relation, they are this whole relation itself." (*EHP*, p. 194) This illustrates that merely breaking through the cognitive model of the subject/object dichotomy and entering ontology from epistemology is still far from enough to secure the development of contemporary aesthetics, which furthermore requires breaking through the philosophical perspective of "anthropocentrism," and reaching the philosophical height of "eco-holism."

(2) **Methodological Ground: Phenomenology**

Heidegger's contemporary ontology is first of all the critique of traditional epistemological philosophy. He believes the predominant illness of traditional epistemology is complying with the cognitive model of the "subject/object dichotomy," and separating beings from Being, thereby resulting in the "concealment" of the truth of things. He specifically analyzed Descartes' traditional proposition of rationalist epistemology, namely "Cogito, ergo sum," insisting that "[w]ith the *cogito sum'* Descartes had claimed that he was putting philosophy on a new and firm footing. But what he left undetermined when he began in this 'radical' way, was the kind of Being which belongs to the *res cogitans*, or—more precisely—the *meaning of the Being of the 'sum.'* (*Being and Time*, p. 46) He also states, "Descartes distinguishes the '*ego cogito*' from the *res corporea*. This distinction will thereafter be determinative ontologically for the distinction between 'Nature' and 'spirit'. (p. 123) This is to say, Heidegger thinks that Descartes' greatest error as the traditional theorist of rationalist epistemology is separating "I think" [cogito] from "I am" [sum], and ultimately the inevitable result is separating beings from Being and spirit from Nature as well. In opposition to this, Heidegger insists that beings are only concrete things, while Being is the process of things becoming illuminated [cleared] out of concealment through interpretation, so the two cannot rigidly split apart. This ontological difference of Being and beings can only be successfully crossed with access to the method of phenomenology.

[38] Ibid., p. 180.

Phenomenology first emerged in 20th century Germany, founded by Edmund Husserl (1859–1938). Husserl's *Logical Investigations* represents the emergence of contemporary phenomenology. Its most important contribution was revealing the possibility of a new method of philosophical thought or a more primordial horizon of looking at philosophical problems. By means of "suspending" all substances (objective objects and subjective ideas), phenomenology returns to the most primordial "intentionality" of cognitive activity, making the phenomenon present its essence in the process of intending to it, thereby attaining "eidetic intuition." This is "the phenomenological reduction," during which the intentionality of subjective perception has an enormous constitutive function such that "the subjectivity of constituting" becomes the chief theme of Husserl's phenomenology. In the famous *Cartesian Meditations*, Husserl proposes the important idea of "inter-subjectivity," which is that the "egos" in intentional acts coincide with the phenomena that the ego constitutes, and so all relations found in intentionality become relations "between subjects."

Heidegger added depth to and reformed the "marginal region" within the structure of intentional constitution that Husserl discovered. Heidegger gave it ontological meaning in the discovery of ontological phenomenology. He replaced the phenomenon of consciousness structured by the transcendental subject with Being, and made phenomenology become the search for the meaning of Being. A fact itself is not abstracting some thing out of its connections with human being and with the world, but rather it is going back into the field out of which it originally emerges in order to show the implications of its Being in the horizon of its initially connecting with human being and the world. Thus, so-called "face the facts" becomes "return to Being," and what it places in suspension is beings. It is precisely this long forgotten and unquestioned "Being" that is what phenomenology brings into facts as the theme under consideration. "With regard to its subject-matter, phenomenology is the science of the Being of entities—ontology." (*Being and Time,* p. 61) "Phenomenology is our way of access to what is to be the theme of ontology, and it is our way of giving it demonstrative precision. *Only as phenomenology, is ontology possible.*" (Ibid., p. 60) "Ontology and phenomenology are not two distinct philosophical disciplines among others. These terms characterize philosophy itself with regard to its object and its way of treating that object." This is to say, for Heidegger, investigating the problem of Being demands the means of phenomenology, and the principle of phenomenology itself demands that the problem of Being is its origin and aim, the two produce one another, glorify one another and together make up philosophy itself. Thus, with reference to the phenomenological methodology as the ground, "the return to the human's Being" is but the return to the human being's beginning and the genuine starting point of aesthetics.

(3) **Resources of Thought: Meditations about Technology**

In the context of the technological age, modern technology is the obstacle of thought and is destined to be overcome and sublated by thought; it is also clear evidence of a philosophical turn and the ultimate possibility of a philosophical turn. People's reflection on technology is actually the manifestation of humanity's current existential predicament in thought, and is thus unavoidable. Heidegger's meditation on technology is not opposition to technology, but rather the attempt to peer into the essence of technology.

1. The origin of technology: as one of the ways of the disclosing of Being

In the eyes of ancient Greeks, technology was not only a means, but was rather a way of showing that enables something to naturally emerge as itself. In this showing or disclosing, the holistic linkages of human being, Nature and world are still there. It uncovers Being as "presence." But this "presence" is the presence of temporalized emergence, and not by any means an eternally enduring presence. "Technology is the phenomenon in the ontological sense; it is essentially more appropriative than human being as subject, and hence more powerfully and deeply participates in shaping human being's realm of historical appropriation," (*Commentary on Heidegger*, p. 283).

2. The Essence of Modern Technology: En-frame (Ge-stell)

Heidegger's eco-ontological aesthetics contains extremely rich significations. He first ruthlessly critiques the Western modern age's anthropocentrism that took shape out of technology's "provocation," pointing out: "The men of this earth are provoked by the absolute domination of the essence of modern technology, together with technology itself, into developing a final world-formula which would once and for all secure the totality of the world as a uniform sameness, and thus make it available to us as a calculable resource. The provocation to such making-available orders everything into a single design, The togetherness of the four 'voices of destiny' no longer rings out."[39] Modern technology's so-called "provocation" is the dominating violence or powerful oppression that humanity abusing technology to limitlessly plunder and destroy Nature gives shape to, the result of which "levels the harmony of the infinite relation" between humanity and Nature, and "the togetherness of the four 'voices of destiny' no longer rings out," namely the fourfold game of sky, earth, gods and man. This necessarily results in the "anthropocentrism" of humanity attempting to "order" the space of the universe. Heidegger points out: "Europe's technological-industrial domination has already covered the entire earth. On the other hand, the earth, as a planet, has already been included in the interstellar-cosmic space which is placed at man's disposal by the planned projects of man. The earth and heaven of the poem have vanished. Who would dare say whither? The in-finite relation of earth and heaven, man and God, seems to have been destroyed."[40]

What the enframing of modern technology meets with is none other than human being's fragments and signs, and every material being is abstracted into a calculable market-price. Human being becomes the subject of the world, while the world devolves into an objectified object. Unbounded exploitation not only deprives Nature of the protection it deserves, but also deprives human beings of the protection they deserve as they are eventually replaced in the well-ordered working of modern technology. Ecological aesthetics is dedicated to improving the continuously worsening

[39]Heidegger and Hoeller [16], p. 202.

[40]Ibid., p. 200.

relationship of tension between humans and Nature, and so must thoroughly reflect on modern technology. The goal of reflection is not to oppose science and oppose technology, but to delve deeper into its origins and peer into the essence and mystery of technology in order to provide some possibility of transcending and overcoming modern technology. Heidegger's meditation on technology is undoubtedly one of the most important resources of thought for ecological aesthetics.

(4) **The Horizon of Time**

1. Invoking time was Heidegger's most important contribution to theory and had epoch-making significance.
2. In his later thought on Being, Heidegger places linguistic, poetic and artistic disclosing in opposition with technological disclosing, because the closeness that language, poetry and art bring about includes the original primordial time; and the reason why authentic poetry and art is so important is because reserved in what they uncover is the chance, destiny and primordial experience of the emergence of Being;
3. Metaphysics forgets and covers up primordial time, and technology becomes pre-fabricated technology, which does not mean that it does not have its own view of time, but rather that metaphysics presupposes the temporality of forgetting the emergence of Being (primordial time) or abstracts from the primordial time of Being and rigidly sticks to the chanceless present of the now. Modern technology cannot return to the temporal realm of the emergence of Being, due to forgetting Being and abstracting from the primordial realm of time, but on the contrary, by means of this forgetting and abstracting, takes beings as Being itself, rigidly holds onto the ready-made without temporality, while covering up the connection of temporality between humanity and Nature and continuously consuming the essence of beings to exhaustion through the grasping (researching and probing) of beings.
4. Returning to conditional and temporalized primordial time is one of the most important paths on which ecological aesthetics has made breakthroughs in research and building.

The actual foundation that ecological aesthetics proposes is this present condition of global ecological crisis brought about by modern technology's ruthless destruction of ecology.

Forgetting or dispossessing original time and submerging into the sequence of presents of ordinary time (the time of physics), while viewing everything (affairs, things) as a lifeless, ready-made state is an extremely important characteristic of traditional metaphysics and modern technology. Therefore, any reflection on them must begin with the time that they forget or cover up. Only in this way can we fundamentally overcome or twist out the defects and insufficiencies that they engender with respect to beauty and aesthetic problems.

Many facts indicate that the formation of Heidegger's perspective of ontological philosophy and especially his later fundamental turn from anthropocentricism to eco-holism and Eastern culture was closely interrelated with the influence of Chinese

Daoist thought. From the 1930s onward Heidegger was already rather fluent in applying the Daoist thought of Laozi and Zhuangzi. Heidegger's ecological thought about "the fourfold game of sky, earth, gods and man" can be traced back to section 25 of the *Laozi*, "*Dao* is great, heaven is great, the earth is great, and the [sage] king is also great; there are four greats in the world and the [sage] king reigns as one of them." He also used section 28 of the *Laozi*, "know the clear, yet safeguard the dark," in order to explain his thought of "moving from concealment to clearing," and used section 11 of the *Laozi*, "[t]he thirty spokes converge at one hub, but the utility of the cart is a function of the nothingness of the hub," in order to explain his distinction between "beings" and "Being." In other words, he uses the spokes of a cart's wheel converging at the central emptiness of the hub, which alone enables it to turn as a metaphor for Being, whose absence alone is what enables it be useful. In addition, Heidegger also uses the first section of the *Laozi*, "[t]he *dao* that can be spoken is not the eternal *dao*," in order to explain his "[s]aying is not speaking"; he uses "the greater use of the useless" to explain his thought that "dwelling" is without utility; he uses Zhuangzi's dialogue with Huizi while wandering over a bridge, where they argue about "the joy of fish," in order to explain how it is impossible from an ordinary standpoint to understand the joy of fish swimming freely in water and that one can only get a feeling for this point from the perspective of ontology, by means of which he illuminates the distinction between ontology and epistemology. He says in *A Dialogue on Language*, "a nature of language can reach the thinking experience, a nature which would offer the assurance that European-Western saying and East Asian saying will enter into dialogue such that in it there sings something that wells up from a single source [*einzigen Quelle*]." (*DL* 8/93–94) Chinese philosophers also speak of the ecological idea in Heidegger's aesthetics as "an explanation of *language*."

References

1. [U.S.] Merchant, C. (1999). *The death of nature—Women, ecology and the scientific revolution*, translated by Wu Guosheng et al. Jilin People's Publishing House, "Foreword".
2. [French] Foucault, M. (2001). *Words and objects—Humanities and archaeology*, translated by Mo Weimin. Shanghai Sanlian Bookstore.
3. [French] Derrida, J. (2001). *Writing and differences*, translated by Zhang Ning. Sanlian Bookstore.
4. Huxiang, F. (1996). *Human, nature and culture: A comparative study of Chinese and Western environmental philosophy*. People's Literature Publishing House.
5. Marx and Engels. (1972). *Selected works*. People's Publishing House (《马克思恩格斯选集》,人民出版社1972年版).
6. Nash, R. F. (1999). *The rights of nature: A history of environmental ethics*, translated by Yang Tongjin. Qingdao Press.
7. Leopold, A. (1997). A Sand County Almanac, translated by Hou Wenxuan. Jilin People's Publishing House.
8. Nuo's, W. (2011). *European and American Ecology literature*. Peking University Press.
9. Yi, L. (2001). *Deep ecological thought research*. Tsinghua University Press (雷毅:《深层生态学研究》,清华大学出版社2001年版).

10. [U.S.] Rolston, H. (2000). *Environmental ethics: The value of nature and human obligations to nature*, translated by Yang Tongjin. China Social Sciences Press.
11. (English) Alexander, S. (2000). *Art, value and nature*, translated by Han Donghui et al. Huaxia Publishing House.
12. [U.S.] Carson, R. (1979). *Silent spring*, Lu Ruilan, Li Changsheng. Science Press.
13. [U.S.]Ray Griffin, D. (1998). *Postmodern spirituality and society,* translated by Wang Chengbing. Central Compiler Publishing House.
14. Heidegger, M. (1962). *Being and time*, translated by John Macquarrie & Edward Robinson. Oxford, Blackwell Publishers Ltd.
15. [U.S.] Fromm, H. (1996). *From transcendence to degeneration: A road map*. Containing in *Ecological Criticism Reader*, University of Georgia Press.
16. Heidegger, M., & Hoeller, K. (Eds.). (2000). *Elucidations of Hölderlin's poetry*. Amherst: Humanity Books.
17. Heidegger, M. (2011). *Basic writings*. Routledge Classics.

Chapter 3
The Literary Context of the Birth of Ecological Aesthetics

In addition to the birth of ecological aesthetics sharing a relationship with the modern social economy and modern cultural philosophy, it is also very much intimately tied to the development of ecological literature. The rise and development of modern ecological literary critique not only promoted the birth of ecological aesthetics, but also injected an abundance of theoretical resources and practical experiences into the founding of ecological aesthetics. Literature is a form of aesthetics that is unique to humanity due to relying on language, and the extension of ecological theory into literature greatly propelled the birth and development of ecological literature. In this chapter, we sort through the birth, intension and extension of ecological literary critique in the modern West.

3.1 The Literary Foundation of the Birth of Modern Ecological Critique

The entirety of Western literature has been guided by "anthropocentrism," and especially since 18th century enlightenment thinking, the principle of humanism has gradually become the most basic principle of Western literature as the concentrated manifestation of "anthropocentrism." But at the same time, there has been a branching of ecologism in Western literature since the 18th century.

Literary romanticism in the late 19th century West sang the praise of Nature and glorified a plain and peaceful way of life, manifesting a distinct ecological tendency. The poet of the British Romanticism and the representative of the Lake Poets, William Wordsworth (1770–1850), lived reclusively among the natural hills and waters of the western central region of Great Britain, and created many ballads praising Nature and spring. For instance, he wrote in the poem "The Tables Turned":

F. Zeng, *Introduction to Ecological Aesthetics*,
https://doi.org/10.1007/978-981-13-8984-9_3

One impulse from a vernal wood
May teach you more of man,
Of moral evil and of good,
Than all the sages can.

Sweet is the lore which Nature brings;
Our meddling intellect
Mis-shapes the beauteous form of things:–
We murder to dissect.

Another of the Lake Poets, Coleridge (1772–1834) tells a story in his famous longer narrative poem entitled *The Rime of an Ancient Mariner* about a mariner killing an auspicious bird, an albatross, and punished by nature, the ship runs into unthinkable hardship, the crew dies off, and with no food and water but unbearable suffering, he prays to God, and is only saved after blaming himself in deep remorse. Here, this mariner sails around, heeding humanity to respect Nature through the personal experience of it. This poem has been declared the English literary world's "greatest ecological fable" by the eco-literary researchers Neil Roberts and Terry Gifford.[1]

Here, we should specifically introduce two American ecological literary artists and their literary works on ecology. Their achievements have become the direct sources of Western ecocriticism and especially of the birth of American ecocriticism.

Henry David Thoreau (1817–1868) is the famous American forerunner of ecological theory and ecological literature. His work, *Walden*, became the inspiring work behind modern ecological philosophy, ecological ethics and ecological literature, and Thoreau himself became "the greatest ecological writer of the romanticism age." Thoreau's history is quite simple. He studied at Harvard, and after graduating went to his hometown middle school to teach for two years, after which he also became a follower and assistant of the famous writer, Emerson. From the March of 1846 to the May of 1847, he went to his hometown's Walden lake carrying an axe in hand, and lived in solitude for 26 months through the simplest forms of work. During this period, he contemplated human life, came to realizations about Nature, and wrote out *Walden*, which interweaves prose, philosophy and scientific observations in extraordinary fashion. After this book was published in 1854, it was met with massive feedback. Thoreau passed away in 1862 from tuberculosis at the age of 44.

Walden is an extraordinarily interesting book. By the middle of the 19th century, the industrial revolution was advancing full speed ahead, and people were fully immersing themselves in material pleasures, but Thoreau uniquely had the prescience to discern the disastrous destruction of Nature brought about by the industrial revolution, and insightfully probed the authentic relationship of human being to Nature in advocating a pure and primitive life of partnership with Nature. He experienced Nature and entered into dialogue with Nature with the unique profundity of a philosopher, the sensitivity and loving passion of a poet, and a delicate stroke of the pen, leaving behind a series of eternal adages that caution following generations, while

[1]Murphy et al. [1].

enriching humanity. The renowned American eco-critic Lawrence Buell is a specialist on Thoreau, and wrote many investigative treatises with many discoveries in them. Based on our reading experience, The true inspiration that Thoreau gives us in *Walden* may be summarized as: humanity should always live in close partnership with Nature.

Thoreau's time was still the heyday of American capitalist industrialization, people had an inexhaustible desire for material pleasures, and heading to cities and civilization in the search for money became a fashion and trend. Thoreau graduated in the prestigious Harvard University, so although he had the conditions to seek material pleasures, he on the contrary abandoned all of them, headed to the expansive Walden pond resolutely with axe in hand and felled trees to make a shelter, planted legumes and vegetables, and fished on boat. He remarked: "It was a pleasant hillside where I worked, covered with pine woods, through which I looked out on the pond, and a small open field in the woods where pines and hickories were springing up."[2]

The reason why he did this came out of resistance to the materially decadent life of the times; he thus chose a primitive and pure way of healthy life that cherishes Nature. He remarked: "It would be some advantage to live a primitive and frontier life, though in the midst of an outward civilization."[3] Thoreau believed that humanity—so caught up in the inflation of material desires and the fetishistic worship of money—"is sick." He attempted to cure humanity through this primitive and authentic life, "If, then, we would indeed restore mankind by truly Indian, botanic, magnetic, or natural means, let us first be as simple and well as Nature ourselves, dispel the clouds which hang over our own brows, and take up a little life into our pores."[4] What kind of life ultimately would his simple life be like? That would be working no longer than 6 weeks a year, but making enough to cover all of life's expenses with the remainder of time open to read and experience Nature while relying solely on one's two hands, a knife, an axe, a hoe, a push cart and physical effort. Thoreau insists: "Most of the luxuries, and many of the so-called comforts of life, are not only not indispensable, but positive hindrances to the elevation of mankind. With respect to luxuries and comforts, the wisest have ever lived a more simple and meagre life than the poor."[5] Moreover, extraordinarily noble was that Thoreau did away with any private property in such a society in which the goal of life was money. He fervently loved the beautiful garden scenery, but did not however take the garden bought as his own private property. He remarked: "It makes but little difference whether you are committed to a farm or the county jail."[6] Such a state of mind in those times was far from ordinary.

---

[2]David Thoreau [2], p. 36.

[3]David Thoreau [2], p. 10.

[4]Ibid., p. 72.

[5]Ibid., p. 12.

[6]Ibid., p. 78.

3.2 Treating Nature as Your Neighbor and Friend, Experiencing Nature and Speaking with Nature

Thoreau delicately described the lakeside scenery of Walden Pond in *Walden*; he lived in a wooden shack and treated Nature as a neighbor, spoke with Nature and directly experienced many profound realizations arising from Nature, relating his fervent love for Nature and his neighborly and friendly concern for Nature. He wrote about several of his neighbors. "Such was not my abode, for I found myself suddenly neighbor to the birds."[7] He also wrote about other neighbors, moles, plants, deer, bears, etc. Of course, for him, his most important neighbor was Walden Pond itself. He lived beside the pond, lived there and grew there, and was filled with deep passion for Walden Pond. He said, "This small lake was of most value as a neighbor in the intervals of a gentle rain storm in August."[8] With a pen soaking with emotion, he made a deeply passionate description of Walden lake. "Even then it had commenced to rise and fall, and had clarified its waters and colored them of the hue they now wear, and obtained a patent of Heaven to be the only Walden Pond in the world and distiller of celestial dews. Who knows in how many unremembered nations' literatures this has been the Castalian Fountain? Or what nymphspresided over it in the Golden Age? It is a gem of the first water which Concord wears in her coronet." (The Ponds) [9] "It is earth's eye; looking into which the beholder measures the depth of his own nature."[10] He considers Nature to be the mother, and in accepting the gift of Nature's selfless giving, he always felt thankful of Nature. "[The sand cherry] in the fall weighed down with good sized and handsome cherries, fell over in wreaths like rays on every sode. I tasted them out of compliment to Nature, though they were scarcely palatable."[11] Thoreau experienced Nature and spoke to Nature precisely with such a sense of thanks, friendship and equality. In the stillness of dusk and night, he focused his ear to listen to all of the sounds of Nature, and considered such sounds to be of incomparable beauty, the "flutes of Heaven" unheard of in the human world. He distinguished all kinds of sounds with delicate hearing and feeling, and enjoyed this incomparably beautiful music. When describing the call of the nightingales, he remarks: "Sometimes I heard four or five at once in different parts of the wood, by accident one a bar behind another, and so near me that I distinguished not only the cluck after each note, but often that singular buzzing like a fly in a spider's web, only proportionally louder."[12] He was even capable of making out the meaning of the nightingale's call: "*Oh-o-o-o-o* that I never had been *bor-r-r-r-n!* sighs one on this side of the pond, and circles with the restlessness of despair to some new perch on

[7]Ibid., p. 80.
[8]Ibid., p. 81.
[9]Ibid., p. 169.
[10]Ibid., p. 174.
[11]David Thoreau [2], p. 106.
[12]Ibid., 116.

the gray oaks."[13] Thoreau became a friend of Nature who can discern the language of birds, "I was suddenly sensible of such sweet and beneficent society in Nature."[14]

In summary, Thoreau immersed himself completely in Nature, and considered himself to be a part of Nature, remarking: "I go and come with a strange liberty in Nature, a part of herself." Precisely because of this, he became aware of a sweet, warm, healthy and uplifting feeling in Nature: "There can be no very black melancholy to him who lives in the midst of Nature and has his senses still. There was never yet such a storm but it was Aeolian music to a healthy and innocent ear."[15]

3.3 The Critique of the Over-industrialization and Anthropocentricism that is Hostile to Humanity and Nature

In the middle of the 19th century, America experienced the brutal primitive accumulation of capital, and industry rapidly developed, but Thoreau sharply saw the unrestrained stripping away of Nature as well as the hostility to humanity and Nature behind industrialization, remarking that life is too luxurious, "Of a life of luxury the fruit is luxury, whether in agriculture, or commerce, or literature, or art." An important manifestation of this life of unrestrained luxury is hostility to Nature, the unrestrained robbing of Nature, abusively felling trees for wood, and destroying the great earth, while at the same time exploiting the blood and sweat of human beings. He cites the example of the railway cutting across the United States, which he thought was actually convenient transportation at the cost of the fresh blood and life of uncountable working men, "Did you ever think what those sleepers are that underlie the railroad? Each one is a man, an Irish-man, or a Yankee man."[16] Thoreau himself saw through to the fact that the rich minority representing the capitalist class exploits the majority, and so intentionally disobeyed and refused to pay taxes. He wrote: "I did not pay a tax to, or recognize the authority of, the state which buys and sells men, women, and children, like cattle at the door of its senate-house."[17] Of course, due to going overboard on ecologism, Thoreau did not understand many aspects of modernization as the trend of historical development, which is perhaps where his one-sidedness and limitations are to be found. He stated: "Men think that it is essential that the *Nation* have commerce, and export ice and talk through a telegraph, and ride thirty miles an hour, without a doubt, whether *they* do or not."[18] ("Where I Lived, and What I Lived For") But as the forunner of ecological theory and ecological literature, Thoreau's affinity with Nature, his fiercely defended way of life of "simplicity, simplicity, aye

[13]Ibid., pp. 116–117.

[14]Ibid., p. 123.

[15]Ibid., pp. 122, 124.

[16]Ibid., p. 87.

[17]Ibid., p. 161.

[18]David Thoreau [2], p. 87.

simplicity,"[19] and his aphorisms singing the glory of Nature have eternally become our prized spiritual heritage, holding inspiring significance for us humans today who life in extreme opposition to Nature.

3.4 Lighting the Way with the Ancient Chinese Confucian Thought of "Humane Love," and Illuminating the Path of Partnership between Humanity and Nature

Thoreau borrowed the light of Eastern ancient wisdom, and especially the ancient Chinese Confucian thought of "humane love" to elucidate his viewpoint that humanity should befriend and neighbor Nature. In the chapter "Spring," he applies Mencius' "theory that human nature is good" to show that the beauty of Nature is the beauty of the original nature of humanity, and that treating nature with care is the original virtuous nature of human being, citing "Gaozi I" to elucidate the thought that the beauty of Nature as well as the humane treatment of Nature are determined by the originally good tendency of human nature. "A return to goodness produced each day in the tranquil and beneficent breath of the morning, causes that in respect to the love of virtue and the hatred of vice, one approaches a little the primitive nature of man, as the sprouts of the forest which have been felled. In like manner the evil which one does in the interval of a day prevents the germs of virtues which began to spring up again from developing themselves and destroys them." "After the germs of virtue have thus been prevented many times from developing themselves, then the beneficent breath of evening does not suffice to preserve them. As soon as the breath of evening does not suffice longer to preserve them, then the nature of man does not differ much from that of the brute. Men seeing the nature of this man like that of the brute, think that he has never possessed the innate faculty of reason. Are those the true and natural sentiments of man?" After this, Thoreau took care to write a poem to elucidate Mencius' thought:

"Not yet the pine felled on its mountains had descended
To the liquid waves that it might see a foreign world,
And mortals knew no shores but their own.
There was Eternal spring, and placid zephyrs with warm
Blasts soothed the flowers born without seed."

This is a time of the early creation of the human age, spring gusts ripple, gentle breezes stir, and the flowers of Nature freely bud forth, human being and Nature harmoniously mingle, the age of friendship with Nature is Thoreau's ideal age, the utopia in his bosom.

Thoreau also used Confucius' thought of humane love to illuminate the ideal of harmonious co-existence between humans and nature as well as that between humans

[19]Ibid., p. 86.

themselves. He quotes the *Analects* 12.19, which records Confucius speaking to Ji Kang, "You who govern public affairs, what need have you to employ punishments? Love virtue, and the people will be virtuous. The virtues of a superior man are like the wind; the virtues of a common man are like the grass; the grass, when the wind passes over it, bends." ("The Village") Using this to demonstrate, Thoreau suggests that a simply life of partnership and friendship with natural society ought to stem from the humane governance arising from the affective humane core of humane virtue, which corresponds precisely with the Homeric Hymn:

Nor wars did men molest,
When only beechen bowls were in request.

Thoreau's thought of humanity neighboring Nature is an important source for contemporary thought on the aesthetics of ecological ontology. Thoreau's thought about humanity in friendly partnership with nature is advocating a completely new way of life and existence unlike that of the age of industrialization, in which humanity dichotomizes and opposes subject and object, but what Thoreau was advocating was a way of life and existence wherein human being is there in the friendly neighborhood of Nature intimately bonded to the world. The former is the epistemological way of life and existence and the latter is the existential way of life and existence; the former necessarily results in the opposition of humanity and nature, thereby gradually worsening humanity's way of existing, while the latter involves gradually getting a better life in the friendly co-existence between humanity and Nature. This way of existence of "being there closely bonded to the world" is what Heidegger called "Being-in-the-world." Here, "Being-in" does not refer to something being inside something in the spatial sense, but means rather "'to reside,' '*habitare*', 'to dwell' [sich auf halter.]. '*An*' signifies 'I am accustomed', 'I am familiar with', '1 look after something'[20] This shows that the relationship of human being to Nature is not that of objectivity, but rather of inhabiting; human being and Nature are a whole and inseparable. Human being succeeds in existing and gradually unfolds her own authenticity in the direction of poetic dwelling precisely in this inhabiting relationship, which is an eco-existential kind of aesthetic existence. Thoreau was undoubtedly the forerunner in advocating this ecological aesthetic way of existing and idea of beauty.

(1) **Leopold and *A Sand County Almanac***

Aldo Leopold (1887–1948), the founder of modern American ecological theory and ecological literature. His famous work, *A Sand County Almanac*, put forward such famous ideas as "thinking like a mountain," an "ecological community" and "earth-ethics." Leopold was born In Iowa, United States, in the November of 1887, and received a master's degree in forestry at Yale, after which, he once served as the vice-principle of the National Forestry Forest Laboratory, and then went to Wisconsin University to serve as professor of Wildlife Management until he passed away. He fervently loved Nature, researching it and coming to deep realizations of it. In 1935, Leopold bought a piece of discarded farmland on the bank of the Wisconsin River, and

[20]Heidegger [3], p. 64.

in the following several decades, he, his wife and several children utilized weekends and break time to live in an extremely beaten-up wooden shed on that farmland, devoting themselves to farming, observation and experiments. He wrote down his own work and realizations, namely in the book *Sand County Almanac*. In April of 1948, Leopold died of heart disease in the process of organizing and conducting a brigade to extinguish a fire disaster, living a total of 61 years. In 1949 *Sand County Almanac* was published by Oxford University Press, and became an important classic of ecological theory and ecological literature.

1. The shift from capitalist conservationism toward ecological holism. As an official representative of the state government forestry bureau, Leopold held the same position as the government's capitalist conservationism, but here the so-called conservation of capital still treats Nature, deals with natural resources, and makes the distinction between "useful" and "useless" natural resources from the position of anthropocentrism and even from the temporary and narrow profit of humankind. All of this betrays the "ecological laws' of reason for being of any natural form, the result of which is necessarily bringing about ecological imbalance and even the more serious consequence of destruction of the ecosystem. For example, hunting and killing a high quantity of wolves in order to eat the deer meat that wolves eat results in the overpopulation of deer, which ultimately makes it impossible for dear to survive. Another example, conserving "Nature" and "wilderness" for the purpose of human "recreation" still upholds the dichotomized opposition between Nature and humanity, the result of which is necessarily bringing about the destruction of Nature and wilderness.[21] Leopold points out: "To him who seeks something more, recreation has become a self-destructive process of seeking but never quite finding, a major frustration of mechanized society."[22] Leopold shows from the height of an ecological scientist and ecological philosopher that a plant species is a "history book." Because this plant species lives through a long history growing even longer than humankind, it not only testifies that the earth and human history but also that itself also varies in symbiosis and competition with other species. A species of animal or plan being destroyed is equivalent to burning this content rich history book until nothing but ash remains. Yet, modern industrialization along with the people who seek material pleasure do not care about such aspects, ripping to shreds every such precious book of history without even wincing at the annihilation of species after species of animal for the sake of humanity's own so-called progress. Leopold angrily states: "Mechanized man, oblivious of floras, is proud of progress in cleaning up the landscape on which, willy-nilly, he must live out his days."[23]
2. The "land ethic" of the "ecological community." In ecological theory, Leopold both criticizes "anthropocentrism" while distancing himself from "eco-centrism," putting forward the thought of "the land ethic" of "ecological community." He states: "protectionism is only possible when people assume the role

[21] Leopold [4], p. 156.

[22] Ibid., p. 157.

[23] Ibid., p. 43.

of a citizen in a community where soil, water, plants and animals are all part of each other; in this community, each member is interdependent, each Members are eligible to occupy a position in the sun." [24]This is actually a new extension of contemporary ethics. Since traditional ethics only touched on human to human relations, and later extended to human to society relations, it now must extend to human and land (nature) relations. Leopold points out: "The land ethic simply enlarges the boundaries of the community to include soils, waters, plants, and animals, or collectively: the land."[25] Leopold never denies humanity's right to utilize the land (nature), but he insists that we must consider the being of the land (nature) from the perspective of the "community," that is "whole." He states: "A land ethic of course cannot prevent the alteration, management, and use of these 'resources,' but it does affirm their right to continued existence, and, at least in spots, their continued existence in a natural state."[26] Of particular value, Leopold raised the important concept of "the health of the land" and healthy rivers, which gives the biotic intension of life to the land (Nature), just as we speak of the "healthy Yellow River" and "healthy Yangtze." Leopold believed: "A land ethic, then, reflects the existence of an ecological conscience, and this in turn, reflects a conviction of an individual responsibility for the health of the land."[27] What is important is that the "community" of Leopold's idea of "land ethic" is established on the theoretical basis of the "land pyramid" of the "food chain." He states: "The image commonly employed in ecology; the biotic pyramid. I shall first sketch the pyramid as a symbol of land, and later develop some of its implications in terms of land-use." [28] This pyramid tells us that plants absorb energy from the sun, the layer underneath plant life is the soil, the layer growing on top of plants is that of insects with horses and rodents forming another layer on top of insects, predatory animals on top of them, and humans living at the apex of the pyramid. Every substratum provides food to the stratum above it, but the higher the stratum the shorter the food chain of that stratum is, and thus the more fragile it is. This pyramid makes up a recycling community. The earth, soil and plants are the base of the pyramid, the base of energy production and availability. From this perspective, is humanity truly separable from the earth? Is it really possible to put the being of the earth out of consideration? This is a scientific judgment, and also a humanistic judgment, a judgment that relates to the long-term better survival of humankind.

3. The problem of the unique value of "wilderness." Leopold brings up the problem of the value of "wilderness" when criticizing "anthropocentrism" and "environmental conservationism." He states, "Hence the wilderness that he cannot personally see has no value to him. Hence the universal assumption that an unused hinterland is rendering no service to society. To those devoid of imagination, a

[24]Ibid., p. 216.

[25]Ibid., p. 193.

[26]Leopold [4], p. 194.

[27]Ibid., p. 209.

[28]Ibid., p. 203

blank space on the map is a useless waste; to others, the most valuable part."[29] The chapter of his that specifically deals with wilderness, namely "Wilderness" details the recreational value, scientific value and wildlife conservation value of wilderness, setting up the theoretical foundation for Rolston's book *Philosophy Gone Wild*.

4. Bringing up the unique way of thought from the perspective of ecological holism—"Thinking like a mountain." Practice proves that the emergence of ecological problems and environmental pollution is a problem of cultural attitude, namely a problem of whether we should proceed from the cultural attitude of "anthropocentrism" or rather from that of "ecological holism." Answering this, Leopold brings up the unique way of "thinking like a mountain" according to ecological holism, but he does not have a specific chapter devoted to "Thinking like a mountain" in *A Sand County Almanac*. Leopold brings up this thesis when criticizing the erroneous viewpoint that "the less wolves there are, the more deer there are." He argues, when looking at the eyes of that "old wolf" shot near to death by the hunter, he deeply senses that whether it is a wolf or a mountain, none would agree with this idea.[30] Since this idea that "the less wolves there are, the more deer there are" betrays the law of ecological holism, and necessarily results in ecological destruction and environmental degradation. Because of this, Leopold warns humanity: "He has not learnt to think like a mountain. Hence we have dustbowls, and rivers washing the future into the sea."[31] Leopold's "Thinking like a mountain" also insists that humanity ought to engage in dialogue with Nature and experience Nature on an equal footing. When describing his own experience of a sand county, he vividly writes about listening to animals congregating in the marshes on an April night, when he heard the sound of owls hooting, the sound of the isabelline wheatearflapping its wings, the clucking of long-billed dowitchers, the sharp ear-splitting calls of wild geese, and finally the sound of large geese mingling. He felt that if he were truly a part of Nature, he could hear even more sounds, "Once again, I would I were a muskrat!"[32] (p. 22) The description of the Draba's experience of needing a little warmth, the appreciation of two sandhill cranes dancing midair, the consideration of each and every plant and animal in sand county as "tenants" and the focus on the "birthday" of each and every plant all show Leopold treating every being in Nature on an equal footing, and living "symbiotically" with them.

[29] Ibid., p. 166.

[30] Ibid., p. 123.

[31] Leopold [4], pp. 123–124.

[32] Ibid., pp. 21–22.

3.5 The Historical Emergence and Development of Ecological Literary Criticism

The scholarly world generally recognizes the publication of Rachel Carson's *Silent Spring* in 1962 as the commencing of literary ecological criticism in the modern West. But the true beginning of ecological literary criticism should be the publication of Joseph Meeker's *The Comedy of Survival: Studies in Literary Ecology* along with the proposition of the main idea that "human beings are the earth's only literary creatures." Literary "ecological criticism" was proposed by the American ecological critic William Rueckert. He published his thesis entitled *Literature and Ecology: an Experiment in Ecocriticism* in Iowa Review 9.1 (Winter 1978), where the term "ecocriticism was first used, naming it as an ecologically conscious literary activity; in 1985, Fred Waage edited, compiled and published *Teaching Environmental Literature: Materials, Methods, Resources*, which includes the class content of 19 scholars on "environmental literature," thereby opening the first river of cooperative research on ecocriticism; in 1989, the American Nature Writing Newsletter was founded, consisting of theses and book reviews about nature and the environment; in 1990, the University of Nevada set up the academic position of "literature and environmental research"; in 1991, the ecocritic Harold Fromm organized a special seminar entitled *Ecocriticism: The Greening of Literary Studies* in the MLA, which enabled American ecocriticism to gain widespread notice at a specialized academic conference; in 1992, the ecocritic Glen Love held a special talk entitled *American Nature Writing: New Contexts, New Approaches* at the American Literature Association. In the same year, the Association for the Study of Literature and Environment (ASLE) was founded, chaired by Scott Slovic. The establishment of this association was the mark that ecocriticism had been founded as a new school of literary research; in 1993, Patrick Murphy created the journal Interdisciplinary Studies in Literature and Environment; in 1995, the ASLE, the Association of Studies in Literature and Environment was held for the first time in the US, and Harvard University's Lawrence Buel published the work *The Environmental Imagination: Thoreau, Nature Writing and the Formation of American Culture*; in 1996, Cheryll Glotfelty and Harold From co-edited the collection on ecocriticism *The Ecocriticism Reader*; in 1998, the first American collected works on ecocriticism, *Writing the Environment: Ecocriticism and Literature*, was published; in 1999, *New Literary History*, Summer No., published a special edition on ecocriticism; in June of 2000, University College Cork of Ireland held an international seminar on ecocriticism. In October of the same year, Tamkang University of Taiwan organized the international conference on "Ecological Discourse"; in 2001, David Mazel edited and published *A Century of Early Ecocriticism*; in 2002, University of Virginia published *Explorations in Ecocriticism Series*. In March of the same year, the ASLE held the National Conference on New Bearings in Ecocriticism; in 2003, the ASLE held the fifth academic conference in Boston on the theme of "how ecological literature promotes environmental protection."[33]

[33]Gloverti [5], pp. 17–18.

The previous paragraph shows a brief history of the development of ecocriticism. Now the problem is, why is it that ecocriticism as a literary theory only truly emerged in 1978, and why it is still relatively immature as a field up till today. On this problem, I think, the forces of "anthropocentrism" are still too strong to break through in the field of literary theory and literary criticism; another reason is, the ecological view of literature itself involves more complications than ecological philosophy and ecological ethics. Since aside from the relationship of ecology to the humanities, it also involves the problems of literature and aesthetics, which makes things much more complicated. Because of this, no mature theoretical system has emerged yet up till now in this respect, and there is even the phenomenon of conflating ecological literature with realism in literature and sensationalism in psychology, which is precisely a manifestation of the immaturity of the critical theory of ecological literature. Regardless of how, ecocriticism from 1978 till now has taken a twisted path for 30 years, and in displaying overflowing vitality as well as a strong propensity for continuous growth and maturation, it has theoretically made many achievements indeed. However, there are still many problems that need to be solved, and at present what most especially needs summarization is the achievements of past research on the foundation of which new creations spring forth.

3.6 Principles and Characteristics of Ecocriticism

The famous American ecological literary theorist, Jim Schwab, asserts that ecocriticism is a cultural criticism. This proposition is both right and wrong. Ecocriticism actually represents a massive shift in cultural standpoints, but this shift brought about a change in aesthetic principles. Sharply dividing the problem of literature into external laws and internal laws is still a practice of dichotomized thought, which is unscientific. Ecocriticism is still in the midst of continuous growth at present, and we can only summarize its principles and characteristics on the basis of the materials currently at hand. When explaining the implications of ecological aesthetics, we already pointed out that the basic principles of ecological aesthetics are generally one and the same with those of ecocriticism. In order to avoid repetition, some content like consciousness of "place" require positioning it in the theoretical treatment of the implications of ecological aesthetics in order to give an account of it. On the basis of materials currently at hand, the principles of ecocriticism may be summed up into the following six aspects:

(1) **Ecocriticism is a Literary Critique that Envelops the Ecological Dimension**

When ecocriticism was first brought up, some theorists summarized it as a combination of ecology and literature. Of course, the "ecology" spoken of here should refer to "deep ecology" namely ecological philosophy. This summary is not nonsense, but a little to simple, and even runs the risk of rigidly wedding "ecology" as a natural science to a branch of humanities. Actually, saying that ecological literary criticism is a combination of ecological philosophy and literature is unproblematic. When ecocriticism was first brought up, scholars had no experience with it and unavoidably came up with opinions that were not entirely comprehensive, which is understandable. The first one to bring up the concept of "ecocriticism" was Rueckert who believed, "As should be clear by now, I am not just interested in transferring ecological concepts to the study of literature, but in attempting to see literature inside the context of an ecological vision in ways which restrict neither and do not lead merely to proselytizing based upon a few simple generalizations and perceptions…" [34] Rueckert emphasizes that "ecocriticism" is not a "proselytizing," namely an attempt to majorly shift the principle of critique. Of course, Rueckert also asserts that ecocriticism cannot restrict itself simply to making this shift, but also must put it into practice. Later, William Howarth would go onto define the meaning of the word "ecocritic," stating, " 'A person who judges the merits and faults of writings that depict the effects of culture upon nature, with a view toward celebrating nature, berating its despoilers, and reversing their harm through political action.' So the *oikos* is nature, a place Edward Hoagland calls 'our widest home', and the *kritos* is an arbiter of taste who wants the house kept in good order, no boots or dishes strewn about to ruin the original décor."[35] Lawrence Buell argues [in the Chinese edition of *Writings for an Endangered Earth*] that ecocriticism generally unfolds in the practical spirit of the environmental movement. In other words, the ecocritics do not only see themselves as people engaging in an academic activity, but more deeply concern themselves with the environmental crisis today, and many—although not all—also join different movements acting in the betterment of the environment. They also believe that the humanities and especially literary and cultural studies may make contributions to understanding and rescuing the environmental crisis.[36] In summary, ecocriticism differs from all of the past forms of criticism most importantly in that it includes the ecological dimension. Just as the ecocritic, Harold Fromm, put it: "Thus, 'the problem of the environment,' which many people persist in viewing as a peripheral arabesque drawn around the 'important' concerns of human life, must ultimately be seen as a central philosophic and ontological question about the self-definition of contemporary humanity." [37]

From this we can see that the shift of the ecological dimension into literary critique is a "change in belief," an important question involving philosophy and ontology. Previous to this, all models of literary critique that had flourished in the history of human literature and literary theory should generally be considered "anthropocentric." For instance, so-called social historical critique finds its orientation of values in traditional "humanism;" aesthetic critique also finds its criteria in such aesthetic ideas centered on the "theory of subject" as the traditional "theory of Ideas" and "theory of copying"; the new critique is a "text-centrism," but the "text" here is also human language and human text lacking all dialogue with non-human species, which have always been absent and aphasic; even though psychoanalytic critique

[34]Ruckert [6], p. 115.

[35]Howers [7], p. 69.

[36]Bury [8], p. 1.

[37]Fromm [9], p. 38.

emphasizes the irrational "libido," it ultimately can only enter the sphere of human control by undergoing "sublimation"; the critique of archetypes has an "anthropological" flavor, emphasizing the "collective unconscious," but it is still a description of the genesis of human consciousness. Even contemporary phenomenological and hermeneutic critique still emphasizes the "constitutive function of the subject." In sum, past forms of literary critique lack the "ecological dimension," and only ecocriticism is finally the first to enfold the ecological dimension into literary theory, and make it become the most basic cultural standpoint. This cultural standpoint is precisely the standpoint of contemporary ecological philosophy or rather the standpoint of contemporary ecological holism. Just as Leopold put it in *A Sand County Almanac*, "A thing is right when it tends to preserve the integrity, stability, and beauty of the biotic community. It is wrong when it tends otherwise."[38] This is a principle of contemporary ecological philosophy and ecological theory as well as a principle of contemporary ecocriticism.

(2) **Ecocriticism is the Manifestation of Contemporary Literary Workers' Moral Responsibility to Ecology**

Considering the production and practice of ecocriticism, it is a form of literary criticism that embraces contemporary literary workers' passionate sense of moral responsibility for ecology. The unity of literary criticism and moral responsibility for ecology is undoubtedly its most important principle. In this sense, ecocriticism is clearly demarcated from pure aestheticism's "art for art's sake" and "form for form's sake" spirit of critique. As is widely known, ecocriticism was born in the 1970s when environmental pollution was so severe and the ecological crisis was intensifying. Clearly, the emergence of ecocriticism is the manifestation of the ecological moral responsibility of literary workers who were filled with a sense of justice and "ultimately concerned" with humanity's end in the plight to "save the earth" and "save humankind." Just as the chief advocate of ecocriticism, William Luckhart put it, "[t]he problem now, as most ecologists agree, is to find ways of keeping the human community from destroying the natural community, and with it the human community. This is what ecologists like to call the self-destructive or suicidal motive that is inherent in our prevailing and paradoxical attitude toward nature. The conceptual and practical problem is to find the grounds upon which the two communities—the human, the natural—can coexist, cooperate, and flourish in the biosphere."[39] Clearly, the proposing of ecocritcism is for the sake of stopping the human community's destruction of the natural community from ultimately running into the grave aftermath of self annihilation. Another critic, Lynn White Jr reflected on the ways humanity has intensified its influence about the natural environment since the literary renaissance of the 14th century: using dynamite to extract minerals and burning to make charcoal brings about the erosion of soil and elimination of forests; the threatening force of the hydrogen bomb is enough to exterminate every biotic gene on earth; the growing population, chaotic urban development and increasing quantity

[38]Leopold [4], p. 213.
[39]Ruckert [6], p. 107.

of sewage waste even more so give the earth and human population unprecedented harm. In sum, "there are no other living things like human beings who mess up their own dwelling space to such a degree."[40] But how can this problem be solved? White answers: "What shall we do? No one yet knows. Unless we think about fundamentals, our specific measures may produce new backlashes more serious than those they are designed to remedy."[41] Eco-literary criticism is an important alternative founded on changing human being's cultural position and attitude toward Nature after thinking about fundamental problems; it is an entirely new path that differs from past attempts to influence the natural environment with civilization and restore the natural environment by civilizing. Its unique responsibility and contribution, as Glen Love put it, "is a sense of responsibility toward ecology."[42]

(3) **Ecocriticism is a Green Reading of Literature that "Reconstructs Values"**

Early on in the later part of the 19th century, Nietzsche announced the end of the cultural thought of rationalist philosophy, and proposed the world shaking and convention shattering "reevaluation of values" in his famous work *The Birth of Tragedy*. But today, beginning from the later part of the 20th century, the rise of ecological literature and ecocriticism has actually precipitated another new "reevaluation of values" in the field of literary arts. This reevaluation of values is of course a reconstruction of values as well, a reconstruction with "ecological" and "green" characteristics, so we call it "green reading." Lawrence Buell insisted that "we have no way to talk about green literature without green thought and green reading".[43] Glen Love wrote an article specifically for the rise of ecocriticism entitled *Revaluing Nature*, revaluing the literary description and expression of Nature, where he explicitly posed the problem of revaluing some literary and critical texts. He states: "our profession must soon direct its attention to that literature which recognizes and dramatizes the integration of human with natural cycles of life. The time cannot be far off when an ecological perspective will swim into our ken. Just as we now deal with issues of racism or sexism in our pedagogy and our theory, in the books which we canonize, so must it happen that our critical and aesthetic faculties will come to reassess those texts—literary and critical—which ignore any values save for an earth-denying and ultimately destructive anthropocentrism."[44] For this, he spoke of three points aimed at the future effect of the Western Literature Association: Western Literature will rank as the frontier of the redirection of criticism in accordance with expectations; revaluing Nature will be accompanied by the rearrangement of important literary genres; Western literature is not the only component within its own ecological perspective. In Glen Love's view, if we are to do away with anthropocentric values, we must adhere to the basic starting point of the cultural position and "green reading" of the ecological whole. Such a "green reading" will need to reevaluate the classics

[40]White [10], p. 5.

[41]Ibid., p. 5.

[42]White [10], p. 230.

[43]Bury [11], p. 1.

[44]Love [12], p. 235.

and authors. Of course, such revaluating and reading does imply negating history, but rather implies giving renewed interpretations to its contemporary value. For instance, in *The Comedy of Survival*, Joseph Meeker proceeds from contemporary ecological theory to give an entirely new explanation of the traditional theory of comedy. In the traditional theory of theatre, tragedy is a praise of higher spirits and heroic figures, while comedy focuses on minor characters. Because of this, tragedy ranks higher than comedy. But Meeker draws the conclusion that comedy ranks higher than tragedy from the contemporary perspective of ecological theory. He states, "Tragedy demands that choices be made among alternatives; comedy assumes that all choice is likely to be in error and that survival depends upon finding accommodations that will permit all parties to endure." Therefore, "comedy is essentially ecological".[45] Again, if we were to face the famous tragedies emitting unique auras in the history of literature during the literary renaissance starting with Shakespeare, we would say that we must make a "reevaluation of values" to give it a green reading. For instance, as the famous praise of human spirit from *Hamlet* goes: "What a piece of work is a man! How noble in reason, how infinite in faculty! In form and moving how express and admirable! In action how like an angel, in apprehension how like a god! The beauty of the world. The paragon of animals." This is clearly championing anthropocentrism. In today's world marked by humans wantonly destroying nature in the rapid expansion of material desires, we can no longer continue to affirm such anthropocentric praises, but we also cannot deny its great effect in liberating the human spirit from the religious repression of the Middle Ages and feudal despotism because of this. Reevaluating classics does not equal negating their significance. Green reading is a reading that aims at "symbiosis," "the whole" and "life," a reading that envelops all kinds of reading and models of critique. When explaining his own deep ecology, Naess says his ecological philosophy is only ecosophy T and there still needs to be an ecosophy A, B, C, etc., and need to supplement one another. Of course, to embrace is not to strangle, but it must not jettison "green" and "ecological" values.

(4) Ecocriticism Advocates an "Environmental Imagination" that Adheres to the Ecological Standpoint

Literature is an artistic form that appeals to the imagination, and creating images through imagination is precisely the mission and function of literature. The Famous American ecocritic Lawrence Buell forcefully advocated a theoretical idea of "environmental imagination," constituting one of the important theoretical principles of contemporary Western literary ecocriticism. His first work of ecocriticism was *The Environmental Imagination: Thoreau, Nature Writing and the Formation of American Culture*. Buell established four indicative elements of environmental imagination: 1. The presence of non-human environments does not merely act as a means of framing the background; 2. Human interests are not understood as the only legitimate interests; 3. Human responsibility for the environment is a component part of the ethical orientation of the text; 4. Nature is not a constant thing or presupposed

[45]Mick [13], p. 164.

fact.[46] Based on these four indicative elements, Buell's environmental imagination obviously tend to be colored by the values of ecological holism. But "environmental imagination" is not ultimately a neutral concept. It encompasses the two valuing orientations of "literary imagination," anthropocentrism and ecocentrism. With respect to environmental imagination proceeding from anthropocentrism, the poet Shelley's famous adage puts it, "poets are the unacknowledged legislators of the world." Buell insists that we have to think about anthropocentrism from the environmental perspective. Buell does not completely agree with "ecocentrism" either; he uses deep ecology as an example, and frames the ecocritic Jonathan Bate's words as the ecological dream that may never be completely realized on earth. He insists that for those living in society, what is of the highest priority is the concern of health, safety and subsistence, which is understandable. Clearly, what Buell holds to is still striking the medium between anthropocentrism and ecocentrism in "eco-humanism," which is perfectly noble. He also talked on the problem of the role of environmental imagination, insisting that it could foster connectivity between reader and natural environment in the following several ways: foster connections between reader, and human and non-human struggles and bitter experiences; incite readers to imagine different kinds of futures: promote their liberation from material desires, and so on.[47] Of course, Buell's theory of environmental imagination is contested in the scholarly world. Because of this, the word "environment" originally envelops anthropocentric implications in contrast to the word "Nature," so "environmental imagination" may possibly lead writers down the old beaten track of anthropocentrism.

(5) **The effect of ecocriticism is to make Nature's "negative entropy" become possible through "green reading"**

Researchers and practicers of eco-literary criticism are determined by the specific object of the earth and Nature that they face, they are also determined by attempting to resolve the specific mission of "environmental pollution" and again determined by their early aim of trying to "integrate literature and ecology," so in the theoretical construction and practice of eco-literary criticism, they unavoidably make use of concepts in the natural sciences such as "energy," "the ecosphere," "balance" and so on, which could be the distinctive feature of contemporary eco-literary criticism. The first to propose eco-literary criticism, Rueckert, applied the concept of "entropy" unique to the study of physics, when discussing the effect of eco-literary criticism, stating: "McHarg says that symbiosis makes negentropy possible; he identifies negentropy as the creative principle and process at work in the biosphere which keeps everything moving in the evolutionary direction which has characterized the development of all life in the biosphere."[48] As is widely known, in physics, "entropy" refers to the instability intrinsic to physical structures, but "negentropy" refers to overcoming this instability and making it tend toward stability. Rueckert uses the concept of "energy" in physics, insisting that the creation, education, reading and transmission of litera-

[46]Bury [11], pp. 7–8.

[47]Bury [8], p. 1.

[48]Ruckert [6], p. 120.

ture are all processes of energy transfer. He states: "The central endeavor, then, of any ecological poetics would have to be a working model for the processes of transformation which occur as one moves from the stored creative energy of the poem, to its release by reading, teaching, or writing, to its transmutation into meaning, and finally to its application, in an ecological value system, to what McHarg variously calls "fitness and fitting," and to "health"—which he defines as "creative fitting" and by which he means to suggest our creation of a fit environment. This work could transform culture and help bring our destruction of the biosphere to an end.[49] He describes this process as literature translating into actions that clean and rescue the biosphere. Here, "energy" is nothing more than a metaphor, and ultimately, the role that ecological literature and ecocriticism plays still that of transforming cultural attitude. Therefore, today's ecological problem is at bottom a problem of selecting a way of life and the means of production, which is a problem of cultural attitude and position. Changing cultural positions and cultural attitudes and selecting ways of production and life in symbiosis with Nature through ecological literature and ecocriticism is not only the path of humanity rescuing itself, but also what the role of eco-literary criticism is.

(6) **The Construction of Ecological Poetics**

When Rueckert puts forward the concept of ecocriticism, he proposes the conceiving of the construction of contemporary ecological poetics. He states, "I am going to try to discover something about the ecology of literature or try to develop an ecological poetics by applying ecological concepts to the reading, teaching, and writing about literature."[50] The ecological poetics that Rueckert is talking about is the theory of ecological literature or ecological arts, which is a new literary theory encompassing the ecological dimension. Based on our understanding of the contemporary condition of ecocriticism, this ecological poetics is still in the process of being created, the pathways forward for which could include the two principles of establishing new principles of poetics and utilizing and rehabilitating the original principles of poetics. For example, in *Revaluing Nature* Glen Love clearly insists that the "land ethics" that Leopold proposes in *A Sand County Almanac* "might well be the litmus test for the new pastoralism."[51] Next is the utilization and rehabilitation of the original principles of poetics, and further "greening" them on the foundation of the originals. Eco-literary criticism's utilization and rehabilitation of the theory of poetics made by the pre-Soviet theorist, Bakhtin is clear evidence of this. As is widely known, Bakhtin's theories of communicative dialogue, space and time, polyphony, openness and unfinalizability all have thick scents of ecological meaning, and contemporary Western ecological literary artists even say this theory is a literary form of ecology

[49]Ibid.

[50]Ruckert [6], p. 107.

[51]Love [12], p. 234.

founded on relational sciences.[52] Because of this, some ecocritics in the contemporary West try to use ecological theory to give a renewed explanation of Bakhtin's understanding of poetics in the contemporary context, and make it "green" into a new precious resource of ecological poetics.

3.7 Ecofeminism and Ecofeminist Literary Criticism

Contemporary ecological theory is grounded in pluralism and symbiosis, and every feminism and anti-racism that aims at dissolving monism and domination is necessarily embraced in it. Ecofeminism and its literary criticism was born in the rising tide of contemporary ecological theory's lively development.

(1) **Ecological Feminism**

Feminism welled up in the 1880s of Europe with the main purpose of women's liberation as the cultural trend. Its first wave began in the 1960s centered around opposition to gender discrimination; after the 1970s, it became characterized by engaging in dialogue with all the other cultural theories, from which derived post-structuralist feminist theory, psychoanalytic feminism, post-colonial feminism and ecological feminism, among others, which became the feminist theory in the post-modern context. In 1974, the French feminist Francoise d' Eaubonne first proposed the term "ecofeminism," whose purpose was to combine the feminist movement and the ecological movement and promote the deeper development of both. In 1980, professor Caroline Merchant Ph.D. of the University of California Berkeley published *The Death of Nature: Women, Ecology and the Scientific Revolution* in the attempt to evaluate the scientific revolution from both feminist and ecological perspectives, which became one of the most important works of ecofeminism.

1. Organicism and Nature as Mother

The basic intension of ecofeminism is opposing anthropocentrism and phallocentricism, which are intimately tied to one another; phallocentricism could be considered the derivative of anthropocentricism. Their philosophical foundation is the the mechanistic philosophy characterized by the dichotomized opposition between man and nature, subject and object, body and mind, sensibility and rationality along with the ossified forms of thought related to it. The ecological holism and ecofeminism opposed to this is based on the philosophical foundation of organicism which eliminates all of such dichotomies. Therefore, organicism is the most basic theoretical origin of ecofeminism. Merchant's *The Death of Nature* unfolds the relationship between organicist philosophy and natural ecology with great concentration; it includes the flesh and blood connection between man and nature in humankind's early pre-scientific organicism along with the loss of organicism in the Enlightenment period and its contemporary recovery, but its most basic viewpoint is "Nature as the mother of humanity" that organicism necessarily leads to.

---

[52]McDowell [14], p. 372.

First of all, Merchant insists that early on humans continuously depended on immediate organic connections with the natural order for sustenance. Humans come from Nature and are intimately tied to Nature, which is the most basic fact, due to which organicism is the most authentic realistic connection between man and nature. Merchant states, "From the obscure origins of our species, human beings have lived in daily, immediate, organic relation with the natural order for their sustenance."[53] Merchant reflects on the ancient Western philosophers' treatments of organism, starting with Plato's *Timaeus*, which grants life to the entire world and compares this world to an animal, believing that the diety "framed one visible animal comprehending within itself all other animals of a kindred nature." Aristotle then underscores the primacy of internal growth and development within Nature on the basis of organic theory. In *Metaphysics*, he defines Nature as "the source of movement of natural objects, being present in them either potentially or in complete reality." The Neo-Platonism of Plotinus integrates Platonism with Christian philosophy, dividing the feminine soul into two components, the higher part shaped by divine God and the lower part (nature) then generates the phenomenal world.[54]

Second, Nature as female and Mother. The necessary result of organic theory is metaphorically referring to Nature as female and Mother, because, organic theory is always accompanied with the characteristic that Nature has of nurturing human beings and all other beings. Just as Merchant says, "Central to the organic theory was the identification of nature, especially the earth, with a nurturing mother: a kindly beneficent female who provided for the needs of mankind in an ordered, planned universe."[55] Merchant explains by listing some of the artistic works of the renaissance period. Under the pen of writers like Chaucer of the English literary Renaissance, Nature was "a kindly and caring motherly provider, a manifestation of the God who imprinted a designed, planned order on the world." In Shakespeare's tragedy *King Leer*, King Leer's daughter represents utopian nature and the Nature of ideal unity as the opposite; The pastoral poems of the Renaissance period represented another image of Nature as female, the yearning of past ages for the motherly loving embrace; the German painter Lucas Cranach's "The Nymph of The Spring" has the representation of "the female earth nymph resting in a bed of flowers while doves, symbols of peace, feed near the edge of a trickling stream and deer water on its farther bank."[56] Merchant mentions that the atmospheric chemist James Lovelock identifies the earth as Gaia, and states that the human pollution of Nature disturbs the earth mother Gaia's life, and destroys her body.[57]

The Gaia Theory was first proposed by Timothy Zell in 1970, and James Lovelock refined it in 1972, giving it a more profound explanation. Lovelock identifies the earth as the earth god "Gaia" in ancient Greek mythology; she not only gives birth to the earth, but also nurtures all beings with her milk, and thus is living and alive. This was

[53]Merchant [15], p. 1.

[54]Ibid., pp. 11–15.

[55]Ibid., p. 2.

[56]Merchant [15], pp. 6–10.

[57]Ibid., p. 1.

an imagistic and new ecological idea, and became an important theoretical resource for ecological theory and especially ecofeminism. Later, there were also those who called it the "Gaia rule" which powerfully demonstrates the motherly character of earth and Nature, and specifically the organic nature and vitality intrinsic to earth and nature.

Finally, it is the call for the production of a new organic ecological worldview and the attempt to bring Gaia back to health. In *The Death of Nature*, Merchant powerfully criticizes the severe ecological destruction and death of nature brought about by the mechanistic world view, and at the same time presages the immanent birth of a new organic ecological world view. She insists that relativism, quantum theory, process physics, neo-thermodynamics and chaos theory are all challenges and break-throughs against mechanistic theory that signal the emergence of a new organic worldview. She states, "The machine image that has dominated Western culture for the past 300 years seems to be giving way to something new. Some call the transformation a 'new paradigm'; others call 'deep ecology'; still others call for a postmodern ecological world view."[58] This new ecological world view is a non-mechanistic science and an ecological ethics as well as an organic world view that has been baptized by the new era of Scientific Revolution. This new organic world view retains the values of those flesh and blood bonds between man and nature in the pre-modern organic world view, but at the same time fills it with the content of modern science and technology, giving it unprecedented value as it plays an unprecedented role. Merchant argues that it "must support a new economic order grounded in the recycling of renewable resources, the conversation of nonrenewable resources, and the restoration of sustainable ecosystems that fulfill basic human physical and spiritual needs. Perhaps Gaia will then be healed." [59]

2. Mechanistic Theory and the Death of Nature

Since the middle of the 17th century, the mechanistic world view gradually gave way to the mechanistic world view, and it directly resulted in the "Death of Nature" and the severe destruction of the ecological environment. This is precisely what Merchant's book focuses on explaining. As she put it in her book, "the new conceptual framework of the Scientific Revolution—mechanism—carried with it norms quite different from the norms of organicism."[60] The order of this new mechanism along with the values of power and control connected to it inevitably led Nature to death.

First of all, She insists that at the verge between the 16th and 17th centuries, an image of the organic universe with vitality and feminine earth as its core gave way to a mechanical world view.[61] As is widely known, at the intersection of the 16th and 17th centuries, Europe experienced an overwhelmingly powerful scientific and industrial revolution, the important hallmark of which was the traditional organic world view giving way to the mechanistic world view. This was a major upheaval

[58]Ibid., p. 3.

[59]Ibid., p. 5.

[60]Ibid., p. 210.

[61]Merchant [15], p. 3.

in the fields of philosophy and thought, the representatives of which were Bacon and Hobbes from Britain and Descartes from France. In *The Death of Nature*, Merchant explains their mechanistic philosophical ideas in detail; she states that Bacon is universally considered the originator of the concept of the modern research institute, the philosopher of industrial science and the main inspiration behind the Royal Society well as the founder of the inductive method.[62] Bacon believed that Nature existed in the three states of freedom, error or bondage, the third of which, bondage, meant "she is put in constraint, molded, and made as it were new by art and the hand of man; as in things artificial."[63] Here, Nature becomes a completely passive entity constrained and molded by man. Bacon's mechanistic world view is centrally embodied in his utopian work of 1624, *New Atlantis*. He imagined a forest island in a utopian society, that was a capitalist industrial society, "hierarchical and patriarchal, modeled on the early modern patriarchal family." The highest controlling institution of this utopian society, he called, the scientists from Salomon's House, and the governor of Salomon's House was a scientist who became the ruler, promoting a Baconion program reinforcing aggressive attitudes toward nature, speaking out in favor of 'mastering' and 'managing' the earth.[64] In accordance with this program, human beings begin engaging large scale reconstruction and destruction of nature. Merchant remarks, "In the *New Atlantis* lay the intellectual origins of the modern planned environments initiated by the technocratic movement of the late 1920s and 1930s, which envisioned totally artificial environments created by and for humans. Too often these have been created by the mechanistic style of problem solving, which pays little regard to the whole ecosystem of which people are only one part. The antithesis of holistic thinking, mechanism neglects the environmental consequences of synthetic products and the human consequences of artificial environments."[65] Merchant specifically makes her own analysis of the Bensalem model, "Bacon's mechanistic utopia was fully compatible with the mechanical philosophy of nature that developed during the seventeenth century. Mechanism divided nature into atomic particles, which, like the civil citizens of Bensalem, were passive and inert. Motion and change were externally caused; in nature, the ultimate source was God, the seventeenth century's divine father, clock-maker, and engineer; in Bensalem, it was the patriarchal scientific administration of Salomon's House."[66] In France, the promotor of mechanistic philosophy was the famous philosopher Descartes, who proposed the infamous "Cogito, ergo sum," and placed reason and sensibility in opposition to one another. In *On Method*, Descartes indicates that he himself desired to master and possess nature. In *Treatise on Man* written in 1622, Descartes clearly describes the human body as a machine, and in his 1644 work *Principles of Philosophy*, he reconstructs the universe as a mechanical apparatus, "a mechanism, based on the motion of inert material corpuscles that transmitted motion consecutively from part

[62]Ibid., p. 181.

[63]Ibid., p. 188.

[64]Ibid., p. 207.

[65]Ibid., p. 205.

[66]Ibid., p. 204.

to part through efficient causation."[67] Another British philosopher, Hobbes, applied this mechanistic philosophy to the social sphere, believing that "The body politic was composed of equal atomistic beings united by contract out of fear and governed from above by a powerful sovereign."[68] In summary, Merchant illustrates the overall countenance of mechanistic philosophy since the 16th and 17th centuries along with its intrinsic rigidity and absurdity in *The Death of Nature*.

Next, Merchant rather comprehensively explains the ecological crisis of globalization and its grave threat to the health of mother earth brought about by mechanistic philosophy. In *The Death of Nature*, Merchant proposes an extremely important view of historicism, namely that every idea far from abstract and transcendental comes into being through a certain history, including the concepts of nature and woman. She argues that "the concept of nature like the concept of woman is a socio historical construct".[69] This is an extremely important and valuable theoretical position to take, showing that every theoretical idea grows out of a certain political and social economic background, and understanding the implications of such ideas is only possible under such a perspective. This shows the scientific nature of Merchant's ecofeminism; she establishes her own ecological and feminist perspective from such a perspective, insisting that nature only devolves into the slave of man and is plundered when she finds herself in the background of the capitalist market economy and economic culture of mechanistic philosophy. Because the market economy parades under the pretext of progress, concentrating more wealth in the hands of businessmen, clothing merchants, entrepreneurs and independent farm owners, which intensifies the gulf between the lower and higher stratums of society, and ultimately results in severe ecological crisis. She states, "Today, a global ecological crisis that goes beyond the environmental crisis of the 1970s threatens the health of the entire planet. Ozone depletion, carbon dioxide buildup, cholorflurocarbon emissions, and acid rain upset the respiration and clog the pores and lungs of the ancient Earth Mother, rechristened 'Gaia,' by atmospheric chemist James Lovelock. Toxic wastes, pesticides, and herbicides seep into ground water, marshes, bays, and oceans, polluting Gaia's circulatory system. Tropical rainforests and northern old-growth forests disappear at alarming rates as lumberers shear Gaia of her tresses. Entire species of plants and animals become extinct each day. A new partnership between humans and the earth is urgently needed."[70] She exposes the reality of humans ravishing the earth mother at multiple levels, and with respect to agriculture, "Although the early effects of agricultural improvement were healthy for the soil, built into the emerging capitalist market economy was an inexorably accelerating force of expansion and accumulation, achieved, over the long term, at the expense of the environment and the village community—the natural and human resource bases."[71] In addition, fertilizer and pesticide use have resulted in unanticipated side-effects, monoculture grains have

---

[67] Merchant [15], p. 224.

[68] Ibid., pp. 229–230.

[69] Ibid., p. 3.

[70] Ibid., p. 1.

[71] Ibid., p. 60.

wreaked the havoc of blooming pests and diseases, and the continuous bringing of new virgin lands into cultivation have all but destroyed ecosystem balances. "The marsh—drained for pasture, blamed for disease, and exploited for its wildfowl—has today become a *cause célèbre* of ecology." (p. 61) Capitalist values "irrevocably altered the English fenlands...," and "capitalist modes of production accentuated human impact on the forests over and above effects attributable solely to population pressure."[72]

In addition, Merchant also exposed that as capitalism fleeced Nature, it also ruthlessly exploited women, forcefully advocating that women should in all docility and forthrightness remain in the position of following orders. Mechanism brings about anthropocentricism but not without instituting phallocentrism at the very same time; as capitalism fleeces Nature, it does not do so without exploiting women in the very same movement. Merchant points out, "[s]ymbolically associated with unruly nature was the dark side of woman. Although the Renaissance Platonic lover had embodied her with true beauty and the good, and the Virgin Mary had been worshipped as mother of the Savior, women were also seen as closer to nature than men, subordinate in the social hierarchy to the men of their class, and imbued with a far greater sexual passion. The upheavals of the Reformation and the witch trials of the sixteenth century heightened these perceptions. Like wild chaotic nature, women needed to be subdued and kept in their place."[73] Content demonizing femininity flooded the artistic works and thoughts of the Renaissance, characterizing women as lascivious and insolent, as figures who brutalize and cheat husbands, as witches intoxicated and lustfully sexual. The physician Johann Weyer's *Histoires Disputes et Discours* (1579) argues that women are fragile in mind and spirit, and could very easily be depraved in reason due to the melancholic humors that burdened their brain. Anthropologists at the time argued that both Nature and women are on a lower level than culture. The Scottish Protestant Reformer John Knox argued that since the flesh is subordinate to the spirit, the woman's position is below that of the man's, and that natural law determines that the man should command the woman. Some physicians even absurdly argued that the female only provides the material in reproduction, while the active principle comes from the male sperm, and that the soul is transmitted through the male system. When summarizing the repression of women in the capitalist age brought about the "Scientific Revolution" marked by mechanism, Merchant points out, "[f]or women, this aspect of the Scientific Revolution did not bring about the presumed intellectual enlightenment, objectivity, and liberation from ancient assumptions traditionally accorded it."[74]

3. The Ideal of Ecological Society—the Simultaneous Expression of Nature's Rights and Women's Rights

Merchant clearly proposed the conception of an ideal ecological society that simultaneously expresses nature's rights and women's rights in *The Death of Nature*. She

[72]Merchant [15], p. 70.

[73]Ibid., p. 146.

[74]Ibid., p. 180.

first largely affirmed the two ecological utopian social forms born in the early 17th century along with the resources that they offer us to think with. These two utopias include the historically famous Tommaso Campanella's City of the Sun in 1602 and Johan Valentin Andrea's Christianopolis in 1619. They tried to put these utopian ideal states into practice, and Campanella organized an uprising against the Spanish authorities, ultimately ending in failure. *City of the Sun* is the work he wrote in prison. Andrea tried to effectuate his Christianopolis plan, but of course ultimately failed. These two ecological utopian ideals and their actualization in the construction and development of human societies offer extremely important wisdom and inspiration for theory. Merchant argues that these two utopias forcefully championed organicist thought, "articulated a philosophy of communal sharing that responded to the interests of artisans and the poor for a more egalitarian distribution of wealth based on an original harmony between people and nature."[75] This is to say that these two utopian societies carried forward the organic thought of original organic harmony between people and Nature, and used it to guide the construction of societies. The "builders" of these two utopias were followers of the naturalist philosophy of the Renaissance, and in their belief, God is immanent in all creatures, and all matter is alive. Under the guidance of this organicism, these two utopias adhered to holism, which is effectively the concrete manifestation of organic theory. When Merchant discusses this holism, she points out "a second variation of the organic theory, growing out of peasant experience and village culture, was based on the leveling of differences and stressed, instead, the primacy of community, the collective will of the people, and the idea of internal self-regulation and consent. Here the communal whole was still greater and more important than the sum of the parts, but the parts were of equal or nearly equal value."[76] This holistic ecological perspective closely tied to the organic theory developed later into one of the most basic theoretical supports of contemporary ecological philosophy and aesthetics, and are of extremely important value. As Merchant puts it, "Recognized today as keys to viable ecosystems in nature are the interrelationships and organic unity among a system's parts, and the maintenance of ecological diversity."[77] This shows that the organic holism of ecology is the most important characteristic of ecosystems. Next is the realization of women's liberation and women's rights. Merchant argues that the two aforementioned "utopian communities postulated a more egalitarian view of woman and man, artisan and master."[78] "In both Christianopolis and the City of the Sun, women were more liberated than in the real sixteenth-century society."[79]

Merchant also introduced the ecological utopias conceived by other contemporary ecological theorists. In her view, the two utopias that appeared in the early 17th century only represented "the pre-industrial form of social revolution." In other words, they were thoughts that never underwent the industrial revolution. When

[75]Merchant [15], p. 89.

[76]Ibid., p. 85.

[77]Ibid., p. 93.

[78]Ibid., p. 89.

[79]Ibid., p. 101.

facing the severe ecological crisis of today, some ecological theorists have proposed their own ecological utopias, which embody two new characteristics: one is they attempt to unify "is" and "ought," namely morality and science; another is they try to unify the continued enjoyment of modern technology with the harmonious coexistence between humans and environment. Merchant mentioned that in Ernest Callenbach's *Ectopia*, northern California, Oregon, and Washington secede from the United States in 1980 and develop in isolation. In this stable state ecological utopia, society is based on and guided by natural ecological philosophy, and women become the leaders of the main political party, and with private property abolished, people live in small village communities or micro-cities and modern life is highly consistent with ecological protection. Merchant argues that "it reflects the need for social changes and the most idealistic aspirations of contemporary society."[80]

Finally, Merchant argues that faced with the increasingly serious ecological crisis, health can ultimately only be restored after reversing modern values and a revolution in economic priorities, which means, that "the world must once again turn upside down".[81] Merchant points out that contemporary society's polluting of nature and ecology has already reached an extremely grave magnitude, and she especially points to the Three Mile Island Nuclear Reactor Accident in Harrisburg, Pennsylvania as even more powerful evidence that "the death of Nature" will result from the earth's sickness caused by radioactive waste, pesticides, plastics, photochemical smog and fluorocarbons. The graveness of the problem is that the mechanistic world view still plays a repressive role in the economic conditions of capitalist markets where the situation of Nature and women being exploited fundamentally cannot change. "The mechanistic view of Nature that developed out of 17th century natural philosophy based on the Western mathematical tradition traceable back to Plato still remains dominant in science today."[82] Merchant argues that "[m]echanistic assumptions about nature push us increasingly in the direction of artificial environments, mechanized control over more and more aspects of human life, and a loss of the quality of life itself."[83] Even more important is that the current capitalist system resists the liberation of nature and women. Merchant is an historicist, who forcefully advocates that every idea is an historical and social construct. Because of this, she recognizes that the liberation of nature and women is not realized in the social institution of capitalism, and that the conjunction of conservation and ecology movements women's rights and liberty have already gone in the opposite direction reversing the subjugation of both nature and women.[84] Because of this, her statement that this world must once again turn upside down not only refers to turning mainstream values around but also to reforming the capitalist system that generates profit through consuming nature at the expense of working people, while establishing "a new social

[80]Ibid., p. 109.

[81]Ibid., p. 327.

[82]Merchant [15], p. 322.

[83]Ibid., p. 323.

[84]Ibid., p. 326.

model of fittingness",[85] which raises the problem of social revolution to the forefront of the ecological movement and ecofeminist movement, and is revolutionary. This is the distinctive feature of the ecofeminist theory Merchant details in *The Death of Nature*.

4. Eco-feminist Literary Criticism

After ecofeminism came ecofeminist literary criticism. The "Introduction" to the American ecocritic Glotfelty's first book *The Ecocriticism Reader* (1996) summarized the three stages of development of feminist ecocriticism: The first stage is the unearthing of the themes and works of feminist literature; the second stage is the tracing of the tradition of feminist literature and uncovering its meaning; the third stage is the examination of the inner structure of ecofeminist literature including canonical texts.[86]

Contemporary ecofeminist literary criticism is rich in content. It applies the variety of viewpoints from ecofeminism such as those on organic theory and mother nature along with the championing of ecologically ideal social constructs as well as the criticism of mechanism and the death of nature to the critique of literary arts. At the same time, it also includes the following contents: first, it fiercely criticizes phallocentrism in light of the ecological literary critique of anthropocentrism. Just as Rueckert said when analyzing Adrianne Rich's book *Diving into the Wreck*, "there is a demonstrable relationship between the ways in which men treat and destroy women and the ways in which men treat and destroy nature."[87] Second is unearthing the rich content about the image of the earth mother in literary works and other works. For example, the British scientist Lovelock's uncovering and elucidation of the implications of the earth mother Gaia, from which he develops an extremely important ecological law—"Gaia's law." Third is describing and explaining a living place of a woman's own. For instance, William Howarth states in *Some Principles of Ecocriticism*: "Ecocriticism finds its strongest advocates today in feminist and gender critics, who focus on the idea of place as defining social status. Of particular interest is "a woman's place," often described as an attic or closet, that contains yet sustains individuals until they locate a congenial environs."[88] Fourth is the revaluation of feminist literature and feminist writers. This point has the nature of re-establishing values. In the history of literature feminist literature and feminist writers have been easily looked over and distorted, and must be reappraised from the right feminist position. Fifth is encouraging enlarging the active feminine participation in the literary creations of ecofeminism, and joining other ecological movements, and make them truly become driving forces of protection for female rights, earth conservation and child protection. Ecofeminist criticis argue that due to the closeness of femininity and natural nature, in a certain sense we could say they care for the earth, nature and the naturalness of humanity's future.

[85] Ibid., p. 327.

[86] Gloverti [5], p. 215.

[87] Ruckert [6], p. 117.

[88] Howers [7], p. 82.

References

1. Murphy, P., Gifford, T., & Yamazato, K. (1998). *Literature of nature: An international sourcebook*. Taylor & Francis.
2. [U.S.] David Thoreau, H. (2004). *Walden*, translated by Xu Chi, Shanghai Translation Press.
3. [Germany] Heidegger, M. (2006). *Being and time* (revised translation), translated by Chen Jiaying, Wang Qingjie, Sanlian Bookstore.
4. [U.S.] Leopold, A. (1997). *A Sand County Almanac*, translated by Hou Wenhui. Jilin People's Publishing House.
5. [U.S.] Gloverti, C. (1996). *Foreword: Literary studies in the age of environmental crisis*, Containing in *The Ecocriticism Reader*, University of Georgia Press.
6. [U.S.] Ruckert, W. (1996). *Literature and ecology: An experiment of ecological criticism*. Containing in *The Ecocriticism Reader*. University of Georgia Press.
7. [U.S.] Howers, W. (1996). *Some principles of ecological criticism*, Containing in *The Ecocriticism Reader*. University of Georgia Press.
8. [U.S.] Bury, L. (2001). *Writing for the dangerous earth*. Harvard University, Bernard Press.
9. [U.S.] Fromm, H. (1996). *From transcendence to degeneration: A road map*, Containing in *The Ecocriticism Reader*. University of Georgia Press.
10. [U.S.] White, L. (1996). *The historical roots of our ecological crisis*, Containing in *The Ecocriticism Reader*. University of Georgia Press.
11. [U.S.] Bury, L. (2001). *Imagination of the environment: Thoreau, the formation of nature and American culture*. Harvard University, Bernard Press.
12. [U.S.] Love, G. (1996). *Reevaluating nature: Facing an ecological criticism*, Containing in *The Ecocriticism Reader*. University of Georgia Press.
13. [U.S.]·Mick, J. (1996). *The mode of comedy*, Containing in *The Ecocriticism Reader*. University of Georgia Press.
14. [U.S.] McDowell, M. (1996). *The Bachkin way to ecological criticism*, Containing in *The Ecocriticism Reader*. University of Georgia Press.
15. [U.S.] Merchant, C. (1999). *The death of nature—Women, ecology and the scientific revolution*, translated by Wu Guosheng et al. Jilin People's Publishing House.

Part II
The Theoretical Guidance of Ecological Aesthetics

Chapter 4
The Ecological Theory of Marxism

Ecological theory results from human beings rethinking history in the context of the intensifying conflict between human beings and nature due to the extreme expansion of capitalism since the 1970s. In fact, such reflections started to emerge from the great theories of Marx and Engels in the 19th century. As mentors and great theorists of the contemporary human spirit, they have analyzed and predicted the aesthetic relationship between human beings and nature with profound insight and sharpness. According to our current research, the profoundness and foresight of their analysis are very impressive. It is an extremely valuable theoretical guide and an important ideological resource for us to think deeply and to explore ecological aesthetic issues in the new century. The discourse of Marxist ecological theory in this chapter focuses on the ecological theory of Marx and Engels, the theory of ecological civilization of China's contemporary theory of socialist innovation, and the ecological outlook of contemporary Western Marxism represented by David Pepper.

4.1 Marx and Engels' Ecological Theory of Materialist Existence Theory

As we all know, after the death of Hegel in 1831, three transitions took place in the fields of western thought, culture, and philosophy: from epistemology to existentialism, from the subject/object dichotomy to intersubjectivity, and from "anthropocentrism" to "ecological holism." Until now, the debate about these transitions continues. Marxism is the most advanced world outlook in the history of human society. It not only reflects the developmental trend of philosophy-culture transformation, but also plays a guiding role.

F. Zeng, *Introduction to Ecological Aesthetics*,
https://doi.org/10.1007/978-981-13-8984-9_4

(1) The common subject for Marx and Engels—the creation of a rich, ecological, and aesthetic sense of materialist practice

The common subject for Marx and Engels is the creation of a rich, ecological, and aesthetic sense of materialist practice. This is the core of the materialist theory's world outlook and ecology. In the relationship between human beings and nature, this kind of practical outlook with a rich, ecological, and aesthetic sense not only broke through the metaphysical view of the subject/object dichotomy, but also broke through the anthropocentric outlook. Its core is the harmony and equality between master and nature, or a common symbiosis. The ecological theory of Marx and Engels is a philosophical view of contemporary significance and value. The concept of practice created by Marx and Engels contains a strong ecological awareness and can serve as a theoretical guide and an important resource for us to build a contemporary ecological outlook and ecological aesthetics today.

1. On the basis of criticizing idealism and old materialism, Marx and Engels created a new worldview characterized by a breakthrough in metaphysics. This new worldview is the materialist view of practice with which we are familiar. It is a materialist worldview that is different from all old materialism, characterized by subjective and active practice. That is, it is an existential worldview of materialism. Marx's *On Feuerbach's Syllabus* has a well-known remark: "The main defect of all hitherto-existing materialism—that of Feuerbach included—is that the Object [*der Gegenstand*], actuality, sensuousness, are conceived only in the form of the object [*Objekts*], or of contemplation [*Anschauung*], but not as human sensuous activity, practice [*Praxis*], not subjectively."[1] Regarding the significance of materialist practice in breaking from the drawbacks of traditional metaphysics, Marx has a more concrete explanation in *the Economic and Philosophical Manuscripts of 1844*, wherein we find the familiar discussion about, "the unity of complete naturalism and complete humanitarianism." Marx pointed out: "Here we see how consistent naturalism or humanism is distinct from both idealism and materialism, and constitutes at the same time the unifying truth of both. We see also how only naturalism is capable of comprehending the action of world history."[2] What Marx refers to here as "naturalism" and "humanitarianism" borrows the basic concepts of Feuerbach. What Feuerbach called "naturalism" and "humanitarianism" are both based on the abstract concept of "human nature," and will inevitably lead to the division between human beings and nature, and will therefore be incomplete. Marx believes that complete "naturalism" and "humanitarianism" should be the unification of human beings and nature in social practice. In this way, Marx achieved two important breakthroughs. First, this unification is not one based on abstract human nature, but is rather a unification based on social practice. Furthermore, this unification can only be achieved by a communist society after the true elimination of class. In this way, we can truly combine naturalism and humanitarianism. Thus, it can be seen that Marx's con-

[1]Marx and Engels [1], (Vol. 1), p. 16.

[2]Marx and Engels [2], (Vol. 42), p. 167.

cept of materialism contains the extremely important respect for nature in the concept of "thorough naturalism," and the ecological awareness that nature is a factor in the development of human society. What needs to be clarified here is that although the concept of "nature" is borrowed from Feuerbach, there is a clear difference between Marx's "nature" and "nature created by people" which Feuerbach mentioned. Marx's "nature" is a "nature" that contains ecological significance. We believe that this materialistic world view of practice is at the same time also a world view of the existentialism of materialism. This is because, first of all, Marx's materialist practice viewpoint is based on the liberation of individuals and the good life. Secondly, it is based on taking the entire proletariat and humankind's liberation and flourishing life as their ideals and goals. Finally, it is based on taking social practice as the most important way to bring about social revolution and revolutionize production practices (that is, to overthrow the capitalist system). Only in this way can we truly overcome the contradiction between human beings and nature. The unification of human beings and nature can only be achieved on the basis of Marxist practical ontology and social practice. Of course, there is an essential difference between Marx's existentialism of materialism and the existentialism of Western existentialism. The core difference is that firstly, Marx emphasizes social practice. Secondly, he emphasizes social revolution. Thirdly, he not only emphasizes the existence of individuals, but also emphasizes the existence of the general public as well as the people. One of the basic propositions of existentialism is that "existence precedes essence." This "existence" is an individual existence. Marx's materialism of existence not only emphasizes the individual, but also emphasized the liberation and good life of the entire human race, especially the proletariat and the broad masses of people. This is the most basic feature of the Marxist theory of materialist practical existence.

2. Marx's concept of materialistic practice contains not only an obvious ecological sense, but also an obvious ecological, aesthetic sense. At this point, we can understand from Marx's famous exposition of "the law of beauty." In Marx's *Economic and Philosophical Manuscripts of 1844*, referring to the difference between human production and animal production, he pointed out: "An animal produces only itself, whilst man reproduces the whole of nature. An animal's product belongs immediately to its physical body, whilst man freely confronts his product. An animal forms only in accordance with the standard and the need of the species to which it belongs, whilst man knows how to produce in accordance with the standard of every species, and knows how to apply everywhere the inherent standard to the object. Man therefore also forms objects in accordance with the laws of beauty."[3] As for the "species scale" and "internal scale," there is much ambiguity in the academic understanding. Some people think that these two scales are the same thing, all referring to "human scale." However, "the laws of beauty" are completely the result of anthropocentrism. "Constructing according to the laws of beauty" means building the society according to human standards, implying that nature is also completely humanized nature. In my opinion, this

[3]Marx and Engels [2], (Vol. 42), p. 97.

is not in line with Marx's original intention. "Species scale" refers to the scale of species, while the "internal scale" refers to the human scale. Therefore, the proposition that "man is also built according to the laws of beauty" contains a clear and profound ecological sense. In other words, the so-called "law of beauty" is the harmony of the laws of nature and the laws of human beings. What Marx refers to here as "standards" means "standards, specifications, and norms." When combined with a contextual understanding, it also contains the meaning of "basic needs." The so-called "scale of any one species" means that the vast natural world contains the basic needs of various animals and plants. The "law of beauty" must include these basic needs, and cannot "alienate" it and turn it into the opposite of human beings. This already includes the recognition of the value of nature. In recognizing the "basic needs" of natural things, we must recognize its independent value, or at least a relatively independent value. The so-called "inherent Standard" literally is "inherent, and born of standards and specifications." That is, it is a human-specific, conscious, comprehensive, and liberating need beyond the instinctual needs of narrow species. However, this conscious need of transcendence should be "built according to the laws of beauty" and under the premise of recognizing the basic needs of nature. It should be built within the context of a combination of naturalism, humanism, and harmonious unity between human beings and nature.

3. Marx's "construction in accordance with the law of beauty" as stated in *The Economic and Philosophical Manuscripts of 1844* is a brand-new worldview. It is an indispensable and important element of practical materialism, which is of extremely important theoretical value and the significance. In his *On Feuerbach's Syllabus*, Marx said: "Philosophers have hitherto only *interpreted* the world in various ways; the point is to *change* it."[4] This "outline" was written by Marx in notes in Brussels in the spring of 1845, and was not ready for publication, so it cannot be expanded in context. In connection with *The Economic and Philosophical Manuscripts of 1844*, we think that what Marx refers to here as "changing the world" should include the meaning of generating "construction according to the laws of beauty." Therefore, a more complete statement by Marx in this passage should be as follows: Philosophers interpret the world in various ways, but the point is to change it. They should change it by building the world according to the laws of beauty. In this way, the concept of materialism contains a rich sense of ecological aesthetic.
4. Marx and Engels' have other, related discourses on the concept of ecology. In his *Economic and Philosophical Manuscripts of 1844*, Marx discussed many times that human beings are part of nature, which contains the ecological concept of equality between human beings and nature. He said: "Man *lives* on nature,"[5] as well as, "Man is directly a natural being."[6] Engels has repeatedly discussed the relationship between human beings and nature in his *Dialectics of Nature*,

[4]Marx and Engels [1], (Vol. 1), p. 19.
[5]Marx and Engels [2], (Vol. 42), p. 95.
[6]Ibid., p. 167.

eloquently illustrating the great role of labor in the transition from ape to man, revealed in the truth that human beings originate in nature. Engels also pointed out that, "The whole of nature accessible to us forms a system, an interconnected totality of bodies."[7] This already includes the ecological chain of thinking of ecology, which indeed has a value.

(2) Marx: The Sublation of Alienation—Reconstruction of the Harmonious Relationship between Human beings and Nature

This is actually a critique of the natural destruction of the capitalist system and capitalist industrialization. The emergence of the concept of ecological aesthetics has a profound and realistic foundation, which is the serious confrontation between human beings and nature caused by deforestation and the destruction of nature because of capitalism's blind pursuit of economic interests. Marx classifies the phenomenon of "opposition" as "alienation," deeply discusses its connotation and solutions, and gives profound revelation to contemporary, ecological, aesthetic construction. Marx said, "*Man's species-being,* both nature and his spiritual species-property, into a being *alien* to him, into a *means* of his *individual existence*. It estranges from man his own body, as well as external nature and his spiritual aspect, his *human* aspect."[8] The "alienation" here includes the human body, nature, spiritual nature, and human nature etc. Nature is one of its important aspects.

First of all, nature, as an organic part of producing products, is in a different, opposite state from that of workers in a state of alienated labor. Marx pointed out, "It is true that labor produces for the rich wonderful things—but for the worker it produces privation. It produces palaces—but for the worker, hovels. It produces beauty—but for the worker, deformity. It replaces labor by machines, but it throws one section of the workers back into barbarous types of labor and it turns the other section into a machine. It produces intelligence—but for the worker, stupidity, cretinism."[9] Workers have created wealth and beauty in the work of transforming nature, but these are far away from themselves, as works go on living a poor, ugly, unnatural, and non-beautiful life.

Second, the alienation of nature and human beings in social labor is also manifested in the grave destruction and pollution of nature throughout the labor process. Originally, social labor should be activity "constructed in accordance with the laws of beauty," and a harmonious unity of human beings and nature. To contrast, alienated labor pollutes and destroys nature. In criticizing Feuerbach's intuitive materialism, in particular the so-called "outside world unrelated to human beings," Marx said that all of the nature is "humanized nature," but excessive "humanization" has led to pollution. The blind development of industry pollutes nature and even the fish have lost the essence of their existence–clean water. In a letter, he pointed out:"Every time a new invention is invented, every time the industry goes further, there is a new site that will be drawn out of this field……The 'essence' of a fish is its 'existence', that is,

---

[7]Marx and Engels [1], (Vol. 3), p. 492.

[8]Marx and Engels [2], (Vol. 42), p. 97.

[9]Ibid., p. 93.

water. The 'essence' of river fish is river water, However, once the river is under the control of industry, once it is contaminated with dyes and other wastes, there are ships in the river. Once the river water is introduced into the aqueduct, as long as the water is drained, the fish can lose the channels of the living environment. The water of this river is no longer the 'essence" of the fish. It has become an environment unsuitable for the survival of fish."[10] This shows that the development of modern industry has seriously polluted the natural environment, and the polluted river no longer becomes the essence of the existence of fish. Instead, it has become its antithesis. Of course, it is alienated and opposed to human beings. In light of the cruel deprivation of laborers in alienated labor, the unprecedented confrontation between human beings and nature, and the deteriorating living conditions of human beings, Marx clearly proposed the sublation of alienation, the abandonment of capitalist private ownership, the construction of a communist society, and the beautiful ideal of reconstructing of harmonious relationship between human beings and nature. Marx put forward the very important idea of "abolition of private property" in *Economic and Philosophical Manuscripts of 1844*. He said: "The abolition [*Aufhebung*] of private property is therefore the complete *emancipation* of all human senses and qualities,"[11] He also said: "*Communism* as the *positive* transcendence of *private property* as *human self-estrangement,* and therefore as the real *appropriation* of the *human* essence by and for man;"[12] This is a very profound theory. It was the profound thinking of the alienation of nature and human beings, of the alienation of human beings from each other, and of how to solve these issues. These ideas in Marx were caused by his ideas about private ownership and capitalism over a hundred years ago, which has strong theoretical and practical significance. If we say that contemporary deep ecology is a "deeper questioning" of ecological issues at the philosophical level, then Marx has carried out a sociological deep meditation on the relationship between human beings and nature in the *Economic and Philosophical Manuscripts in 1844.* Furthermore, his meditations are closely linked with the social and political system.

Finally, Marx profoundly revealed the nature of the infinite proliferation and expansion of "capital." In *Capital*, he explained the nature of capitalism's brutal looting of both man and nature. Marx profoundly exposed the infinite expansion of "capital" and pointed out: "As a producer of the activity of others, as a pumper-out of surplus labour and exploiter of labour-power, it surpasses in energy, disregard of bounds, recklessness and efficiency, all earlier systems of production based on directly compulsory labour."[13] In discussing the relationship between the big industry and agriculture under the capitalist system, he pointedly pointed out that "Capitalist production, therefore, develops technology, and the combining together of various processes into a social whole, only by sapping the original sources of all wealth—the soil and the labourer."[14] That is to say, capitalism proceeds by sapping land and

---

[10]Marx and Engels [2], (Vol. 42), p. 369.

[11]Ibid., p. 124.

[12]Ibid., p. 120.

[13]Marx [3], (Vol. 1), p. 344.

[14]Ibid., p. 553.

workers. Here, he profoundly exposed the double destruction of capitalist production to workers and nature.

(3) Engels: The Founding of Dialectical Materialism's View of Nature—The Philosophical Dimension of the Unity of Human and Nature

Engels founded the view of nature in dialectical materialism, which includes critiquing "anthropocentrism," criticizing idealism, emphasizing the connection between human beings and nature, and emphasizing that people's scientific and technological capabilities are limited in the face of nature, and so on. As we all know, Engels once studied a great deal of literature on the important issues of natural science from 1873–1886, and completed about ten related papers, leaving more than 170 notes and fragments, which comprise the *Dialectics of Nature,* published later. The book contains the profound Marxist view of nature.

First of all, Engels focuses on the connection between human beings and nature, emphasizing the unity of human beings and nature and criticizing the view of idealism that "mankind is superior to other animals," However, he also makes a profound exposition of the limitation of human labor, science, technological ability, and nature's inviolability. These theoretical views have very strong realistic and theoretical value for the current criticism of the traditional concept of "anthropocentrism." With respect to dialectics, Engels makes it clear that "The general nature of dialectics to be developed as the science of interconnections, in contrast to metaphysics."[15] In speaking of nature, Engels believed that "the whole of nature accessible to us forms a system, an interconnected totality of bodies",[16] which includes all the species that exist in nature. Engels, by means of the theory of cells, demonstrates the consistency between human beings and nature, and criticizes the traditional viewpoint that humankind is superior to animals on the basis of the fact that human beings and animals and plants alike are composed of cells with the same basic structure. He said: "This much is certain—comparative physiology gives one a healthy contempt for man's idealistic arrogance in regard to other animals. At every step it is forcibly brought home to one how completely his structure corresponds to that of other mammals; he has basic features in common with all vertebrates and even—if less distinctly—with insects, crustaceans, tapeworms, etc."[17] Engels views the concept of "human beings are superior to other animals" as "idealist exaggeration," giving it "extreme contempt," which is already a powerful critique of "anthropocentrism." This criticism that all living things, including humans, are made up of cells is critically powerful from a comparative scientific perspective of physiology. Additionally, Engels further demonstrates the homology between human beings and nature from the perspective of the evolution of humankind from apes. He said: "Almost certainly as an immediate consequence of their mode of life, for in climbing the hands fulfill quite different functions from the feet, these apes when moving on level ground began to drop the habit of using their hands and to adopt a more and more erect posture

[15]Marx and Engels [1], (Vol. 3), p. 484.

[16]Ibid., p. 492.

[17]Ibid., pp. 337–338.

in walking."[18] He also argues for the similarity between human beings and nature from the similarity between children's actions and animal actions, stating: "Among our domestic animals, more highly developed thanks to association with man, every day one can note acts of cunning on exactly the same level as those of children. So the mental development of the human child is only a still more abbreviated repetition of the intellectual development of these same ancestors, at least of the later ones."[19] Starting from this, Engels expounds the relationship between human beings and animal from the perspective of philosophy—"This is also true"—thus criticizing the metaphysical separatism between human beings and nature. It is precisely this kind of "also" that is due to the "middle stage" between things and ought to be integrated and transitioned. Thus it can be seen that "anthropocentrism," which confronts human beings and nature, is precisely the metaphysics that Engels criticized as "non-one" contrary to dialectics. However, after all, humans have a qualitative difference from animals. That is, animals can only passively adapt to nature while humans can carry out purposeful and creative work. Engels pointed out: "And what do we find once more as the characteristic difference between the band of monkeys and human society? *Labor*."[20] Because of this, it is impossible for animals to stamp their will upon nature, as only human beings can change nature through purposive labor, "and brings about changes in it simply by his presence; man by his changes makes it serve his ends".[21]

Second, Engels criticized the blind confidence of human beings in their ability to transform the environment as well as the growing severity of humankind's destruction of the environment. Engels pointed out: "When an individual manufacturer or merchant sells a manufactured or purchased commodity with only the usual small profit, he is satisfied, and he is not concerned as to what becomes of the commodity afterwards or who are its purchasers. What did the Spanish planters in Cuba, who burned down forests the slopes of the mountains and obtained from the ashes sufficient fertiliser for *one* generation of very highly profitable coffee trees, care that the tropical rainfall afterwards washed away the now unprotected upper stratum of the soil, leaving behind only bare rock?"[22] This closely links the destruction of the natural environment with the pursuit of profit under the capitalist system. It not only shows that the destruction of the environment is closely linked with the capitalist political system, but also is closely linked with the mode of production, lifestyles, and modes of thinking, in which people blindly pursue economic interests. At the same time, the destruction of the environment and the development of science and technology together have led people to overconfidence in their abilities, which leads to the concept and behavior of indiscriminate and predatory nature. While he portrays that religion progressively narrows its territory as science advances, Engels wrote, "But what God has had to suffer at the hands of his defenders! In the history

[18]Marx and Engels [1], (Vol. 3), p. 508.

[19]Ibid., p. 517.

[20]Ibid., p. 513.

[21]Ibid., p. 517.

[22]Marx and Engels [1], (Vol. 3), p. 520.

of modern natural science, God is treated by his defenders as Frederick William III was treated by his generals and officials in the Jena campaign. One division of the army after another lays down its arms, one fortress after another capitulates before the march of science, until at last the whole infinite realm of nature is conquered by science, and there is no place left in it for the Creator."[23] Science and religion contested the realm of nature, and finally, with great satisfaction, the "infinite realms of nature" were conquered by them. However, Engels is keenly aware that humans are overly intoxicated and overly optimistic about their own so-called conquering of nature by virtue of their pursuit of economic interests and their technological capabilities. From the philosophical perspective of the universal contact between man and nature, he said: "Let us not, however, flatter ourselves overmuch on account of our human conquest over nature. For each such conquest takes its revenge on us. Each of them, it is true, has in the first place the consequences on which we counted, but in the second and third places it has quite different, unforeseen effects which only too often cancel out the first."[24] This is a very well-known and often quoted phrase that is very profound and wonderful. It not only points out that humankind should not overindulge in its own abilities, but also emphasizes that the so-called triumph of humankind's conquest of nature will surely be retaliated and eventually its outcome will be canceled. This idea foresees both the intensification of the relationship between human beings and nature as well as the emergence of the ecological crisis.

Elsewhere, Engels also set off the necessary warnings to humankind: "Thus at every step we are reminded that we by no means rule over nature like a conqueror over a foreign people, like someone standing outside nature—but that we, with flesh, blood, and brain, belong to nature, and exist in its midst,"[25] This means that the destruction of nature by humankind ultimately amounts to the destruction of humanity itself. Engels attacked the anti-natural, cultural tradition of Europe developed from ancient times and in Christianity. From the perspective of dialectical materialism, he further elaborated the viewpoint that people "agree with themselves and with nature." He said: "the more this happens, the more will men not only feel, but also know, their unity with nature, and thus the more impossible will become the senseless and antinatural idea of a contradiction between mind and matter, man and nature, soul and body, such as arose in Europe after the decline of classic antiquity and which obtained its highest elaboration in Christianity."[26] This is a very philosophical and scientific concept of nature. However, it is also a scientific ecological outlook. Even today, it is still very inspiring. While attacking people's over-superstitious scientific and technological capabilities, Engels did not completely deny the role of science and technology, he believes that the development of science will enable people to correctly understand the laws of nature so as to learn to govern and overcome the far-reaching natural influences caused by productive activities.

---

[23]Ibid.., p. 529.

[24]Ibid., p. 517.

[25]Ibid., p. 518.

[26]Marx and Engels [1], (Vol. 3), p. 518.

Engels believes that the solution to the fundamental antagonism between human beings and nature is to establish "conscious organization of social production, in which production and distribution are carried on in a planned way"[27] through the socialist revolution. Human beings are blindly pursuing economic interests, causing imbalances in the relationship between human beings and nature and between people, which is actually a kind of alienation of human free and conscious nature akin to a retreat to the animals. Only this "conscious organization of social production, in which production and distribution are carried on in a planned way" is the beginning of a new historical period of socialism as the restoration of humankind from an animal state and the return of human essence. Today, as China has entered into such a historical period, it possesses the possibility of eliminating the blind pursuit of economic interests. As long as we further perfect the "conscious organization of social production, in which production and distribution are carried on in a planned way", we will surely make the relationship between people and nature, and between people themselves further harmoniously coordinate so as to realize the survival of human aesthetics.

Finally, it is the question of limitations of the times and history about the ecological outlook of Marx and Engels. Engels once pointed out: "Taken historically the thing would have a certain meaning: we can only know under the conditions of our epoch and as far as these allow."[28] As a Marxist historicist, Engels speaks very profoundly. Therefore, we can also recognize the inevitable historical limitations of the ecological aesthetic outlook of Marx and Engels. In the mid-19th century, capitalism was still in the flourishing period of historical development. The contradiction between human beings and nature was yet to be highlighted. By the middle of the 20th century, the contradictions between human beings and nature had become increasingly prominent, and the environmental issue had become extremely acute. In human society, not only the economic crisis as revealed by Marx and Engels but also the ecological crises they have never seen before have emerged. Therefore, Marx and Engels certainly do not have enough arguments on the acuteness of environmental issues. Although they foresaw the historical trend that socialism will inevitably replace capitalism, they had not yet anticipated the trend of ecological civilization replacing industrial civilization. The full understanding of the unique value of nature and ecology needs to be further deepened. For example, Marx believed that there is value for air and water only when used by mankind, and this notion needs to be further deepened. Facts have proved that natural resources such as air and water are not inexhaustible. Moreover, they not only have their unique value, but also have their bearing capacity. However, Marx's and Engels's expositions on ecological issues from the perspective of the fundamental causes of social economy still have extremely important values in the present age, and they shine with immortality.

[27]Ibid., p. 458.

[28]Ibid., p. 562.

4.2 Discourse of China's Contemporary Theory of Socialist Innovation on Ecological Civilization

In October 2007, after China's socialist material civilization, spiritual civilization, and political civilization, China clearly put forward the important measure of building "ecological civilization" for the first time. It will serve as a new goal and a new requirement for building a well-to-do society in our country and will give a series of in-depth expositions on the connotation of "ecological civilization." This is undoubtedly of great significance to the construction and development of our country's contemporary ecological theories including ecological aesthetics and ecological literature. It will also open up a vast space for development. We should learn well and understand the profound theoretical connotations of the discussion on "building an ecological civilization" in the theory of socialist innovation in contemporary China. We should also use it to guide the construction of ecological aesthetics and ecological literature. Here, we will collectively refer to the theory of socialist ecological civilization with Chinese characteristics.

(1) **Proposition of Socialist Ecological Civilization with Chinese Characteristics and Its Rich Connotation**

Facts have proved that the theory of ecological civilization with Chinese characteristics is the inheritance and development of the Marxist theory of ecology. It is an important part of the Marxist theory with Chinese characteristics and contains an extremely rich connotation.

We begin with a reflection on the traditional mode of development. It is put forward in the background of concluding the "difficulties and problems" of our country's contemporary construction and development. One of the important issues is "the resources and environment of economic growth are too expensive." It is under the premise of such deep reflection that the emergence of extremely important theoretical understanding and transcendence of the traditional model has become the realistic background and reason for "building an ecological civilization." In this sense, the theory of "building an ecological civilization" in our country also carries a positive and constructive, "postmodern" nature, which is a reflection and transcendence over the development model of industrialization of capitalism and "governance after pollution first" of China's past industrialization. Second, we discuss the two footholds. According to the theory of socialist ecological civilization with Chinese characteristics, the construction of ecological civilization "relates the fundamental interests of the people and the survival and development of the Chinese nation." This clearly proposes two footholds of ecological civilization construction: The fundamental interests of the people and the survival and development of the Chinese nation. In other words, the construction of contemporary ecological civilization is based on the survival and development of the people and the nation. Its purpose is the well-being of the people, the nation, and humanity, which is the embodiment of the "people-oriented" principle in the scientific concept of development as the contemporary development of socialist humanism. Third, there are the three relations. The contemporary theory of ecological civilization with Chinese characteristics involves three extremely important relations in the construction of contemporary ecological civilization. The first is the relationship between ecological civilization and economic development. On the one hand, the modernization of our country regards development as the "top priority for governing and rejuvenating the country." At the same time, we regard "protecting the environment" as the "foundation" of economic development and clearly put forward the idea of "sound and rapid development." This unifies the construction of ecological civilization with economic development in the process of modernization, and makes it an integral part of socialist modernization. The other is the relationship between ecological civilization and science and technology. China's socialist modernization explicitly shifts economic and social development from relying on resource and environmental consumption to relying on scientific and technological progress. It also points out that ecological civilization must make good use of scientific and technological means, and "develop and promote advanced and practical technology of conservation, substitution, recycling, and pollution control." Ecological civilization construction and science and technology will be well unified. The theory of ecological civilization with Chinese characteristics also involves the relationship between China and the world in the construction of ecological civilization, and proposes "to strengthen the capacity building to address climate change and make new contributions to the protection of the global climate." This shows the global character of the contemporary ecological environment and the international ecological responsibility that China should bear as a responsible, big country. Fourth, there are three important measures for the construction of ecological civilization. The theory of ecological civilization with Chinese characteristics emphasizes that the construction of ecological civilization should start with three aspects: legal policy, institutional mechanisms, and work responsibility systems. In terms of laws and policies, the transition from previous administrative penalties to tax channels and scientific proofs should be made. In terms of institutional mechanisms and work responsibility systems, environmental protection should be closely integrated with the performance evaluation of local and cadres' work. Fifth, there are five major changes involved in the construction of contemporary ecological civilization. First, on the economic level of industrial structure, we should shift from, "relying mainly on resource consumption to relying mainly on scientific and technological progress, improving the quality of workers, and management innovation." In terms of the mode of growth, we should shift from the blind pursuit of economic benefits to "sustainable" and "sound and fast" methods. Regarding consumption, we should shift from the blind pursuit of material needs, extravagance, waste, and comparing with each other to saving and moderation. In theory, we should make "the concept of ecological civilization firmly established in the whole society." Here, "the concept of ecological civilization" should include contemporary ecological philosophy, ecological ethics, ecological aesthetics, and ecological literature, which have a very rich contemporary ecological and cultural form with the advanced nature of the times and Chinese characteristics. Finally, regarding development goals, we should make our country a "country with a good ecological environment."

(2) The Great Significance of the Proposal of "Ecological Civilization Construction" Theory

The theory of socialist ecological civilization construction in our country is of great significance to the contemporary ecological theory construction, including ecological aesthetics, in our country. The theory of "ecological civilization construction" provides a solid, realistic basis for the construction of contemporary ecological theory in our country. For a long time, the theory of ecology in our country has been basically marginalized. It has been due to the incomplete nature of ecological theory itself and the fact that theorists are not unified in their understanding of the contemporary social and theoretical transformation. This leaves ecological theory is basically in question. Regarding the proposal of the contemporary theory of socialist ecological civilization, the construction of contemporary ecological theory is an important part of the theory of socialism with Chinese characteristics. This is in line with the direction for the construction of the advanced socialist culture in our country, and precisely reflects the inevitable trend of the transition from industrial civilization to ecological civilization, as well as the theoretical "human-centered" to "friendly transformation of people and the environment" in our contemporary society. Under the broad topic of "ecological civilization," a series of theories such as economic concepts, development concepts, consumption concepts, political outlooks, philosophy concepts, values, aesthetic concepts, and literary concepts are re-established between human beings and nature, manifesting in the concept of "win-win" in economics, the concept of "symbiosis" in philosophy, the concept of "ecological value" in ethics, the concept of "poetic dwelling" in aesthetics and the concept of "human and nature friendly" in literature. Each of these will be the social and ecological concepts that need to be developed. Of course, this good situation at present only provides a good realistic environment and premise for the development of ecological theory in our country. It does not mean that the theory itself is complete. Rather, we also need our ecological theorists to redouble their efforts in this unprecedented good environment for better theoretical innovation, as theoretical work should play a positive role. It ought to make its due contribution by being imported into the torrent of "socialist ecological civilization" of our contemporary socialist construction. What needs to be particularly explained here is that the proposal of the theory of socialist ecological civilization in our country has pointed out the direction for creating an ecological theory with Chinese characteristics, including ecological aesthetics. That is, we must first proceed from China's national conditions. The construction of ecological theory in China must closely integrate with China's economic, social, and ecological conditions and cannot copy Western theories. At the same time, the construction of ecological theory in China should draw lessons from and absorb the rich ecological wisdom of traditional Chinese culture, so that it has the Chinese style.

4.3 The Basic Contribution of Contemporary Western Marxism in the Construction of Ecological Theory

The ecological theory of contemporary Western Marxism inherits the ecological concept of traditional Marxism and argues that capitalism will inevitably lead to ecological crisis. It also conducts reflection and criticism on the alienation of the relationship between human beings and nature in today's society, and advocates an ecological awareness and ecological civilization. On this basis, it puts forward the theoretical conception of ecological socialism, which has important enlightenment and reference significance for analyzing and solving the ecological problems facing humankind today. The basic characteristics of its theory is its attempt to use historical materialism to analyze and explore the root causes of the contemporary ecological crisis and its solution, and put forward the political ideology of ecological socialism. This chapter focuses on the ecological socialism theory of British scholar, David Pepper.

(1) **The Background and Nature of Contemporary Eco-socialism**

Contemporary ecological socialism and ecological Marxism is something that some left-wing intellectuals attempt to use. They treat Marxism as their weapon to demonstrate their attitude to the serious contemporary ecological problems and controversy on ecological issues, to describe their own point of view, and to put forward solutions to the problem. They flourished under increasingly serious ecological circumstances after the middle of the 20th century, amid the booming of various ecological theories. Judging from the international perspective, there are two major contradictions facing the current theoretical field. The first is the contradiction between Marxism and capitalism, which we understand as the contradiction between Marxist theory and the contemporary reality of capitalism, the contradiction between serious environmental pollution and environmental pollution transfer, the theory of capitalist industrialization, the contradiction between the theory of infinite expansion, and the theory of Marxist ecological practice. Second, there is the so-called "red and green dispute," mentioned by David Pepper in *Eco-socialism from Deep Ecology to Social Justice*. "Red" refers to Marxism, and "green" refers to the green theory, that is, ecocentrism, which even includes the "Green Party." Under such circumstances, eco-Marxism and eco-socialism inevitably come into existence and take part in the controversy and the solution of the problems with their own opinions. The nature of Western modern ecological socialism belongs to the general trend of thought of contemporary Western Marxism, and holds a radical leftist standpoint.

(2) **Today's Ecological Socialism Is to Give a Powerful Counterattack to Various Attempts to Misinterpret Social Contradictions in the Increasingly Serious Ecological Crisis Situation**

In the current situation of increasingly serious ecological crisis, some with ulterior motives have tried to misinterpret social contradictions, thinking that class contradictions have been replaced by environmental problems and other risks and are no longer the major social contradictions. Contemporary ecological socialism has criticized this theory, arguing that "although the above arguments contain some elements of truth, they are generally based on an extremely wrong and exaggerated position."[29] In fact, the emergence of these problems proves that "the theories and practices of socialism and communism have become more in demand than ever before."[30] They insisted on the basic Marxist position.

(3) Full Confirmation of Ecological Theory of Marx and Engels by Contemporary Ecological Socialism

Under the current situation of serious ecological crisis, some theorists have denied that the theory of ecology is contained in Marx and Engels' theory, and that there is no ecological school of Marxism. Contemporary ecological socialism has criticized this view and held that "Marxism does already contain enough in the way of a meaningful- albeit mostly implicit—perspective on ecology",[31] "Marx and Engels were forerunners of human, political and social ecology ".[32]

(4) Efforts to Interpret the Current Serious Ecological Crisis in Marxist Historical Materialism Way

Contemporary Western Marxists try to explain the causes and solutions to the current ecological crisis by using the social existence and social awareness of Marxism, as well as historical materialism as the economic basis and the superstructure. They argue that "A capitalist mode of production implies 'capitalistic' relations with nature and with each other—capitalist relations of production, to which correspond particular political and legal arrangements and particular 'forms of social consciousness'". And also said "if we want to change social and society-nature relationships we must seek such changes not simply in the minds of people—their insights or philosophies, i.e. their 'forms of social consciousness'—but also in their material, economic, life."[33] They also argue that alienation may be controlled by "reappropriating collective control over our relationship with nature, via common ownership of the means of production:".[34]

(5) Adhere to the Basic View That Ecological Socialism is "Anthropocentrism."

David Pepper pointed out in the "Foreward" of Eco-socialism: From Ecology to Social Justice, that "The aim is to outline an eco-socialist analysis that offers a radical, socially just, environmentally benign—but fundamentally anthropocentric—perspective on green issues."[35] He believes "that this is what the green movement now needs, rather than its current 'biocentric' and politically diffuse approach, in

---

[29]Pepper [4], "Preface", p. 1.

[30]Ibid.

[31]Ibid., p. 91.

[32]Ibid., p. 93.

[33]Pepper [4], p. 101.

[34]Ibid., p. 355.

[35]Ibid., p. *Forward* p. 2.

order to appeal to the concerns of the many who are still alienated by or indifferent to it. Furthermore, and pragmatism aside, I think it important not to allow our concern for nonhuman nature to become a substitute for, or a priority over, concern about people. Some greens believe that we should protect and respect nature for its 'intrinsic worth', whatever that is, rather than its worth for (all) people. I am not comfortable about this. Social justice, I think, or the increasingly global lack of it, is the most pressing of all environmental problems. And the Summit showed clearly that attaining more social justice is the prerequisite for combating ozone depletion, global warming and the rest."[36] He puts social justice first and negates the intrinsic value of nature. In Ecological Socialism's Summary of *Eco-socialism: From Ecology to Social Justice*, Pepper argues that "Eco-socialism is anthropocentric and humanist. It rejects the bioethic and nature mystification, and any anti-humanism that these may spawn, though it does attach importance to human spirituality and the need for this to be satisfied partly by non-material interaction with the rest of nature. But humans are not a pollutant, neither are they 'guilty' of hubris, greed, aggression, over-competitiveness or other savageries".[37]

Therefore, the limitations of Pepper's theory of ecological socialism are obvious. The first is his improved position. He attempts to change the existing social system through a modified approach, through unionism and utopian socialism, returning to land movement, and at the same time change the ecology. The second is the reservation of anthropocentrism. This is a retreat from the ecological outlook of Marx and Engels. Facts have proved that the "anthropocentrism" accompanied by the industrial revolution has shown itself to lag behind the times, and should be replaced by the "ecological holism" and "ecologic humanism" of the gradual development of ecological civilization. The theory of "symbiosis," "win-win," and "environment-friendly society building" advocated by the theory of socialist ecological civilization with Chinese characteristics inherits the Marxist ecological concept, which also draws on the relevant points of view found in Western Marxist ecological concepts. Of course, it is based on the existing experience of contemporary Chinese socialism.

Here, we have to mention an important ecological socialist theorist, Andre Gorz of the present age in France. In his later life, he published a series of important essays on ecological socialism such as *Ecology as Politics* (1975), *Ecology and Freedom* (1977), *Critique of Economic Reason* (1988) *Capitalism, Socialism and Ecology* (1991) etc. In these treatises, Gorz put forward a series of viewpoints in the vein of ecological socialism. The points are as follows: The profit motive of capitalism must destroy the environment. The crisis of capitalism is essentially an ecological crisis. The best choice to protect the ecological environment is to take the socialist road, and so on. It is important that he closely linked the critique of capitalist society with the critique of the unlimited expansion of productive forces and science and technology. He thus avoided falling into the trap of "anthropocentrism." In his later years, facing the complex and ever-changing information society and consumption society, he felt that the "utopian" ideal of eco-socialism was difficult to achieve. At the age of 84,

[36]Ibid., p. 1.

[37]Ibid., p. 354.

Gorz chose to commit suicide. His series of essays on ecological socialism have become a valuable spiritual legacy for humankind to truly move toward ecological civilization.[38]

In short, the theory of western ecological socialism criticizes the capitalist system's dual plunder of nature and humankind, as well as adheres to a certain cultural standpoint in order to promote the harmonious development of human beings and nature. The approach it proposes to solve the ecological crisis through the socialist system are very valuable spiritual assets for the construction of contemporary socialist civilization.

References

1. Marx and Engels. (1972). *Selected works.* People's Publishing House (《马克思恩格斯选集》,人民出版社1972年版).
2. Marx and Engels. (1979). *Collected works.* People's Publishing House.
3. Marx. (1975). *Capital.* People's Publishing House.
4. [English] Pepper, D. (2005). *Ecological socialism: From deep ecology to social justice*, translated by Liu Ying. Shandong University Press.

[38]See *Journal of Social Sciences*, 7th edition, October 11, 2007.

Part III
The Western Resources of Eco-aesthetics

Chapter 5
Ecological Aesthetics in Western Countries

5.1 Western Ecological Aesthetics Resources Since the 18th Century

Although the West had begun the Enlightenment and the Industrial Revolution in the 18th century, it has generated "anthropocentrism" marked by humanism. Although its philosophical mode of thinking is mainly in terms of the subject/object dichotomy, there are also some ideological and theoretical resources on natural and ecological aesthetics that are worth learning from.

(1) **Vico's Original Poetic Thinking**

Giambattista Vico (1668–1744) was an Italian jurist, historian, linguist and aesthetician, former professor of rhetoric at the University of Naples, and historian of the Kingdom of Naples. He published the book *New Science* in 1752, which covered many disciplines. The goal was to establish an all-encompassing social science and to explore the law of the development of human society's history and culture. This book relates to the original poetic thinking, which is closely related to aesthetics. We can call it a pioneering work of modern aesthetics. What is particularly valuable here is that Vico's primitive, poetic thinking is an original and natural aesthetic thinking. Thus, he also became the direct pioneer of ecological aesthetics. According to the *New Science*, primitive, natural thinking has the following characteristics:

1. Imaginative Thinking with Human Senses

The human body has always been a sensory, natural entity. According to Vico, the naturalness of human primitive thinking first manifests itself in this kind of thinking that relies on the human senses. He said: "These primitive people did not have the ability to reason, but had strong feeling and vivid imagination. This metaphysics is their poem, and poetry is a function they have when born (because they are born with these senses and imagination)."[1] In Vico's view, the primitive humans did not have

[1] Vico [1], pp. 161–162.

F. Zeng, *Introduction to Ecological Aesthetics*,
https://doi.org/10.1007/978-981-13-8984-9_5

the ability to think abstractly. The spirit and the flesh were not completely separated, and the subject and the object were not separated. Therefore, their thinking relies upon the natural imagination, and is a sense of the natural form of thinking.

2. This is a way of thinking of the unity of human beings and nature according to which "all things have spirit"

It is precisely because of the era in which the primitive humans existed. People still had no abstract thinking ability. They lacked scientific knowledge, which therefore led to the "animism of all things". Any magical things were attached to a natural entity. For example, there is thunder and there are thunder gods, rains and rain gods, flowers and flower gods, and so on. Human beings have only awe, reverence, and dialogue with these spiritual things. As Vico points out, primitive humans "by their own point of view, things that amaze themselves each have a physical presence, just as children hold inanimate things in their hands and talk to them as if they were living beings."[2] This is indeed "pantheism", but to some extent also demonstrates a "naturalistic" tendency. This is the defining feature of the era of primitive ignorance.

3. Metonymy and metaphor in primitive poetic thinking: human beings and nature have obtained the status of equality

An important feature of poetic thinking is the use of various natural phenomena to act as a metaphor for reality. As Vico points out, in primitive thinking "it is noteworthy that most expressions of inanimate things in all languages are formed by metaphors of the human body and its parts and of the sensual and erotic nature of the human being. For example, use the 'head' to express the top or the beginning, use the 'forehead' or 'shoulder' to express a mountain site, the needle and potatoes can have eyes, cups or pots can have 'mouth', rake, saw or combs can all have 'teeth', any void or hole can be called 'mouth', wheat's beard, 'tongue' of shoes, 'throat' of river, 'neck' of ground, etc …".[3] Metaphor in primitive thinking is to use a part of the human body to describe natural things. There is another kind of "substitution" or "metonymy" that replaces the part by the whole or vice verse. For example, Vico wrote of "ugly poverty, miserable old age and pale death"[4] etc. In short, no matter what kind of metaphor is based on the friendship between human beings and nature, the starting point is that human beings and nature are in an equal position.

In short, Vico's primitive poetic thinking is consistent with Leopold's "thinking like a mountain" in *A Sand County Almanac*. It has an important reference for us to build the artistic thinking mode of ecological aesthetics.

(2) Naturalistic Aesthetic Thoughts of Santayana of the United States

George Santayana (1863–1952), a famous philosopher, esthetician, poet and literary critic in the United States, was a professor at Harvard University and later went to Europe to live in various countries. His aesthetics is mainly found in his first book

[2] Vico [1], p. 162.

[3] Ibid., p. 180.

[4] Ibid., p. 181.

on aesthetics, *The Sense of Beauty* (1896). The connotation of nature and ecology in his naturalistic aesthetics is mainly reflected in the following aspects:

1. Emphasized that the sense of beauty is the satisfaction of one's "natural function". Ecological aesthetics not only treats people as rational, but also treats them as living beings who form a whole with nature. It is in this sense that ecological aesthetics never denies the natural characteristics of human beings. Of course, it does not exaggerate the natural characteristics of human beings. In *The Sense of Beauty*, Santayana emphasizes that "beauty is a value," in addition to the value of good. However, it also contains the value of satisfying people's "natural functions". "Beauty is one of the highest merit, it satisfies a natural function, and some of our basic needs or abilities in mind. Therefore, beauty is an intrinsic positive value and a pleasure."[5] This kind of aesthetic value that satisfies the natural functions should contain some kind of ecological aesthetics.
2. Stressed that all human functions are involved in the aesthetic. Santayana's *The Sense of Beauty* emphasizes that all human functions are involved in the aesthetic: "all human functions may contribute to the sense of beauty".[6] In fact, to a certain extent, this is a refutation of the participation of traditional audiovisual and sensory organs in the aesthetic. It is the forerunner of Berleant's "Aesthetics of Engagement" in later generations. However, the role of Santayana in emphasizing human sexual instincts is somewhat over-emphasized, which is a defect.
3. Emphasizes aesthetic utilitarianism and criticizes the traditional aesthetic "disinterestedness", which opens the way for the establishment of ecological principles in ecological aesthetics. As we all know, ecological aesthetics does not agree with Kant as the representative of the aesthetics of "disinterestedness", emphasizing that aesthetic contains the ethical principles of ecology and the value of life. Santayana clearly opposes the traditional aesthetics of "disinterestedness", and advocates that aesthetics contain utilitarian interests. He said: "The characteristics of aesthetic pleasure is not its disinterestedness"[7] and added that "it is impossible to separate the aesthetic function of things from the practical and moral function of things in the history of art, it is also impossible in the reasonable judgment of artistic value."[8] What Santayana calls "the practical function of things" embodies the function of ecological life, which is in line with his philosophical aesthetics of naturalism.
4. Put forward the important aesthetic ideas "that beauty is a harmony of life," as the prelude to aesthetic praise and interpretation of ecological harmony of life.

[5]Santayona [2], pp. 30, 34.

[6]Ibid., p. 36.

[7]Santayona [2], p. 25.

[8]quoted from Liyuan [3], p. 218.

As we all know, eco-aesthetics is a tribute and aesthetic appreciation of the harmony of life as a whole. Santayana later proposed a new statement about beauty: "beauty is a harmony of life, an image that is felt and dissolved in the form of eternal life."[9] It is here that an important conclusion, that "beauty is a harmony of life," is actually an elaboration of the chords, resonance, and harmony between human beings and nature, which is just an ecological aesthetic state.

(3) **Dewey's Ecological Aesthetic Thought on "Living Creatures" of the United States**

John Dewey (1859–1952), a famous philosopher, educator, and esthetician in the United States, taught at the University of Michigan, the University of Chicago, and Columbia University, and formed a pragmatic and important faction with his disciples: the Chicago School, which wrote quite good works, and has a wide range of international influence. His main aesthetic work is *Art as Experience* was published in 1934. This is a very special treatise of aesthetics, as it contains an extremely rich aesthetic sense of ecology in the era of soaring industrialization at that time.

1. Puts forward the important ecological concept that "nature is the mother of mankind"

Dewey pointed out: "Nature is the mother and the habitat of man, even if sometimes a stepmother and an unfriendly home. The fact that civilization endures and culture continues—and sometimes advances—is evidence that human hopes and purposes find a basis and support in nature."[10]

2. Strives to break through the traditional materialism and spirit and the binary opposition between man and nature

In the early 1920s, Dewey conducted his own academic activities based on the traditional antagonism of the subject/object dichotomy in order to overcome their efforts. He named his philosophical treatise *The Transformation of Philosophy*, that is, his attempt to change the traditional dualistic opposition and replace it with the performance of a viable organic world outlook. In *Art as Experience*, He clearly defined his own aesthetics as a breakthrough in the traditional dichotomy between subject and object. He said: "A complete answer to the question would involve the writing of a history of morals that would set forth the conditions that have brought about contempt for the body, fear of the senses, and the opposition of flesh to spirit."[11] He went on to further analyze: "Oppositions of mind and body, soul and matter, spirit and flesh all have their origin, fundamentally, in fear of what life may bring forth. They are marks of contraction and withdrawal. They are marks of contraction and withdrawal."[12]

---

[9]Ibid., pp. 220–221.

[10]Dewey [4], pp. 31–32.

[11]Dewey [4], p. 22.

[12]Ibid., p. 26.

3. To restore the continuity of human beings and animals as a starting point to build their own empirical aesthetics

Dewey believes, "To grasp the sources of esthetic experience it is, therefore, necessary to have recourse to animal life below the human scale."[13] To this end, he created a concept of "living creatures" as a starting point for the definition of human beings and their entire empirical aesthetics. It is through the relationship between this "living creature" and the environment that the so-called aesthetic experience is formed. Dewey fully recognized the consistency of human and animal, but also noted the differences between humans and animals. In his book *The Transformation of Philosophy*, he pointed out that animals do not preserve past experiences, while people can save experiences. This shows that animals are not rational, but people are rational.

4. The artistic experience of aesthetics is "a complete experience" produced by the interaction of human beings as a "living creature" with the environment

Dewey's classic aesthetic is summed up as one of "living creatures" being in conflict and harmony with the surrounding environment, which formed a complete experience. He said: "Unlike these experiences, we have an experience as we experience the fulfillment of the material being accomplished."[14] "The moments when the creature is both most alive and most composed and concentrated are those of fullest intercourse with the environment, in which sensuous material and relations are most completely merged."[15]

5. Puts forward the important ecological aesthetic viewpoint of "We live in the world, and it becomes our home"

When explaining the connotation of aesthetic as "a perfect experience", Dewey pointed out that it is people who have formed a complete experience in the conflict with and integration of the environment so that the environment integrates with people and becomes the world inhabited by people. The world thus becomes the home of people's aesthetic existence. He said: "through habits formed in intercourse with the world, we also in-habit the world. It becomes a home and the home is part of our every experience."[16]

6. The role of artistic, aesthetic experience is to make the whole of life dynamic

In summarizing the role of aesthetic experience and art, Dewey concludes that art implies that people have more powerful and abundant vitality. He said that art should make the whole life more energetic, "allowing him to enjoy his life through appreciation."[17] He further elaborates: "Art is the living and concrete proof that man is capable of restoring consciously, and thus on the plane of meaning, the union of sense, need,

[13]Ibid., p. 21.

[14]Dewey [4], p. 41.

[15]Ibid., p. 119.

[16]Ibid., p. 120.

[17]Ibid., pp. 30–31.

impulse and action characteristic of the live creature."[18] This is the return to the aesthetic starting point from "living creatures". The ultimate effect of the aesthetic and the arts is to restore the vitality of human beings as "living creatures."

(4) **Chernyshevsky's Life—natural Aesthetics**

Nikolai Chernyshevsky (1828–1889), Russian revolutionary and democratic thinker, writer and critics, wrote a novel *How to Do?, Prologue,* and many aesthetics papers. *The Aesthetic Relationship between Art and Reality* was published in 1855 as a dissertation by Chernyshevsky. He strongly criticized what was then popular Hegel idealist aesthetic thinking in the book, advocating the materialist aesthetics of "beauty is life" of. It is of progressive significance to advocate the revolutionary democratic aesthetics of "beauty should be such life" and "art is life's textbook." Today, when we read this important aesthetics works of Chernyshevsky, we find that his proposition, "Beauty is Life," implies an extremely rich ecological and natural aesthetic thought, which deserves our reference in the construction of contemporary ecological aesthetics.

1. Puts forward the naive aesthetic proposition of life that "beauty is life, is healthily alive."

The main purpose of writing this book is to criticize the idealism represented by Hegel on aesthetic concept of, "Beauty is the perceptual manifestation of philosophy". Due to the control of Russian procuratorial authorities at the time, Chernyshevsky could not directly use Hegel's remarks in the text, but could only use Fischer's remark by Hegel's disciple, who was essentially consistent with Hegel's remarks. Fischer's definition of beauty is that, "beauty is a concept in the limited manifestations; beauty is an individual perceptual object that is viewed as a mere representation of the concept".[19] Chernyshevsky pointed out that this kind of aesthetic viewpoint is "too pandemonium" and "too narrow". What is most important is the lack of any "dullness" necessary for any theory to exclude the beauty of life and nature from beauty. Chernyshevsky pointed out: "Here, I think it is necessary to point out that beauty is the definition of the unity of concept and image, and it is not about living natural beauty but aesthetic works of art." In this definition, it already contains the kind of aesthetic tendency of the bud or result that usually regards artistic beauty as better than living reality.[20] He further pointed out in a footnote to the book *Aesthetic Relations between Art and Reality* that this view is in fact only paying attention to the performance of beauty, and neglects the beauty of things. "I mean that is essentially beautiful thing, not because it is beautifully shown in art so beautiful things; I am talking about beautiful things and phenomena, not their beauty in works of art."[21] This critique should be the key to demonstrating Hegelian aesthetics to obliterate the beauty of life and nature. In the end of the book, he concludes: "'beauty is the

[18] Ibid., p. 29.

[19] Chernyshevsky [5], p. 3.

[20] Ibid., p. 5.

[21] Ibid.

complete manifestation of the general idea on individual phenomena', the definition of beauty can't stand the criticism; it is too broad and sets out the tendencies of all forms of human activity."[22]

So, what is beauty? Chernyshevsky put forward his own unique, important proposition, which contains the life and the richness of life such as the, "beauty is life." He said: "So, a definition like this: 'Beauty is life'. Anything wherever we see it in our life as we understand it is beauty; anything that shows life or reminds us of life is beauty."[23] He further elaborated on the definition of "beauty is life." "To the ordinary people, 'a good life' and 'what should be life' is enough to eat, live well, and have enough sleep. But for peasants, the concept of life always includes the concept of labor at the same time. It is impossible to live without labor, and it is bored. It is hard work, but does not make them exhausted as a result of a rich life. That young peasants or peasant girls have a very fresh and ruddy complexion—according to ordinary people's understanding, this is the first condition of beauty. There are sufficient clothing and food and hard work, and so farm girls have a strong physique and strong looks. This is a necessary condition for country beauty. 'Fragile' beauty in high society seems to be categorically 'not beautiful' in the country-folk, and even gives them an unpleasant impression, as they have always thought that 'emaciation' is the result of illness or 'fate'."[24] This interpretation of "beauty" includes a very rich meaning. First of all, "beauty" is a full guarantee of human life, including sufficient food, good living, and adequate sleep, etc. Secondly, "beauty" is a kind of life and health that is strong, energetic, and ruddy, brought by hard work, rather than the "fragile" and "emaciated" of the upper class. In a word, in the view of Chernyshevsky, "beauty" is a beautiful and healthy state of life in which people are in a harmonious relationship with the natural ecology in their work, and are energetic. He said, "Yes, health will never lose its value in the minds of people because if it is unhealthy, it is too wealthy, extreme luxury, and life is not good. So rosy face and full of spirit are still attractive for the upper class people; but morbidity, infirmity, drowsiness, and servitude also have the beauty of value in their minds, as long as it is the result of an extravagant, idle life".[25] What we once criticized the beauty of the human life is a vulgar materialism." Although this critique is not unreasonable, the beauty of a healthy, energetic life still has its rationality on the whole, and is an important form of beauty that accords with the law of natural ecology.

2. Puts forward the important idea of "live beauty is higher than art beauty", criticizes Hegel's "art-centric theory" and maintains the legal status of natural beauty

The important connotation of Hegel's aesthetics is that aesthetics is "beauty of the philosophy of art."[26] In this way, natural beauty is basically excluded from the aesthetic. This viewpoint also determines that the beauty of art is higher than beauty

[22] Ibid., p. 101.

[23] Chernyshevsky [5], p. 6.

[24] Ibid.

[25] Ibid., p. 7.

[26] Hegel [6], (Vol. 1), p. 3.

of nature in life. The above view of Hegel is a typical concept of "art center" and a reflection of "anthropocentrism" in aesthetics and artistry. Because, according to Hegel, humankind is the perfect embodiment of absolute ideas, then the art created by humankind has the supreme status in aesthetics, and nature is marginalized to the pre-aesthetics stage. In our opinion, this view is not in conformity with the facts. Aesthetics should include three parts: art, the aesthetic of nature, and life. However, nature also has a basic place in the spiritual life of mankind, including art. The theme of the book *The Aesthetic Relationship between Art and Reality* is trying to subvert this dominant traditional theory. He said: "The essence of this essay is to defend the reality by comparing reality with imagination with each other, it is an attempt to prove that the work of art must not be compared with the reality of life."[27] His conclusion is: "nature and life are better than art."[28] He vividly compared life and nature to a gold bar without a stamp, and many people refused to take it because it did not have a stamp. Works of art were like banknotes, with little intrinsic value, but nevertheless, everyone value it. He said: "The beauty and greatness of real life is rare for us to reveal the truth, and few people can pay attention to the unnoticed thing and value it."[29]

In order to demonstrate life, the beauty of nature is higher than the beauty of art. Chernyshevsky rejected the Hegelian view denying the beauty of life and nature for eight reasons. First, some people think that "the beauty in nature is unintended" and therefore can't be as good as the beauty in art. Chernyshevsky believes that human beings' power is far weaker than that of nature, and that their works are more crude, clumsy and stupid than their natural counterparts. Therefore, "the merit of a work of art is still outweighed by the advantage of a flaw in production."[30] Second, some people think that "the beauty in nature is not created intentionally, and beauty is seldom seen in reality." Chernyshevsky believes that complaining about the real beauty of the scarcity is not entirely correct. Rather, real beauty is not as scarce as the German aestheticians said. "There are so many beautiful and magnificent scenery that there are many places that can be seen everywhere."[31] Third, some people think that "the beauty of the reality is fleeting." Chernyshevsky criticized this point of view from the aesthetic era, he argues that "the beauty of every generation is and should also exist for that generation: it does not undermine harmony, not violating the beauty of that generation; when the beauty dies with that generation, the next generation will have its own beauty, and no one will complain about the new beauty."[32] Fourth, some people think that "beauty in reality is infrequent." Chernyshevsky believes that this view can't be established. "Beauty is not often, is it hindering the reason why it is beautiful? Is it because the beauty of a landscape will be dim at sunset, is this scenery

[27]Chernyshevsky [5], p. 100.
[28]Ibid., p. 81.
[29]Ibid., p. 83.
[30]Ibid., p. 42.
[31]Chernyshevsky [5], p. 43.
[32]Ibid., p. 45.

less beautiful in the morning?"[33] Fifth, on the "beauty of reality is beautiful, simply because we look at it from a point of view it looks beautiful from that point of view, "Chernyshevsky believes that this is untenable because the aesthetic point of view is the relationship between perspectives. For the aesthetics of life and nature there needs to be a perspective of the perspective. For art, the aesthetic needs the same perspective. Sixth, on the issue of "the reality of beauty it is not linked with a group of objects, but rather is linked with a certain object," "thus undermining the overall beauty and unity." Chernyshevsky believes that this is actually a perfect request. In fact, there is no such problem in real life. People only seek something good rather than perfect. He said, "If there is a landscape, simply because there are three shrubs somewhere in it—if it is better to have two or four clumps,—would anyone want to say that the scenery is not beautiful?"[34] Seventh, some people think that "real things can't be beautiful because it is a living thing, and in it is a realistic process of life with all its rough and not beautiful details." Chernyshevsky believes that the view and beauty is not the thing itself but rather is related to the pure form of things, can't be established. He claims, "idealistic idealism, which is far more absurd than this, can't imagine it."[35] Eighth, regarding the idea that "individual things can't be beautiful, because it is not absolute, but the beauty is absolute", Chernyshevsky thinks, on the contrary, individuality is the fundamental character of beauty, "starting from the idea that individuality is the most fundamental feature of beauty, however, we come to the conclusion that absolute guidelines are beyond the realm of beauty.'"[36]

In a word, Chernyshevsky here strongly criticizes the view of Hegelian School in excluding the "aesthetic center" of natural aesthetics, and maintains the legal status of natural aesthetics. He uses a well-known metaphor in his demonstration, that "tea element is not tea, alcohol is not wine,"[37] to show that all ideas can never replace life and nature itself. His final conclusion is: "natural beauty is indeed beautiful, and it is beautiful from beginning to end."[38]

3. Explains the aesthetic value of nature from the connection between nature and human life

Chernyshevsky made his own interpretation of the aesthetic value of nature. He mainly discusses the aesthetic value of nature from the connection between nature and human life. He said, "anything that shows life or reminds us of life is beautiful."[39] The "life" to which he here refers, of course, is human life, especially the healthy and energetic vitality. For animal aesthetics, he applied the theory of "all things showing life" to discuss it, that is to say that animals that show the vitality of human beings are

[33]Ibid., p. 46.

[34]Ibid., pp. 48–49.

[35]Ibid., p. 49.

[36]Ibid., 52.

[37]Chernyshevsky [5], p. 72.

[38]*Chernyshevsky's Literature* (Volume 2), Translated by Xin Weiai, Shanghai Translation Publishing House, 1979, p. 35.

[39]Chernyshevsky [5], p. 6.

beautiful. He said, "in the eyes of human beings, the beauty of the animal kingdom shows the concept of a man's life about a fresh and healthy life. On mammals—our eyes almost always compare their bodies to human shapes—people think beauty is a round figure, full and strong; the elegance of the movement is beautiful, for only creatures with, beautiful look, that is, creatures that remind us of good-looking people rather than deformities, are graceful."[40] Here, fresh and healthy, plump, active, and elegant are all the vibrant performances of human health and vitality. For plants, however, he is to make us think of life to be discussed. He said, "For plants, we are happy to see the fresh colors, lush and diverse shapes, because it shows the thriving life of strength."[41] Obviously, the freshness of plants, lush and diverse shapes remind us of the thriving life of people, so it is beautiful. He said: "The beauty of nature has its beauty only as an implication of man."[42]

What should be clarified here is that Chernyshevsky and Hegel alike all discuss the aesthetic value of nature according to the connection between nature and human beings. However, there is a fundamental difference between the two. Chernyshevsky discusses the aesthetic value of nature from such aesthetics that "beauty is life", where life is healthy living, and from the relationship between the vitality of human beings and nature. Hegel, on the other hand, starts from his theory of aesthetics and discusses the aesthetic value of nature from the connection between nature and philosophy. Hegel put forward the famous natural aesthetic proposition, "Obscure Forebrain", pointing out that "nature can be called beauty when it is the perceptual performance of specific concepts; that is to say, when we look at the natural images in conformity with the concept, we hazyly presume the conformity of the above-mentioned sensibility and reason, and in the perceptual observation, the inherent inevitability and coordination of all the parts are also sensitive." "The contemplation of the beauty of nature ends with such a vague anticipation of the concept."[43] This is still limited to the relationship between nature and philosophy, and is a copy of a kind of discrimination against the concept of natural aesthetics.

4. Criticizes the lofty view of "anthropocentrism" in Kant's "concept overwhelming form," and put forward the idea that "loftiness is the loftiness of nature itself", admitting that nature has its lofty values

Kant discusses the concept of the "sublime" in *Critique of Judgment*. He states that "sublimity does not exist in nature but only in our mind."[44] Here, Kant denied the lofty aesthetic value of nature itself, which is obviously a kind of "anthropocentrism" to regard sublime as a completely human concept. In his *Aesthetic Relationship between Art and Reality*, Chernyshevsky criticized this view of Kant more deeply, and focused on maintaining the lofty aesthetic value of nature. From multiple perspectives, he discusses the important point of view that the sublime lies in nature and not in human

40 Ibid., 9.

41 Ibid.

42 Ibid., 10.

43 Hegel [6] (Vol. 1), p. 168.

44 Kant [7], p. 89.

beings. He said, "A thing is much greater than anything compared with it, and that is sublime."[45] Around this point of view, he first of all directly criticized the sublime not in nature but in the concept of perspective. He insisted on materialism and held that "in observing an exalted object, various thoughts will happen in our minds and strengthen the impression we get; but the occurrence of these thoughts is an accidental matter, however, that object is still sublime anyway."[46] Here, he explicitly placed the subject before his thoughts, thereby laying the foundation of materialism for the aesthetic category of the "sublime". Then he critiqued the loftiness inflicted by the lofty notion of tradition, that the sublime is often not the infinity but the complete opposite of the infinity. For example, he said, the mountains are magnificent but measurable. The sea seems boundless but has a shore. In fact, it still maintains the lofty view of materialism that lofty ideas are not all their own. Only the lofty sense of philosophy can be infinite, and the loft of things can only be limited.

Chernyshevsky concludes, "therefore, it is very difficult for us to agree that the 'sublime is the overwhelming sense of concept' or that the 'lofty essence is to arouse infinite ideas'".[47] He wrote in the concluding part of the book: "One thing, anything that is much larger than anything human can bring to it, or is much stronger than anything else, is sublime to man."[48]

In short, in his book *The Aesthetic Relationship between Art and Reality,* we find the criticism of Hegel's idealism and the aesthetic theory of "art center," the exposition that beauty is a healthy and vibrant life, the emphasis on the beauty of life and nature and the beauty of art, as well as his exposition of sublime lies in nature itself. These ideas are of great value to us in building the ecological aesthetics of contemporary form. He urged us to further consider the proposition that "art is superior to life", which has long been believed in has actually a clear limitation. It is still the result of Hegel's idealistic aesthetics, and it inevitably leads to the depreciation of the beauty of natural ecology. Of course, Chernyshevsky's absorption of Feuerbach's humanism in the book is also a point of criticism, as well as is his anthropocentric view of "looking at nature from the perspective of the owner" when he expounds on life aesthetics. He also has some historic limitations.

5.2 Heidegger's Ecological Aesthetic Concept

American modern ecological theorists have long regarded Heidegger as a modern "metaphysical theorist with ecological concepts", that is, as an ecological philosopher. This judgment is very accurate. Heidegger is the most important ecological theorist and ecological aesthetician in the contemporary West. "The Fourfold Game of of Sky, Earth, Man and Gods" is contemporary ecological aesthetics. Heidegger's

[45]Chernyshevsky [5], p. 3.
[46]Ibid., p. 14.
[47]Chernyshevsky [5], p. 17.
[48]Ibid., 101.

thinking has changed from his early period to his late period. His philosophical aesthetics have been influenced by the philosophical views of the East, especially the "unity of man and nature" in ancient China. Heidegger's treatise provides a particularly rich source of thought for the contemporary western ecological philosophy and ecological aesthetics and deserves investigation and development.

(1) **In his philosophical thinking, Heidegger uses phenomenological methods to construct a holistic view of ecology as "people in the world"**

Heidegger's philosophical view has already been covered in the first few chapters of this book. What needs to be clearly pointed out here is that Heidegger actually constructed an ecological, holistic view of "man in the world" by means of phenomenology. As we all know, Heidegger's philosophical starting point is that he opposes the dualistic paradoxes of Western traditional philosophy, opposes technological mechanistic epistemology, and opposes "anthropocentrism." He opposed the separation of existence and being in traditional western philosophy, as well as the old philosophical point of view that only saw existence, not being. This kind of philosophical view is actually a subject/object dichotomy, in which the subject reflects the object of the "anthropocentric" point of view. He believes that the "subject and object dichotomy" is a traditional and incorrect expression of the living life model, and the living mode of "life is alive" established by him is a living existential mode of being, namely, the "Dasein and the world" living model. This expression of the living model is "man being-in-the-world", which was proposed in *Being and Time.* He conducted a completely new interpretation by using phenomenology and existentialism. In his view, this "being-in" is not one of the traditional things among the other. For example, Shandong University Center for Literary Aesthetics is in the literature and history building. The literature and history building is in the new school of Shandong University, the new campus of Shandong University is in Li Cheng district of Jinan city and so on. It can be deduced so endlessly, because this is still a metaphysical epistemological approach—human beings and the environment can be separated, but they also can be opposed. For example, even if I leave the literary aesthetics research center immediately, this environment will not be affected. But to Heidegger, "being-in" means "I live in the world, and I regard the world as the place I am so familiar with, living in it, staying in it".[49] Here, we hive the ideas of "living", ""“setting something free". "Staying" refers to the integration of human beings and the natural environment, which can't be separated, such as fish can't do without water, and people can't do without air. In our understanding, it is the harmonious integration of human beings and nature that contains the natural environment. This is a philosophical concept of contemporary, ecological ontology. If we say that there were still some traces of "anthropocentrism" in the earlier arguments about the "dispute between the world and the earth" that the truth was open, then in the later period, from about 1936 as a starting point, until the mid-20th century, he began to learn from the wisdom of the East inspired by wisdom, put forward the famous "The Fourfold Game of Sky, Earth, Man and Gods" We can see this shift in treatises like *Things* and *Languages*. As for *Things*, we will discuss later on the category of

[49]Heidegger [8], p. 64.

ecological aesthetics. Let us first introduce the *Language* written in October 1950. In this article, Heidegger elaborates on Trakker's poem *Winter Night*, and puts forward the famous "The Fourfold Game" theory:

> Winter night
> Snow gently whisk out the window,
> The bell of the evening prayer ring loudly,
> The house is ready
> The table for everyone feast.
>
> Only a small amount of rover,
> From the dark path to the gate.
> Glittering bounty tree
> Breathe the cold dew in the earth
>
> Rover quietly entered;
> The pain has put the threshold into stone.
> In the clear light
> Is the bread and wine on the table.

Heidegger's analyzes this poem as follows: "Falling snow bring the man into the twilight under the vast sky. The bell of the evening prayer beats the deathbed to God. The house and the table combine people with the earth. These named things, that is, called things, gather the four parties of heaven, earth, man and God in themselves. These four parties are a coexistence of primitive unity. Let the das Geviert der Vier inhabit itself. This congregation allows Verweilenlassen to be das Dingen der Dinge. We call the world, the fourfold wholeness of the unification of heaven, earth, human beings and gods, which are in the materialization of things."[50] Heidegger believes that all the named objects in the *Winter Night* poem have formed an inseparable whole through the full play of the physical properties of their own things, so that Sky, Earth, Man and Gods constitute a quadruple world, so that a dead person can be mortified and dwelling. Snow, the evening prayer, the house, the table, the feast, the gate, the cold dew, the tree of grace, the threshold, the bread and the wine, and the rovers in winter night have melted into one that makes up a whole and forms a unique rover that dwells Winter night. It is a unique situation, a unique language, which is very pleasant: snow fluttering, the sound of the night bell, cold dew flashes. Without this unique winter night scene, the language will disappear, and roamers can't become living "here."

(2) **On the Linguistic Conception, Taking the Basic Theory of the Relation between the Earth and Language as the Important Point of View that the Earth is the Foundation of Human Beings**

An important point of Heidegger's view is that "language is the home of being".[51] In other words, in his view, people are generated in language. Language is "being", the

[50] Heidegger [9], p. 13.
[51] Ibid., p. 153.

most basic form of human existence. Why did he say that? One of the very important reasons is that he believes that language is closely related to the natural ecology. Language is natural ecology. It is the earth's special gift to people. As the saying goes, one side soil and water raise, people on one side. In fact, one side of soil and water also nurture one side of language. It is precisely in this unique language that people grow and thrive. In conversations with a Japanese person, Heidegger showed great interest when he mentioned that "language" in Japanese was represented by "Koto ba." He said, "petals from Koto. As the word begins to say, imagination wanders and spans inexperienced areas."[52] He is interested in this because Japanese uses the petals (leaves) that are closely related to the earth and the natural ecology to describe language. In his opinion, such imagination is really wonderful and will bring people to the area of experience that has never existed before. Further, he believes that it is this area that reveals the close relationship between language and the earth and its natural ecology. He said: "The difference between dialects is not the only, and the first is not the difference in the way of movement of language organs. In dialects, each speaks differently from the ground, that is the earth. And the mouth is not just an organ in a body that is represented as an organism, it is the surge and growth of the body and the mouth belonging to the earth—as we die, one grows in the surging and growing of the earth, we have gained our foundation and stability from the earth. Of course, if we lose the earth, we lose the foundation."[53] In Heidegger's view, language is the home of existence, and language is the gift of the earth, so the earth will become the most stable foundation for generations of human beings. Starting from the relationship between language and the earth, this argument about the fundamental relationship between people and the earth can be seen as an extraordinary and profound one. It is also an important aspect of Heidegger's ecological thinking.

(3) In the Aesthetic Sense, to Build a Sense of Homeland for a Game of Equality between Human Beings and Nature with "Art is the Self-introduction of Truth"

In Heidegger's existentialist philosophy and aesthetics, truth, benevolence, and beauty are the foundations of existence and are inherently consistent. Therefore, to Heidegger, the so-called beauty and art are the "self-introduction of truth." In his article, "Origin of Works of Art," he said: "Works of art open the existence of beings in their own way. In the works, there is such an opening, that is, Entbtrgen, that is, the truth of the being. In the work of art, the truth of the being is set in the work by itself. Art is the truth set into the work by itself."[54] The "placement" here is not to put in, is the things (truth) hidden in the depths gradually from the shelter to unmask. Of course, this unmasking differs from the technological world digging and sabotaging of all things by technology. This development seems to be able to achieve the purpose of unmasking and utilizing. However, in the words of Heidegger, it is a kind of

[52] Ibid., p. 137.

[53] Heidegger [9], pp. 198–199.

[54] Ibid., p. 21.

impetus to all things in nature—that is, to destroy and repress. To contrast, aesthetic and artistic removal is a kind of free liberation, a kind of protection and care for the return of all things to the earth. He said: "Works return to the place, what the work has brought about in its own return, which we once called the earth. There is a shelter in the earth. The earth is the things unstoppable, unhindered and unruly, indomitable. Standing above the earth and in the earth, historic humans have established their dwelling in the world."[55] Here, He explains that so-called "placement" is to return all things to the earth so as to obtain the due asylum, and further that they may get unhindered and unruly emergence so that mankind can establish its own dwelling in the world. So, how can the truth be put into effect, and the unhindered and unruly be unmasked? In Heidegger's ontology, there is a key factor: "being.", That is to say that people, through human understanding, interpretation, and actions of friendship with nature, protect nature and love all things, so that the nature of everything from the shelter may understand the environment. Of course, this interpretation of man must be in the world of "The Fourfold Game Heaven, Earth, Man God," or it is impossible. He said: "We call these four parties of homogeneity das Gcuiert. Death-beings exist in the quartet through dwelling. However, the basic feature of dwelling is protection, one death eventually protects the quadruplets in their nature, thereby allowing them to dwell. Correspondingly, the dwelling protection is also quadrupled. A mortal one eventually dwelt because he saved the earth—here's the old meaning of the word "salvation", which Lessing understands. Salvation is not only to get something out of danger; the true meaning of salvation is to release something into its own essence. Saving the earth is far from using the earth, or even exhausting the earth. The saving of the earth is not to control the earth, nor to conquer the earth—the latter is but a step of unrestricted plunder."[56] From this we can see that it is in this world of "The Fourfold Game" that "a mortal one" protects all things, saves the earth and returns to the earth. In the words of Hildin, quoted from Heidegger, it is a "returning home"—a return to your own material and spiritual home. This is the real aesthetic state. Therefore, fundamentally speaking, Heidegger's aesthetic view is a song of a wanderer returning home from far away from his hometown, and it is a true ecological aesthetic.

5.3 Western Environmental Aesthetics and Ecological Aesthetics Rising in the 20th Century

In the 20th century in the West, a new form of environmental aesthetics emerged gradually. Concerning the rise of environmental aesthetics, Carlson, a famous Canadian aesthetician, and Seppama, a Finnish aesthetician, all discuss this issue. In his book *Nature and Landscape,* Carlson argued that environmental aesthetics originated from a theoretical debate surrounding natural aesthetics, mainly published in 1966

---

[55]Heidegger [10], p. 28.

[56]Heidegger [11], p. 158.

by Ronald W. Hepburn, entitled "Contemporary Aesthetics and Neglect of Natural Beauty". This article mainly criticizes the aesthetics despising natural aesthetics and points out that the aesthetic appreciation of nature is different from the appreciation of art. These methods should meet the natural characteristics of uncertainty and pluralism, and should also include multiple feelings of experience and a different understanding of nature. According to Carlson, "his essay laid the foundation for a new model of environmental aesthetic appreciation: This new model is that based on both the openness and importance of the natural environment, the recognition of nature's aesthetic experience is very rich in emotion and cognition, which is entirely comparable with that of art."[57] Seppama also believes that Western environmental aesthetics originated from Hepburn's discussion of the art of contemporary aesthetics, "but the situation has changed as well, and one theme in the Aesthetic Conferences organized in the 1970s and 1980s—and most recently in Montreal in 1984—is environmental aesthetics."[58] The main representatives of environmental aesthetics are Carlson of Canada, Seppama of Finland, and Rolston and Arnold Berleant of the United States.

(1) Canadian environmental esthetician Allen Carlson's Environmental Aesthetics

Allen Carlson is a professor of philosophy at the University of Alberta in Canada. His book, *Environmental Aesthetics*, was published in 1998, and the whole book took more than 20 years to complete. The earliest chapters were written in 1976, while the other chapters were written in the 1980s and 1990s. Most of them were published. Carlson is a famous international environmental aesthetics theoretician today. Seppama studied environmental aesthetics under him. From this perspective, Carlson can be said to be one of the founders of western environmental aesthetics. The book *Aesthetics and the Environment* consists of an introductory volume, two parts, and 14 chapters. The introduction mainly discusses the relationship between aesthetics and the environment. It is a discussion of the definition, essence, orientation, and scope of environmental aesthetics. The first part mainly concerns the appreciation of nature, which begins with a historical review. It also includes discussions of the formal characteristics of the natural environment, appreciation and the natural environment, the natural aesthetic judgment and objectivity, nature and affirmative aesthetics, appreciation of art and appreciation of nature, and so on. The second part concerns landscape art and architecture, including the plight of nature and art, environmental aesthetics and aesthetic education, whether environmental art constitutes a violation of the environment, as well as landscape, agricultural landscape, architectural appreciation, the relationship between landscape and literature, and so on. This book involves more content, mainly on the following topics.

The first aspect is about the nature of environmental aesthetics. Carlson believes that the theme of environmental aesthetics is the aesthetic appreciation of the original environment and man-made environment. The aesthetic object of environmental aesthetics is our environment. The very important characteristic of this object is that

[57] Carlson [12], p. 6.

[58] Sepanmaa [13], p. 198.

"thus, we as appreciators are immersed within the object of our appreciation—what we appreciate is also that from which we appreciate. If we move, we move within the object of our appreciation and thereby change our relationship to it and at the same time change the object itself."[59] The scope of environmental aesthetics is from the birth of the wilderness to the rural landscape, suburbs, urban landscape, the surrounding areas, and more places.

The second aspect is the ontological issue of environmental aesthetics. The first is the appreciation mode of natural objects. Carlson explores the object mode, landscape mode, natural environment mode, participation mode, emotional stimulation mode, mysterious mode, and so on. Carlson takes Leopold's *A Sand County Almanac* as an example and emphasizes "which ties appropriate aesthetic appreciation of nature most closely to scientific knowledge: the natural environmental model."[60] In his opinion, this model is conducive to overcoming the "anthropocentric" tendencies in traditional aesthetic theory. Carlson believes that in the aesthetics of the natural environment, there is a contradiction between form and content. Appreciation of the natural form is the aesthetic appreciation of the traditional landscape painting, which is still human-centered. He said: "The natural environment cannot be appreciated and valued in terms of formal beauty, that is, the beauty of formal qualities; rather, it must be appreciated and valued in terms of its other aesthetic dimensions."[61] This "other aesthetic dimension" mainly refers to the ecological aesthetic dimension of harmony and equality between human beings and nature. This is the mode of appreciation of the natural environment. He believes that this model emphasizes two obvious points: that the natural environment is an environment, and that it is natural. This is, of course, aiming at the traditional landscape pattern characterized by formal appreciation. The object of this landscape pattern is not the environment but rather is the form of the natural environment, and it is not natural but selected by the appreciator. To Carlson, this landscape pattern is "anthropocentric" and indicative of a subject/object dichotomy. He agreed with Spitzer's point of view. "Here I initially agree with some of the comments of Spitzer. He believes that considering something on the environment is mainly based on the relationship between 'self and environment' rather than the relationship between 'subject and object' or 'tourist and scenery'."[62] This reflects that Carlson's approach to environmental aesthetics has been trying to break the traditional "anthropocentrism" of the subject/object dichotomy. In his understanding of what and how to appreciate, he thinks that precisely because human beings and the environment constitute a whole, rather than the landscape of the selective appreciation, this mode of appreciation of the natural environment is from the perspective of human survival to appreciate the environment in its entirety. Further, it results by virtue of the appreciation of all the organs in the living environment being human "smell, touch, taste, even warm and cold, atmospheric pressure and humidity."[63]

---

[59] Carlson [14], p. 5.

[60] Ibid., pp. 27–28.

[61] Carlson [14], p. 64.

[62] Ibid., p. 16.

[63] Ibid., p. 76.

This is the "Aesthetics of Engagement" which was further developed by Berleant afterwards. It too was a breakthrough in the traditional static aesthetic represented by Kant. In this regard, he concluded: "We thus have a model that begins to give answers to the questions of what and how to appreciate in the natural environment and that seems to do so with due regard for the nature of that environment. And this is important not only for aesthetic but also for moral and ecological reasons."[64]

After further comparing the appreciation of art and the appreciation of nature, Carlson claims that there are obvious differences between true appreciation of the natural environment and appreciation of the arts. Appreciation of objects and ways of appreciation are different. Therefore, the category of art can't be applied to nature by itself. "In short, nature does not fit into categories of art."[65] This points out the appreciation of the natural environment, or the relative independence of environmental aesthetics. Many theoretical categories of artistic aesthetics are not suitable for environmental aesthetics, which faces a major issue of category reformation. Under the guidance of this kind of thinking, Carlson proposed the important theoretical view of "all nature is beautiful" which became one of the core arguments of *Aesthetics and the Environment*. He said: "all the natural world is beautiful. According to this view, the natural environment, insofar as it is untouched by man, has mainly positive aesthetic qualities; it is, for example, graceful, delicate, intense, unified, and orderly, rather than bland, dull, insipid, incoherent, and chaotic. All virgin nature, in short, is essentially aesthetically good. The appropriate or correct aesthetic appreciation of the natural world is basically positive and negative aesthetic judgments have little or no place."[66] For this reason, he demonstrated this through several discussions, positive and negative, especially with the aid of contemporary ecological science and ecological philosophy. Finally, he said: In short, the objection is that since there is much in the natural world that we do not find aesthetically good, any justification of the position must be incorrect. I agree that there is much in the natural world that appears to many of us not to be aesthetically good. However, this fact itself does not constitute a conclusive objection, for the justification provides the means of showing how the fact is consistent with the positive aesthetics position. First, as suggested by Rolston's remark that "we do not live in Eden, yet the trend is there,"[67] In other words, to Carlson, "all nature is beauty" is not only a non-artificial "given", but is also a holistic and idealistic one. Here, Carlson inherits some of Darwin's views and at the same time transforms it into a principle of environmental aesthetics. That is, he proposes respecting the original nature and preserving its natural aesthetic value without changing it easily!

The second part of "Aesthetics and the Environment" mainly expounds the appreciation of landscape art and architectural art, the appreciation between nature and man-made creations, or between nature and art. Carlson first discussed the evaluation of environmental art, such as trees made of plastic, the head of a cow constructed with

[64]Ibid., p. 81.

[65]Carlson [15], p. 92.

[66]Ibid., p. 109.

[67]Ibid., p. 150.

bicycle parts by Picasso, sculpture art made up of car parts by Zhang Bolun, as well as Duchamp's well-known "spring" urinals, etc. How do we evaluate them? Starting from ecological holism, Carlson answered with the important criterion of the value of life. He said "I expect it is up to us as a community of individuals. For example, it seems to me that while small family farms along the road may express determination, discarded car bodies do not, and while the skyline of a city may express vision, a strip mine does not;"[68] The "community" here is Leopold's "community of life". As an ecological principle, it becomes the most important aesthetic principle in Carlson's environmental aesthetics. As a result, Carlson affirmed the artist's view on environmental art in Sofitel. It, "does not alter, but only displays, nature's aesthetic qualities one step further—to the point of nature itself as art."[69] In the book's discussion of Japanese gardens, agricultural landscapes, architectural art, and the description of the environment in literature, Carlson followed the most basic ecological aesthetic principle of "living community."

(2) Yrjo Sepanma's environmental aesthetics thinking

Yrjo Sepanma, a professor at Joensuu University in Finland. He is also the Former 13th International Society of Aesthetics chairman. For five consecutive years he was chairman of the International Institute of Environmental Aesthetics. His main book *Beauty of Environment* completed in 1986, published in 1993, is an earlier book on environmental aesthetics. In his "Acknowledgments" at the beginning of the book, he said that he began thinking and researching on environmental aesthetics in 1970 and received an equal fellow-style attitude with his mentor Alan Carlson at the University of Alberta in 1982, where they would, "Discuss, question, explain other possibilities".[70] The Finnish Society of Aesthetics organized a multi-disciplinary series of environmental aesthetics in the spring of 1975. The material of the lecture was chosen and published in 1981. He gave a brief summary of his *Beauty of Environment*. He said: "My goal in this book is to systematically outline the field of environmental aesthetics based on the philosophy of analysis." The outline includes: "The first group of questions is about ontology: what is the environment as an aesthetic object," and the second group is about meta-criticism: how is the environment described? Finally, he discusses discuss "the practice of environmental aesthetics and the benefits it can bring."[71] Obviously, Sepanma studied the beauty of the environment on the basis of traditional analytical aesthetics.

The book's basic point of view can be summarized as follows: The first is what he called "ontology of environmental aesthetics," which is the "core area." He said: "The core area of environmental aesthetics is about aesthetic objects."[72] In other words, he wonders how the environment becomes the aesthetic object? Sepanma argues that "making the environment an aesthetic object is usually based on the

[68] Carlson [15], pp. 214–215.

[69] Ibid., p. 285.

[70] Sepanmaa [13], Acknowledgements page.

[71] Sepanmaa [13], p. *The Beauty of Environment: A General Model of Environmental Aesthetics*.

[72] Sepanmaa [13], p. 36.

audience's choice",[73] meaning that the audience can choose art as well as social things as aesthetic objects. Only by choosing the environment as the aesthetic object can the environmental aesthetic be produced, and the aesthetic relationship between human beings and environment can be established. In this way, the audience is the real artist. Sepanma points out, "So who is an artist? It is the audience, people. Man chooses to think of something as a natural artwork, no matter how it is generated."[74] Obviously, what Sepanma calls "ontology" here is still "anthropocentrism." Because in the end, the beauty of environment is still created by human beings and is the result of human choice. The environment is external to human beings, and human beings and the environment do not form a whole. Of course, with a metaphor, he uses the beauty of environment as the "artwork of nature" and uses nature as an artist to replace the absent creator.[75] This absent creator is somewhat mysterious and deified. Subsequently, Sepanma discusses the difference between environmental beauty and artwork in 12 aspects: Art is artificial, while the environment is given. Art is born in custom, while the environment is not. Art is created for the aesthetic, while the beauty of the environment is by-product. Art is fictional, while the environment is real. Art is omissible, while the environment is itself. Art is defined, while the environment is infinite. Art and its authors have names, while environments have no name. Art is unique, while the environment is repeated. Art is style, while the environment is not. Art is sensory, while the environment is theoretical—the senses. Art is static, while the environment is dynamic. Art is limited, while the environment is free,[76] and so on. Thus, most of the beauty of environment spoken by Sepanma refers to the beauty of the scenery that is similar to the artwork.

The second major point is his so-called "meta-criticism." From analytic aesthetics, the criticism related to utterance is the main part. In this way, Sepanma said: "Environmental aesthetics is the philosophy of environmental criticism."[77] He divided this "meta-criticism" into three tasks: describing, explaining and evaluating. As an analytical esthetician, Sepanma holds that the relationship between human beings and nature is a meta-criticism, a relationship between description and interpretation. He said: "It only makes sense to us when it is naturally seen and interpreted."[78] Moreover, he asserted: "There is no environment beyond perception and description—even the term 'environment' implies the human point of view: human beings are in the center, and everything else surrounds him."[79] Obviously, he is "anthropocentric" here. He also listed four preconditions for environmental assessment: "the evaluators are not physically controlled by natural needs", "in a rational sense, turned from a myth—a religious worldview into a scientific worldview", possess "the knowl-

[73] Ibid., p. 41.

[74] Ibid., p. 58.

[75] Ibid., p. 59

[76] Ibid., pp. 77–100.

[77] Ibid., p. 110.

[78] Sepanmaa [13], p. 136.

[79] Ibid.

edge of nature and culture process", have "ability to classify the object correctly".[80] These four preconditions contain obvious anthropocentric tendencies. However, Sepnama's environmental criticism is not entirely "anthropocentric," but still contains some modern ecological philosophy and ecological aesthetics that respect nature and respect life. He divided his criticism of the environment into affirmative aesthetics and critical aesthetics. The so-called "affirmative aesthetics", he believes, is mainly to evaluate anything that has not been transformed by human beings in its natural state. And "it is decisive to choose a suitable method of acceptance and the validity of the standard." Everything that is in its natural state is beautiful.[81] Obviously, Sepanma was influenced by Carlson's concept of "Nature is all beauty." However, Sepanma's "Nature is all beauty" still has its own value orientation, that is natural self-destructive phenomena that occur as a result of natural disasters, such as forest fires, plant diseases, avalanches, volcanic eruptions, hurricanes, etc.. He believes that these are considered "ugly" as "a few exceptions", "but if you put them in a broader context, you can 'save' them: think of them as a stage in the process in which dramatic changes have occurred in climaxes and valleys."[82] In other words, from a process of natural evolution, these phenomena can be regarded as a stage or a valley in which nature is all beauty, but this does not affect the panorama of natural beauty. The other is "critical aesthetics", which is, "evaluating the results of human activities and even conducting negative evaluations when necessary." In his criticism of such an environment transformed by humans, Sepanma saw a clear value orientation. He said: "Human beings adapt their environment to their own purposes, and these are the ends of all values. But there are ethical restrictions on action: "Earth is not just for human use nor is it just the place where human beings live, animals and plants and even natural structures have their own rights, which can't be compromised."[83] Here, Sepanma uses an ethical guideline, which is the criterion that animal and plant rights can't be harmed. This is the standard of modern ecological ethics. When it comes to the relationship between aesthetic value and the value of life, Sepanma apparently places a greater emphasis on the value of life. Of course, what needs to be explained here is that, the aesthetic value he refers to is also the formalist aesthetic of the symmetrical proportionality of the Western tradition rather than deep life and ecological aesthetic. Therefore, he regards this aesthetic value as a shallow one while viewing the value of life as a deep one; the former can't endanger the latter. He said: "Aesthetic goals can't harm the value of life, therefore, a reckless aesthetic system is excluded."[84] He also added: "in the environment, one can't approve something destructive in the deeper sense or even in the aesthetic. Everything—and even the Auschwitz's corpse—can be viewed from the surface as a constitution and color, but in doing so it would be a deformity that departs from the framework given by the

[80] Ibid., p. 151.
[81] Ibid., p. 148.
[82] Ibid., p. 149.
[83] Sepanmaa [13], p. 149.
[84] Ibid., p. 161.

value of life or ideology."[85] At the same time, Sepanma also talked about the value of the wilderness, in his view, the West, unlike the East (China and Japan), adopted a hiding attitude toward the wild in historical tradition. This is mainly based on the relationship between nature and human livelihood. "When people abandon their solemn praise of fertile land and fertile green areas, the desolation of nature, characterized by simplicity and clarity, draws the attention of people."[86] He also talked about the aesthetic principles of ecological issues. He believes that "in nature, when a process of a natural cycle is continuous and self-sufficient, this system is a healthy system …. The natural balance in this sense is dynamic, not static. An aesthetic principle is realized in the economy of the system and in the harmony of all the parts. In the system, anything is necessary, and nothing is superfluous."[87] In his opinion, the continuity, self-sufficiency, dynamism, balance, harmony and health of the ecosystem are the most important ecological principles of aesthetics. Obviously, Sepanma still fails to fully realize the necessary conversion from the scientific principles of ecology to the philosophical principles of ecology and to the aesthetic principles of ecology. In general, he is still inclined to or limited to ecological scientific principles, which is consistent with the value of life mentioned earlier.

The third major point is his "applied environmental aesthetics." On the one hand, Sepanma emphasized the theoretical qualities of aesthetics and philosophical qualities. However, he also mentioned the application of environmental aesthetics. One of the focuses of his research on environmental aesthetics is "to explore the modes of dissemination of theoretical knowledge and the ways of influencing practice through aesthetics",[88] This mainly involves environmental education, people's impact on the environment, and the concreteness of the outlook. All ecological theories including ecological aesthetics not only have a theoretical character, but also have a practical character. For ecological aesthetics, the most important thing is to promote the principle of ecological aesthetics in real life so that people can master these principles and treat the nature with an aesthetic attitude. Sepanma argues that environmental education works by providing the knowledge, ideals, and goals of the environment, and "creating a foundation for understanding the environment and understanding the environment aesthetically".[89] The so-called "aesthetic understanding of the environment" refers to the aesthetic attitude toward the environment. On the human impact on the environment, Sepanma believes that including environmental ideals, legislation, positive aesthetics, and negative aesthetics, etc. The so-called ideal of the environment refers to the "ethical limitation" and "the requirement of harmony and completeness" to be followed in the practice of transforming the environment; In the so-called legislation, that is, in the transformation of the environment, "we must consider the entire change from an aesthetic point of view and ensure that this aspect is taken into consideration through legislation." The so-called positive aesthetics is

[85] Ibid.

[86] Ibid., 170.

[87] Ibid., p. 180.

[88] Sepanmaa [13], p. 192.

[89] Ibid., p. 193.

an aesthetics that tries to exert a positive influence on the taste and the object, while negative aesthetics is a kind of aesthetics trying to eliminate inherent contradictions. The so-called prospect of concrete aesthetics, refers to the relationship between ethics and environmental aesthetics. The former designed the boundary for the latter, but did not decide the specifics within the boundary. This opens up a precedent for the practical development of environmental aesthetics.

(3) **Rolston's American environmental aesthetics**

Holmes Rolston III (1933–) is distinguished professor of Philosophy at the University of Colorado, and one of the pioneers and founders of the internationally renowned ecological philosophy and ecological ethics. In his 1995 book, *Philosophy Gone Wild*, he put forward the famous proposition of "the wilderness shift of philosophy." As a famous philosopher of ecology and the environment, his series of works involves some basic problems of ecological aesthetics and environmental aesthetics. In 2002, the volume *Environment and the Arts: A Multidimensional Perspective of Environmental Aesthetics*, edited by environmental esthetician Arnold Berleant, included a paper by Rolston, "From Beauty to Responsibility: Natural Aesthetics and Environmental Ethics." Rolston's philosophical position on aesthetics has changed before and after. In *Philosophy Gone Wild* he insisted on the "ecological noumenon" position. However, in the article "From Beauty to Responsibility" he primarily adopts the phenomenological standpoint, and emphasizes the constitutional function of the subject. Both his philosophy of wilderness and phenomenology of environmental aesthetics are the products of an opposition to "anthropocentrism" and are of great value.

1. Proposing the important proposition that nature has aesthetic value, and that wilderness is the root of humankind

Rolston persevered with the philosophical view of natural ontology in his famous book *Philosophy Gone Wild*, emphasizing the ultimate value of nature. He said: "Man arrogantly thinks that 'man is the measure of everything.'" but these natural things existed before mankind. This precious world, a world to which mankind can evaluate, is not without value; on the contrary, it produces value—nothing is closer to its ultimate existence than we can imagine."[90] He believes that nature has its own value beyond human evaluations. These values include economic value, life support value, recreational value, scientific value, life value, diversity and unity value, stability and spontaneous value, dialectical value, religious symbolic value, etc., of course, including the aesthetic value of nature. He said: "and the photographs in Audubon or National Wildlife bring out well this natural aesthetic value."[91] However, what he calls aesthetic value can't be understood from the usual practical value and life supporting value. For practical value is defined by the practicality of nature as it is for human beings, while the value of life support is understood from the perspective of the role of nature in supporting people's lives. In short, the first two perspectives are

[90] Rolston [16], p. 9.

[91] Ibid., pp. 132–133.

starting from people, which is also an "anthropocentric" position. Therefore, Rolston clearly put forward the idea that the aesthetic value of nature should be understood from the above two perspectives. He said, "discovering such aesthetic value, it is critical to separate it from both utility and life support, and only those who recognize this difference can value the desert or the tundra."[92]

So, how to understand the independent aesthetic value that nature has? Rolston opposed the idea of nature as human resources, which starts from the needs of human, because this "resource theory" is still a kind of "anthropocentric" position, a traditional view of "humanization and nature." Rolston believes that "pure wildness can be a good thing. These places change us, not we them."[93] He put forward the important natural ontological point of view which is "Wildness Is a Place of Roots". Starting from the original nature of human existence, he explored the value of nature and the wilderness, claiming that "Wildness Is a Place of Roots in Historical and Ongoing Senses"[94] He also said"Before parents and the sacred, one is not so much looking to resources as to sources, seeking relationships in an elemental stream of being with transcending integrities."[95] "Wilderness is the first legacy, the grand parent, and offers dramatic contacts with ultimacies not found in town."[96] This discusses the values, including aesthetics, in the ultimate and ontological sense.

In the following article, "From Beauty to Responsibility: Natural Aesthetics and Environmental Ethics", Rolston still maintains his "natural ontology" position in some places, and proposes an important proposition that "Aesthetics is moving to the wilderness." In elaborating this argument, he states that people enter the wilderness not to live on it, but to acquire a peculiar experience that frees them from themselves and reshapes themselves, thus giving them a deeper sense of identity. This experience makes us feel as if we are back in the arms of our parents and find the source of our own life. We are all intertwined with the sublime, aesthetic feelings of awe, affection, home, and pleasure. He said, "natural beauty is a pleasure—it is just a pleasure—it seems less urgent to ban it in order to protect it. But this attitude will change as we feel the earth at our feet, the sky in our head, and our home in the world. … This is ecological aesthetics, and ecology is the key, a self at home, in its own world. I define the place where I live as my home. This 'interest' leads me to care about its integrity, stability and beauty."[97] Here, Rolston profoundly expounded the so-called "aesthetics of ecology" as a kind of aesthetics which is from the ultimate meaning of human existence and is the aesthetics of humankind's existence and life in close connection with nature. This should point to the essence of ecological aesthetics.

---

[92]Rolston [16], p. 133.

[93]Ibid., p. 204.

[94]Ibid., p. 210.

[95]Ibid., p. 207.

[96]Ibid., p. 208.

[97][U.S.] Holmes Rolston: *From Beauty to Responsibility: Natural Aesthetics and Environmental Ethics*, Berleant [17], pp. 167–168.

2. The natural aesthetic must have two qualities of aesthetic ability and aesthetic characteristics

Rolston persisted in his point of view of "natural noumenon" in the discussion of the aesthetic value of nature, but in the discussion of specific natural aesthetic experiences, Arnold Burnett adopted the ecological phenomenology of his position and persisted strongly that the natural aesthetic experience is based on the combination of the aesthetic ability of people and the aesthetic characteristics of the object. Of course, according to the perspective of ecological phenomenology, aesthetic ability is the main factor in the above two aspects. Only by virtue of the aesthetic ability of the subject can the aesthetic characteristics of the object be developed to produce the natural aesthetic experience. He said, "There are two sorts of aesthetic qualities: aesthetic capacities, capacities for experience that are only in beholders, and aesthetic properties, which lie objectively in natural things."[98] Only when both are encountered, can the aesthetic experience of the natural ecology be produced. Rolston explained his point of view with the appreciation of the black-backed antelope. "We do enjoy seeing impala leap; there is grace in their motions. The aesthetic experience arise in my encounter with them, but the muscular power driving their locomotion is an evolutionary achievement objectively realized in the embodied animal. My aesthetic capacities track their aesthetic properties."[99] Of course, from the point of view of ecological phenomenology, the aesthetic capacities of the subject are the main one among the two. It is a constituent ability of the subject, by virtue of which the aesthetic characteristics of the subject can be developed, constituted, and form an aesthetic experience. Without the development and composition of the human aesthetic ability, it "has only the possibility left." Rolston vividly compared this aesthetic trait to a cake hidden in a refrigerator, which was still in the dark before the refrigerator was opened. Only after opening was it possible for people to taste the sweetness of the cake and appreciate the beauty of the cake. He compared this "opening" with the shining of the lamp when the refrigerator lamp is on. It is humankind that lights up the beauty of nature. He said, "When we light up the beauty in nature, if we do it right, often we are seeing something already there."[100]

Rolston's discourse on aesthetic capacities abandons the traditional view that aesthetic capacities are only necessary for seeing and hearing two kinds of things, or are capacities needed in a noninterested viewpoint. Rather, he puts forward the idea that natural aesthetics is necessary for immediate "entry," and that a variety of sensory organs are involved in the point of view. He said, "Forests is for access, not for viewing. It is very much in doubt whether a person can experience a forest while he is on a roadside or on television. Forests impact on our senses: sight, hearing, touch, and even taste. Visual experience is crucial, but no forest can also be fully

[98][U.S.] Holmes Rolston: *From Beauty to Responsibility: Natural Aesthetics and Environmental Ethics*, Berleant [17], pp. 158.

[99]Ibid., p. 159.

[100][U.S.] Holmes Rolston: From Beauty to Responsibility: Natural Aesthetics and Environmental Ethics, Berleant [17], pp. 157.

experienced without the smell of pine and wild rose."[101] Here, Rolston adhered to Berleant's "Participatory Aesthetics",[102] in line with the direct nature of natural aesthetics.

3. Believing that beauty is a responsibility, aesthetic experience is one of the basic starting points of environmental ethics

Rolston advocates the unification of aesthetics and environmental ethics, and proposes that beauty is an important responsibility. He said, "If beauty, then duty"[103] This obviously treats ecology and the aesthetics of the environment as the aesthetics of the value theory, which is just an important feature of ecology and environmental aesthetics. To this end, Rolston once again abandoned the traditional non-utilitarian static aesthetics. In his opinion, "Aesthetic experience is among the most common starting points for an environmental ethic."[104] He agrees with Eugene Hargrove's point that "the ultimate historical foundation of nature conservation is aesthetics",[105] and illustrates this in the history of the United States as an example of the protection of the Grand Canyon and Great Tieton. To a large extent, the protection is due to aesthetic needs. Aesthetic value is important in environmental ethics. In this way, he proposed that in the natural aesthetic it is actually transferred from 'is' to 'ought'[106] In other words, the natural aesthetic is not only lawful, but also purposeful. However, this unity of law and purpose is not necessarily the characteristic of natural aesthetics. Art aesthetics can also have such characteristics. What is the special ethical basis in the natural ecological aesthetic? Rolston used the famous "land ethic" by Aldo Leopold, a well-known ecological ethicist: "A thing is right when it tends to preserve the integrity, stability, and beauty of the biotic community. It is wrong when it tends otherwise."[107] Leopold uses the land ethics of his ecological community to connect the aesthetic value of the natural ecology with the stability and harmony of the entire biological community. It is in the sense of an ecological community that there is consistency between the "is" and "ought" of the natural ecological aesthetic. Rolston argues that "Leopold connects the 'beauty of the biotic community' with the continuing existence of its members 'as a matter of biotic right'. That does join beauty with duty——."[108] It is the particularity of the aesthetic value of the natural ecology that is linked with the health and stability of the ecological community. Rolston also has a very important point: this kind of expanded aesthetics contains responsibility, and the so-called responsibility is usually what one owes to others. Specifically, it is human beings "'owing' to fauna, flora, species, ecosystems, moun-

[101]Ibid., p. 166.

[102]Ibid

[103]Ibid., 151.

[104]Ibid.

[105]Ibid.

[106]Ibid., 152.

[107]Leopold [18], p. 213.

[108][U.S.] Holmes Rolston: From Beauty to Responsibility: Natural Aesthetics and Environmental Ethics, Berleant [17], p. 214.

tains and rivers to Earth, is appropriate respect".[109] The reason why human beings owe the natural ecology is that human beings, the only rational creatures on the earth, should bear the responsibility of safeguarding ecological stability. Furthermore, in the past era humankind destroyed the natural ecology, and now owes natural ecology itself a repayment. The ecological ethical value of ecological aesthetic factors are more fully described from many aspects by Rolston.

(4) **American Esthetician Arnold Berleant's Environmental Aesthetics**

Arnold Berleant is a phenomenological esthetician, environmental esthetician, honorary retired professor at Long Island University, former Chairman of the International Society of Aesthetics, Chairman of International Advisory Committee on Applied Aesthetics, Secretary General of International Society of Aesthetics, Secretary General of American Society of Aesthetics. His book, The Aesthetics of the Environment in 1992 and *Environment and Arts; Perspectives on Environmental aesthetics* has been translated into Chinese. Recently Professor Cheng Xiangzhan published his translated article "Aesthetic Ecology and Urban Environment." in the *Academic Monthly* in 2008, volume three. It can be said that Berleant is a well-known esthetician familiar to the Chinese aesthetic community. Berleant's *The Aesthetics of Environment* consists of twelve chapters, including the basic theory of environmental aesthetics, environmental criticism, and urban aesthetics practice areas. On the basic theory, Berleant firstly proposed his own concept about the environment against the traditional dichotomy of "anthropocentrism", which takes the environment as an object external to human beings. He said: "environment, as I want to speak of it, is the natural process as people live it, however they live it. Environment is nature experienced, nature lived."[110] "one must be on guard against the tendency to think of them objectively and dualistically, in the sense of regarding humans as placed in them rather than as continuous with them."[111] To this end, Berleant put forward a very important point of view from the perspective of ecological holism: "there is no sanctuary from the inclusiveness of nature."[112] This is what he said in criticizing the establishment of "nature reserves." The so-called "nature reserve", which was set up to serve the so-called sightseeing and leisure activities of people, still regards nature as external to human beings. Berleant said: "At this point the protective divisions disintegrate entirely, for there is no sanctuary from the inclusiveness of nature. Nature here encompasses everything, all subject to the same standards of being, all manifesting the same processes, all exemplifying the same scientific principles, all evoking the same wonder, the same dismay, the same ultimate acceptance."[113] Here he was influenced by Spinoza's unification of human beings and nature, further distinguishing environmental aesthetics from traditional aesthetics. On the whole, traditional aesthetics is limited to natural beauty, while environmental

[109]Ibid., p. 168.

[110]Berleant [19], p. 11.

[111]Berleant [19], p. 11.

[112]Ibid., p. 9.

[113]Ibid., pp. 9–10.

aesthetics must break the line of defense and admit the whole world. In the aesthetic mode, traditional aesthetics is static aesthetics, with visual and other senses, while environmental aesthetics is the "Aesthetics of engagement". In his opinion, this is the problem that the book needs to focus on.[114] Berleant also described environmental aesthetics as a "cultural aesthetics."[115] That is to say, apart from emphasizing people's aesthetic perception of the environment, he also emphasizes the cultural orientation of environmental aesthetics, and emphasizes the memory, beliefs and social relations mixed in aesthetic perception. However, he argues, "the key point is: how to maintain the loyalty of rational knowledge to current feelings rather than artificially editing them to the traditional wisdom."[116] Berleant stands on phenomenological ground, calling environmental aesthetics a "descriptive aesthetics", an "understanding of the environment and aesthetic experience, and the unity between them".[117] Here, the so-called "environment" has two meanings, that is, the original natural environment and the described environment. This description of the environment is "a fusion of organic awareness, of meanings both conscious and unaware, of geographical location, of physical presence, personal time, pervasive movement".[118] Actually, as Berleant said in another article, the aesthetic appreciation of the natural environment is a combination of the subjective composition and the aesthetic potential of the object. In the following chapters of *Environmental Aesthetics*, Berleant also has made very important discourse on "Aesthetics of Engagement," "Homeland Consciousness," "Site Consciousness," "Environmental Phenomenology," and the philosophy of nature as the core of the future philosophy, ecological philosophy, and aesthetics, We will introduce all related content in the chapters that deal with the category of ecological aesthetics. In the field of environmental aesthetics practice, Berleant put forward the issue of "urban aesthetics." He said, "they may serve as useful guides in cultivating an urban ecology to replace the forces of vulgarity and monotony that depress the modern city, thus helping transform it from a place where one's humanity is constantly threatened into a place where it is continually achieved and enlarged."[119] He also put forward the view that "urban design is a kind of home design", and advocated building a "people-oriented city." Berleant is not eager to put forward the design issues of "space community". He pointed out that when humans enter the space exploration era, a series of concepts related to gravity, such as the matrix, level, severity, upper and lower, will change. Therefore, a new concept of design of extraterrestrial communities should be adopted. "Studies have been going on for several decades on the design of extraterrestrial communities, but this research usually devotes little attention to the aesthetic aspects of such uncustomary environmental conditions. It is important, however, to consider not just how we can carry on human life under such circumstances, but to recognize and direct the qualitative

---

[114]Ibid., p. 12.
[115]Ibid., 21.
[116]Ibid., p. 23.
[117]Ibid., p. 29.
[118]Berleant [19], p. 33.
[119]Ibid., p. 56

conditions of those lives. Such communities place in question the nature and role of art, and more generally the aesthetic, in future human environments, specifically those in outer space."[120] We should note that this argument is very cutting-edge. In addition, Berleant also put forward an opinion on the display of museum exhibits from the perspective of environmental aesthetics, and demanded to, "Replace the aesthetic of objects with an aesthetic of experienced. This requires a clear sense of the museum as an environment. Like any other, the museum environment fulfills itself when it functions as a place of involvement. But it is an environment with a special purpose, the purpose of promoting aesthetic appreciation, the experience of aesthetic engagement."[121] This kind of application of environmental aesthetics to museums is also very inspiring for us.

Berleant, like Seppama, also discusses environmental criticism from the perspective of phenomenology. It summarizes the process of interpretation and evaluation and fully affirms its role. It holds that environmental criticism can make the aesthetic values of the environment obtain the same status as other environmental values and contribute to the formation of the same artistic, aesthetic appreciation. Further, aesthetic criticism can promote the improvement of design level and artificially create the environment, and so on.[122] The final chapter of the book, "Reclaiming the American Landscape," is in fact the vision and blueprint that Berleant tried to apply his environmental aesthetics theory to his homeland. In his opinion, "that aesthetic values, broadly understood, are a necessary part of environmental understanding and action, and that these values must be included in any thoughtful proposal for environmental change".[123] The ultimate practical goal of environmental aesthetics is "to make our presence on the Earth more humane, is to honor the values of home across the entire landscape".[124]

From the above summary, we can see that the development of environmental aesthetics in the West did provide a wealth of resources for the construction of ecological aesthetics since the late 1970s, both in the basic construction of ecological theory and in the special category of ecological aesthetics as well as n the development of ecological criticism. All provide us with a wealth of ideological resources. However, there are still many doubts and confusions regarding the relationship between ecology and environment, ecological aesthetics and environmental aesthetics, anthropocentrism and ecological holism, and traditional aesthetics and ecological aesthetics, which we need to further clarify.

It has been nearly 40 years since the rise of western environmental aesthetics in the 1970s. It was not valued until today when it has been recognized by Western academia, which regards it as an important dimension of contemporary aesthetics, which harmonizes with art aesthetics and daily life aesthetics. This is historical progress. To a certain extent, however, we think that this definition in Western

[120] Ibid., p. 89.

[121] Berleant [19], p. 106.

[122] Ibid., p. 128.

[123] Ibid., p. 161.

[124] Ibid., p. 170.

aesthetic circles is still incomplete because the rise of environmental aesthetics is actually a revolution in the field of aesthetics. Its significance lies in its powerful criticism and counterpoise against the rise of "anthropocentrism" and its related "art-centerism" to the domination of the world since the industrial revolution and the Enlightenment. It is well-known that "art-centerism", which has prevailed for a long time, thoroughly excludes natural ecology from aesthetic disciplines with its "aesthetic philosophy" and "aesthetic subjective principle". However, with the advent of "post-industrial ecological civilization era", from 1966, Hepburn made an important proposition with the idea of "natural aesthetics" based in the idea that "aesthetics is art philosophy". There have been efforts to restore the status of the natural ecology in aesthetics, which led to the rise and prosperity of environmental aesthetics. There arose Leopold's "ecological holism" as an important principle of environmental aesthetics, the introduction of ecological phenomenology into environmental aesthetics, the proposition of "aesthetics of Engagement", and its subversion of Kant's "static aestheticism". The revolution that actually took place in the field of aesthetics has actually gone further and further. In fact, the appearance of environmental aesthetics is more than one dimension of aesthetics. It completely subverts all the important aesthetic paradigms of Western classical aesthetics. In such a situation, can we just call it one of the three pillars of the discipline of aesthetics? Therefore, the current situation of contemporary Western environmental aesthetics is still cold and lonely, which is really not a normal phenomenon, and should arouse our deep reflection! This also shows that the traditional tools of rational thinking and related academic systems are stubborn and conservative!

Echoing Western environmental aesthetics, ecological aesthetics has been gradually flourished since the 90s of the last century, and especially in the new century. Especially under the current "harmonious society", "environment-friendly society" and "ecological civilization", the new situation of building social goals, the development of ecological aesthetics in our country has encountered an unprecedented, excellent opportunity. Under such circumstances, we should learn valuable resources and lessons from the development of Western environmental aesthetics. The first is to absorb its abnormalities that break through the traditional "subjectivity aesthetics" and "aesthetics, art philosophy," which distort the essence of aesthetics and dominate the world. We should learn the spirit of their fight and their strategy of deconstruction. At the same time, we must also learn from the rich academic nutrition provided by environmental aesthetics. For example, their critique of "picturesque landscape" of the traditional natural aesthetic, the advocacy of "nothing outside of nature", proficiency in the use of ecological phenomenology, as well as the advocacy of the "aesthetics of engagement", "landscape aesthetics" and environmental aesthetics practice—all deserve our study and reference. In addition, the fate of Western environmental aesthetics also gives us a major warning. So far, the West still regards environmental aesthetics as the scope of practical application of "applied aesthetics" rather than as the noumenon of aesthetic theory. In our opinion, we can't just call environmental aesthetics and ecological aesthetics a kind of aesthetics or a new branch, but should regard it as a new development and a new extension of the aesthetics discipline. It is a new aesthetic theory that contains ecological dimensions that are different from the contemporary one in the past.

We need to further consider that although western environmental aesthetics has its historical inevitability and academic rationality existing in the 20th century, but because it is the product of western society after all, the soil that it produces is Europe and the United States and other western developed countries. Therefore, we can't fully accept some of its academic positions. For example, "wild philosophy" as a philosophical position, in addition to the affirmative value of "ontology", the concept and action of protecting large areas of wilderness is very difficult to promote in our country. Because, at present, regarding most western developed countries, because of the history of imperialist global aggression, not only do they have a vast territory, save the rich resources, but also they have retained a lot of wilderness. Under such conditions, "wild philosophy" can be promoted, and people can appreciate the beauty of completely original ecology. However, our country has a large population whose economic production and development began very early. It experienced the history of being invaded, and its land and resources are relatively scarce. The ecological footprint is relatively limited and the material needs of the people for development are still very urgent. It's actually very hard to keep and shelter the wilds. Another example is the Western "ecocentric" philosophical position, advocating the extension of human rights to animals such as gorillas. This can only be implemented in a limited manner and in the spirit of eco-humanism and a win-win path between human beings and nature. In addition, western environmental aesthetics insists on the theory that "nature is all beauty" strives to maintain the original natural beauty of the original ecology. In this regard, it is not only difficult for us to agree in theory, but is also difficult to completely accept in practice. Our position is "environment-friendly", where development and protection coexist. From the theoretical resources, western environmental aesthetics rely on limited western resources, and learn from Eastern resources from time to time. As an ancient country with oriental civilization, "harmony between man and nature" is our fine cultural tradition. Confucianism, Buddhism, Taoism and other cultural theories contain a wealth of ecological aesthetic wisdom resources, which is an important theoretical support for us to build contemporary ecological aesthetics, waiting for us at any time to explore and organize and contemporary conversion. We believe that on the basis of the dialogue between China and the West and ancient and modern issues, our country's contemporary ecological aesthetics will surely be able to develop better.

5.4 Ecological Theology and Interpretation of Ecological Aesthetic of *Bible*

Christian culture is a very important cultural tradition in the West.it has a very far-reaching impact on the entire Western civilization. Because of this, we must also make good use of the resources of Western Christian culture, especially its most important classic Bible, in the construction of contemporary ecological culture, especially ecological aesthetics. Among them, it mainly includes reference to the contemporary ecological theology resources flourishing in the West and the ecological aesthetics interpretation of the Bible.

(1) **Ecological Aesthetics Resources of Contemporary "of Ecological Theology"**

1. Lynn White's "ecological theology" that started with a boycott of traditional Christian culture

In 1967, Lynn White (1907–1987), the American historian, published the famous "Historical Roots of Ecological Crisis," which attributed the historical roots of the modern ecological crisis to Christian culture, and set off a series of articles on Christian culture. This led to the beginning of a religious reform of its contemporary form and to the emergence of a new theological theory form of "ecological theology" that has become an important resource for the development of various ecological theories, including ecological aesthetics.

First, Lynn White specifically attributed the root causes of contemporary ecological crisis to Christian culture. He summed up the historical roots of the current serious ecological crisis, and conducted an in-depth critique of the Christian culture, believing that only through deep reflection on the old Christian religion and finding a new religious belief could it be possible to get rid of the ecological crisis. He said: "If we do not seek a new religion or reflect on the old religion, even more science and technology will not get us out of the ecological crisis."[125]

Second, White believes that the most important drawback of Christian culture is its performance as "anthropocentrism." He said: "especially in its Western form, Christianity is the most anthropocentric the world has seen."[126] The fundamental reason is that Christianity replaced the ancient polytheism and animism with monotheism and gave the image of God to people so that people have the right to rule the nature. He said: "Christianity, in absolute contrast to ancient paganism and Asia's religions (except, perhaps, Zorastrianism), not only established a dualism of man and nature but also insisted that it is God's will that man exploit nature for his proper ends."[127]

Third, White argues that the religious ideas of Saint Francis of Assisi within Christianity have a very important connotation of ecological equality and indicate the direction of contemporary religious reform. Saint Francis of Assisi (1182–1226) is the founder of the Christian Saint Francis of Assisi and the Saint Francis of Assisi women's congregations. He prescribed that the monks need to abide by the penance, buy simple clothing, walk barefoot, and walk around the poor to promote the Gospel. They also supported the idea of democracy among all living creatures created by God, becoming the forerunner of contemporary "ecological theology." White fully affirmed this. "Francis tried to depose man from his monarchy over creation and set up a democracy of all God's creatures". therefore, "Francis as a patron saint for ecologists."[128]

---

[125] White [20], p. 8.

[126] Ibid., p. 6.

[127] Ibid., p. 9.

[128] Ibid., p. 10.

2. Moltmann's "*Ecological doctrine of Creation*"

J. Moltmann (1926–), a famous modern Protestant theologian in Germany, wrote *God in Creation: an Ecological doctrine of Creation* from 1981 to 1985, and put forward the famous "*an Ecological doctrine of Creation*". He has become an important representative of contemporary "ecological theology", and has important value and positive meaning.

First, "*an Ecological doctrine of Creation*" is proposed to deal with the growing ecological crisis and the urgent need of carrying out new religious reforms. There is no doubt that Moltmann's "*an Ecological doctrine of Creation*" is to advocate the ecological equality of people and things in the "center of God", God's creation of all things. How does such a theory come about? He clearly pointed out that it is in response to the worsening ecological crisis and the need for a new religious reform in the contemporary era. In the Preface of his book, he gave a full account of the current ecological crisis confronting humankind. "What we call the environmental crisis is not merely a crisis in the natural environment of human beings. It is nothing less than a crisis in human beings themselves. It is a crisis of life on this planet, a crisis so comprehensive and so irreversible that it can't unjustly be described as apocalyptic. It is not a temporary crisis. As far as we can judge, it is the beginning of a life and death struggle for creation on this earth."[129] He put it more thoroughly elsewhere: "This crisis is deadly, and not for human beings alone. For a very long time now it has meant death for other living things and for the natural environment as well. Unless there is a radical reversal in the fundamental orientation of our human societies, and unless we succeed in finding an alternative way of living and dealing with other living things and with nature, this crisis is going to end in a wholesale catastrophe."[130] One of the important ways to solve this crisis is to carry out Christian religious reform. He believes that for this great crisis, traditional Christian culture is hard not to blame. This is the result of misunderstanding and misuse of religious beliefs in the Bible for a long time, such as, "conquering the land" in the Bible as God's command to mankind to conquer nature and rule the world. He said: "The Christian belief in creation as it has been maintained in the European and American Christianity of the Western churches is therefore not guiltless of the crisis in the world today."[131] In order for Christianity to create a creed "it is no longer to be itself one factor in the ecological crisis and the destruction of nature, but is instead to become a ferment working towards the peace with nature which we seek".[132] He tried to "explain and reinterpret" the Christian theology." "His Book of *God in Creation* is the first step in the necessary reform of Christian theology in overcoming our destructive modern life system".[133]

Second, the book proposes God-centered idea of "an Ecological doctrine of Creation" which includes "ecological equality". The core of this book is "an ecological

[129]Moltmann [21], "Foreword", p. 1.

[130]Ibid., p. 31.

[131]Ibid., p. 32.

[132]Ibid., p. 32.

[133]Moltmann [21], p. 22.

doctrine of creation". Moltmann writes in the preface to the Chinese translation: "The purpose of this book is to reveal the true presence of God in all that He has created, in order to see the living Spirit in the created community of nature. This view is not on the other side of the earth, but on the Spirit, the 'spirit' among all things, asking us, as Albert Schweitzer put it, to add respect for the life of any creature to the reverence of God, and, in turn, to revere God's presence in all things."[134] This short sentence basically marks the basic elements of his "an ecological doctrine of creation". The first is "indicating the presence of God in all that He has created," this so-called "presence" is God's "indwelling" in all things. He said, "the inner secret of creation is this indwelling of God".[135] The ideas here contains a wealth of ecological connotations, including the ideas of "creation", "presence", and "indwelling". According to this view, human beings and all things are equal before God. "Indwelling" has" homeland consciousness "of" ecology "and everything becomes God's home. He said, "but in a deeper sense it is also a reference to the symbolism of 'home' and 'dwelling' which I have employed in this book. According to its Greek derivation, the word 'ecology' means 'the doctrine of the house' (oikos) the purpose of the Shekinah is to make the whole creation the house of God"[136]; another very important one is to put forward the "natural community created" which is very eco-conscious concept. Moltmann, in his discussion of "some guiding ideas of ecological creationism," explicitly states that its most basic theoretical foothold is "relationship of community." He said: "we can then no longer, either, conceive his relationship to the world he has created as a one-sided relationship of domination. We are bound to understand it as an intricate relationship of community- many-layered, many-faceted and at many levels. This is the fundamental idea behind non-hierarchical, decentralized, confederate theology."[137] This is a breakthrough in the way in which the traditional mechanical theory opposes the subject and object mode of thinking. It is the choice and persistence of communicative, organic, and integrated ecological thinking. Then there is the advocacy for the concept of "reverence for life" put forward by the ecologist Schweitzer. Schweitzer's "reverence for life" thought contains a series of very important ecological concepts such as "the value of life" and "ecological equality". Moltmann absorbed these ideas into contemporary Christian theology, which is undoubtedly an important step in religious reform. It shows that "an ecological doctrine of Creation" is, to a great extent, connected with contemporary advanced ecological theory.

Third, it discusses the intermediated status of "man in God and nature." The issue of human status is an important part of Christian culture. According to the Bible, on the seventh day before the Sabbath, God created man in his own image and let him "manage all things" and "control the earth." In this way, in later Christian theological hermeneutics, it was believed that God put mankind above all things in nature and so could arbitrarily rule and plunder all things in nature. This is an important source of "anthropocentrism." Moltmann's an Ecological doctrine of Creation has reversed

[134]Ibid.

[135]Ibid., p. 3.

[136]Ibid., "Foreword", pp. 2, 3.

[137]Ibid., p. 8.

this “anthropocentric” theology and has given mankind a “dual role” between God and nature. “then in a Christian doctrine of creation human beings must neither disappear into the community of creation, nor must they be detached from that community. Human beings are at once imago mundi and imago Dei. In this double role they stand before the sabbath of creation in terms of time. They prepare the feast of creation.”[138] In other words, human beings represent the creation before God, and represent God before the creation as the “intermediary” between God and nature. According to the Bible, human beings have the responsibility of “managing” and “controlling” the natural world. According to Moltmann, this “management” and “control” responsibility can only be viewed from the special perspective of “intermediary”, but can’t be seen as human absolute rule of nature. In other words, it can only be understood from the perspective of “representing God” that “it means the administration for God” rather than the conquest of the earth”.[139] He said: “Peoples, races and nations which set themselves up to be masters of the world by no means become God’s image in the process, or his representative, or ‘God present on earth’. They become at most a monster It is only as God’s image that human beings exercise divinely legitimated rule; and in the context of creation that means: only as whole human beings, only as equal human beings, and only in the community of human beings.”[140]

Moltmann criticizes the “anthropocentrism” prevalent in Europe since the Enlightenment precisely based on this “intermediary theory”. He said: “Modern European anthropology uncritically took the modern anthropocentric world picture as its premise. According to this conception, the human being is the centre of the world, and the world was created for his sake and for his use. Modern sciences have put an end to this naive attitude.”[141] He also suggests from the perspective of creationism that because mankind was finally created, he depends upon the natural things that were created before. “Without them his existence would not be possible. So while they are a preparation for him, he is dependent on them.”[142] Even from the perspective of relying on all things, mankind can’t become a “center.” Moltmann also specifically discussed the specific and real relationship between human beings and the natural environment, which is the relationship between labor and residence. Of course, the first is the labor relationship between human beings and nature. “People work on nature so as to acquire food, and in order to build up their own world. Under the aspect of work, the human being is always the active agent, while nature is always passive.”[143] He believes that if human beings seem to be the masters, the strong ones, and nature seems to be slaves and the weak ones, it is only from the point of view of labor. He goes on to say: “Has the human being no other elemental need which inevitably determines his relationship to nature? There is a need of

---

[138]Moltmann [21], p. 259.

[139]Ibid., p. 306.

[140]Moltmann [21], pp. 306–307.

[141]Ibid., p. 253.

[142]Ibid., p. 256.

[143]Ibid., p. 66.

this kind. It has hitherto been theoretically overlooked, and was given a back seat when the huge industrial cities were built up—to the detriment of human beings and nature alike. This concern is the interest of habitation. The interests of habitation are different from the interests of work. We can sum up the interests of habitation under the concept of 'home' or—in Bloch's phrase—'home country'. The idea of a 'home country' is not primarily an appeal to a regressive dream of origin, composed of 'fatherland' and 'mother tongue' and childish security. There can only be a home country in liberty, not in slavery."[144] Here Moltmann offered a brilliant definition of "home country" with "liberty", which contains the equal and free relationship between human beings and nature. It also includes equal and free human relations. We must treat nature in a free and fair manner so that our human beings can enjoy the beautiful feeling of "being at home" in freedom. However, for a long time, human beings have not only neglected the "living" relationship with nature, but also neglect the freedom and equality of nature, which has led to the destruction of the natural environment and the important reason why humankind has become homeless. As Moltmann said "The human being has not merely a right to work; he has a right to habitation. The two interests must be balanced out. This is not merely a requirement in the context of social policy. It is also a demand for a radical change of direction in the relation between human beings and nature. The ecological attitude to the natural environment must surmount the one-sided, pragmatic and utilitarian approach."[145]

Fourth, the "Sabbath' is the real sign of creation. Moltmann explicitly states in his discourse on "Some Guiding Concepts in Ecological Creation" that "The sabbath is the true hallmark of every biblical—every Jewish and also every Christian—doctrine of creation".[146] First, we must clarify the meaning of the Sabbath in Christianity. Moltmann believes that for a long time people have attached great importance to God on the 6th day, the creation of animals and people. However, this interpretation does not take seriously the Sabbath on the seventh day, which is not correct. In fact, in the Christian doctrine, the Sabbath is very important. Moltmann said, "The sabbath commandment given to Israel (Exod. 20.8-11) is the longest of the ten commandments, and it is therefore considered to be the most important. God sanctified the sabbath because on that day he rested from creation; so his people are to sanctify it too. Everyone is to sanctify it, parents and children, masters and servants, human beings and animals, Israel's own people and strangers. The sabbath is an order of peace for everyone. It is impossible to celebrate and enjoy it at the cost of other people. The feast can only be celebrated and enjoyed together with all the others. If human beings are to 'have dominion' over the animals (Gen 1.26), here animals are to enjoy the sabbath too. Later on, the sabbath year is extended to the earth as well: it is to remain free from human cultivation for a year (Lev. 25.11). People sanctify the sabbath by abstaining from every kind of productive work, and by recognizing the whole of reality as God's creation—the creation from which, in face of which, and

[144]Ibid., p. 66.

[145]Moltmann [21], p. 67.

[146]Ibid., p. 13.

within which he himself rested."[147] Thus it can be seen that the Sabbath in Christianity not only means the completion of the Creator's Creation on the 6th day, but also a celebration of God and a blessing of God upon creation. At the same time, more importantly, "the Sabbath" is a kind of rest day for God and all things. It is a kind of day in which persons, human beings, and nature, without interference with each other, live in peace. Therefore, it contains important ecological significance. As Moltmann said, "The Bible's ecological wisdom is embodied in this tradition of rest. And, essentially, the Sabbath, together with its non-intervention in natural leisure, is the echo of inhuman wisdom in the *Tao Te Ching*. Because it is God's Sabbath, not mankind, which is the 'crown of creation', my creation theory is entirely for the purpose of the Sabbath. The result is the doctrine of the Sabbath of the Earth, which includes the harmony of human civilization with nature."[148] At the end of the book, Moltmann concludes his discourse on "an Ecological doctrine of Creation" with his observance of the "Sabbath," saying that "In the ecological crisis of the modern world it is necessary and timely for Christianity too to call to mind the sabbath of creation."[149] Later, he further hoped to put it into action and put forward the idea that "The ecological day of rest should be a day without pollution of the environment—a day when we leave our cars at home, so that nature too can celebrate its Sabbath."[150]

From the above analysis, it is very clear that Moltmann's "an Ecological doctrine of Creation", the ecological theology of modern morphology, is indeed very valuable. It indicates that the ecological transformation of contemporary Christian culture will surely become an important resource in the cultural construction of contemporary ecology. However, as with all theological theories, Moltmann's "an Ecological doctrine of Creation is based on the "God" standing high in the other side of the world as its highest belief and the most important core, including the solution to the current ecological crisis. Therefore, our respect for this belief can hardly dispel the nihility and illusion of its own theory. In the final analysis, the solution of all human problems, including the ecological crisis, has to rely on human beings themselves, on human culture, and on the corresponding practical actions. Moreover, a closer look at Moltmann's an Ecological doctrine of Creation also contains some erroneous notions of historical retrogression. In comparing industrial and pre-industrial societies, he said the following: "The earlier civilizations were by no means 'primitive', let alone 'under-developed'. They were highly complicated systems of equilibrium—equilibrium in the relationship between human beings and nature, equilibrium in the relationship between human being and human being, and equilibrium in the relationship between human beings and 'the gods'.9 It is only modern civilizations which, for the first time, have set their sights on development, expansion and conquest. The acquisition of power, the increase of power, and the

[147] Ibid., pp. 385–386.

[148] Moltmann [21], "Foreword of Chinese Version", p. 23.

[149] Ibid., p. 400.

[150] Ibid.

securing of power: these, together with 'the pursuit of happiness', may be termed the values that actually prevail in modern civilizations."[151]

History proves that as a kind of progress in history, modern civilization has become the historical conclusion. Like all other forms of civilization, civilization has its advantages and disadvantages. We can't deny its "advantage" for promoting social progress because of its "disadvantages" caused by environmental pollution. Furthermore, nor can we improperly promote the "pre-modern" low-level "equilibrium." History can't be reversed. We can only step toward a new "ecological civilization" on the basis of modern civilization and with the help of the power of modern civilization.

(1) A New Interpretations of the Bible from the Perspective of Ecological Existence

Theological aestheticism and theological ontology is based on theological existence theory. It begins from the unique perspective of human beings and everything as God's highest being. The relationship between human beings and all things under the precondition of "the center of God" in the Bible is summed up and explained therein. It opens up the broad prospect of Christian culture taking theological ontology as the starting point, and participating in the construction of contemporary ecological civilization. In the following, we mainly proceed from the Bible and explain the basic connotation of an ecologically aesthetic outlook in the context of theological existence.

1. "The Word was with God": Beyond the Beauty

"the Word was with God" is the starting point of an ecologically aesthetic view of Christian theological ontology. It contains extremely rich content, the most basic of which is to claim that God is the highest existence and the creator of all things. *Bible · Deuteronomy* calls God "God of gods and Lord of lords".[152] *Bible · Psalms* also called the Lord the "The highest heavens belong to the LORD".[153] *Bible · Revelations* says by the mouth of twenty-four elders: "you are worthy, our Lord and God. to receive glory and honor and power, for you created all things, and by your will they were created and have their being."[154] Therefore, the important content of Christian culture, especially the Bible, is the creation of God. The so-called "That visible is created from that invisible"[155] The first "Genesis" of the Bible records the course of God's creation on the 6 days. On the first day God created heaven and earth. On the second day God created the sky. On the third day God created grass, vegetables and trees. On the fourth day God created the sun, the moon and the stars. On the fifth day God created fish and water creatures, birds, insects and beasts. On the sixth day God made man in his own image. The seventh day was the Sabbath. God is the Creator, and human beings and all things are created. Therefore, because

---

[151] Moltmann [21], "Foreword", pp. 38–39.

[152] *Bible* [22].

[153] Ibid., p. 789.

[154] Ibid., p. 1969.

[155] Ibid., p. 1654.

all things in the world were created by God, the relationship between them should be equal. Some scholars have emphasized that God stipulates that people have the function of managing all things, and that people are above all things. Indeed, *Bible Genesis* documented the management of all things by people. The Bible records the word of God: "Let us make man in our image, in our likeness, and let them rule over the fish of the sea and the birds of the air, over the livestock, over all the earth, [Hebrew; Syriac all the wild animals] and over all the creatures that move along the ground." "I give you every seed-bearing plant on the face of the whole earth and every tree that has fruit with seed in it. They will be yours for food. And to all the beasts of the earth and all the birds of the air and all the creatures that move on the ground—everything that has the breath of life in it—I give every green plant for food."[156] The above remarks are considered by many theorists as the main basis for the "human center" of Christian culture. However, humankind does not constitute the center of all things from the same creature's point of view. The function God has entrusted to humankind for all things does not necessarily mean that humankind has become the master of all things. It can mean that humankind has to bear more responsibility for taking care of everything. As the *Bible, Hebrews* said, for humankind, "we have not seen everything and served him."[157] As God gave vegetables and fruits for human consumption, it also gave greens and vegetables to the beasts, birds, and other living animals for food, including the Bible's permission to slaughter and eat meat and rest on the Sabbath and rest in peace during the year. All illustrate, to some extent, the understanding of the ecological rules of the biological cycle of Christian culture. It shows that the equality of human beings and everything as creatures in Christian culture is not absolute equality, but equal to the law of circulation and multiplication of all things. Moreover, both human beings and all others, as being, are also present through the Word of God, becoming through it that specific, peculiar thing here and now. The Bible expounds this with a very figurative metaphor, believing that human beings and everything are like seeds. God gives them to different groups according to his own meaning, and that different groups show God's road with their different glories. *The Bible* says: "When you sow, you do not plant the body that will be, but just a seed, perhaps of wheat or of something else. But God gives it a body as he has determined, and to each kind of seed he gives its own body. All flesh is not the same: Men have one kind of flesh, animals have another, birds another and fish another. There are also heavenly bodies and there are earthly bodies; but the splendour of the heavenly bodies is one kind, and the splendour of the earthly bodies is another. The sun has one kind of splendour, the moon another and the stars another; and star differs from star in splendour."[158] On this basis, the Bible holds that human beings and all things have the same value as the beings who present the Word of God. There is a famous saying in *the Bible Luke*: *"Are not five sparrows sold for two pennies? [Greek: two assaria] Yet not one of them is forgotten by God."*[159] Therefore, even

[156]Ibid., pp. 4–5.

[157]*Bible* [22].

[158]Ibid., p. 1569.

[159]Ibid., p. 1592.

the sparrows, which are not as precious as human beings, have their own value as the beings, embodying the Word of God without being forgotten.

To sum up, from the point of view that human beings and all things are beings "due to the same morality," *the Bible* claims that human benigs and all things are created, live together, and have their value due to the same morality. Such philosophies of human beings and things being together due to the same morality contain a transcendental beauty. Originally, existential aesthetics advocated a transcendental beauty. It is through the "suspension" of the material entity and the spiritual entity, beyond being as presence, and beyond presenting the absence of presence, to reach the open clarity of the truth. The theological existence of aesthetics has its own characteristics in the face of spirit and flesh, sacredness and secularity, the unique contradiction between this shore and the other shore transcends the flesh through the Spirit. Divine transcendence becomes secular, and realizes the transcendence of God's Way to all, presenting the glorious beauty of the Way of God. The *Bible, Galatians*, cites the word of God: "So I say, live by the Spirit, and you will not gratify the desires sinful nature for the sinful nature desires what is contrary to the Spirit, and the spirit what is contrary to the sinful nature. They are in conflict with each other, so that you do not do what you want. But if you are led by the Spirit, you are not under the law." Here, the Bible emphasizes the confrontation between sensuality and the Holy Spirit. Under the guidance of the Holy Spirit, it should surpass the sensual, in order to follow the law of God to reach the realm of truth.... In the Bible, the famous "gate for the sheep" serves as an image of the parable of Jesus leading people beyond materialism, toward life, and truth. The *Bible*, Matthew quotes Jesus' saying: "I tell you the truth, I am the gate for the sheep. All who ever came before me were thieves and robbers, but the sheep did not listen to them. I am the gate; whoever enters through me will be saved. [Or kept safe] He will come in and go out, and find pasture. The thief comes only to steal and kill and destroy; I have come that they may have life, and have it to the full."[160] Here, thieves represent the material desire. Jesus is the Holy Spirit, and entering the gate for the sheep means that the Holy Spirit surpasses the material desire. The *Bible* argues that only through this transcendence can we really cross the darkness into the light of truth. In *Bible · John,* Jesus said to all, "I am the light of the world. Whoever follows me will never walk in darkness, but will have the light of life." Then he said, "If you hold to my teaching, you are really my disciples, then you will know the truth, and the truth will set you free."[161] The transcendent beauty of this kind leads to the belief advocated by the ontology of Christian theology, and provides valuable intellectual resources for the aesthetics of the ultra-utilitarianism of later generations. At the same time, this kind of transcendent beauty also provides thoughtful nutrition for the modest recognition of the "natural charm" in ecological aesthetics. The development of science has indeed made humankind know the mys-

---

[160]*Bible* [22].

[161]Ibid., p. 1466.

teries of nature greatly, but the mystery of nature and the color of the other side of aesthetics are inexhaustible and indispensable factors.

2. "Salvation through Word": the beauty of the tragedy

"Salvation" is the most important content and theme in Christian culture. It is also the most important content of the theological and aesthetic outlook of theology, which constitutes the most characteristic and heart-stirring tragic, aesthetic tone. It consists of four aspects: original sin, suffering, salvation, and tragic beauty. The redemption of God is caused by the punishment of humankind's crime and descent into the unpredictable catastrophe. Therefore, it is necessary to first discuss the theory of sin. The third chapter of Bible Genesis discusses in the original sin committed by human ancestors: the ancestors of human beings were lured by a snake to steal the forbidden fruit, and so committed the original sin. They were then expelled from the beautiful, prosperous, and carefree Eden. So, what is the root cause of the sin committed by mankind? Christian doctrine believes that the main point lies in the greed of human nature. The Bible writes: "When the woman saw that the fruit of the tree was good for food and pleasing to the eye, and also desirable for gaining wisdom, she took some and ate it. She also gave some to her husband, who was with her, and he ate it."[162] Thus, the reason why Eve was tempted to steal forbidden fruit was to satisfy her own appetites, eyes, and cognitive desires. It is this desirability that led to human sin. However, the lust of humanity has not changed because of being expelled from the Garden of Eden. The Bible believes that this lust is human nature and has been exposed repeatedly. In the sixth chapter of *Bible · Genesis* writes: "The LORD saw how great man's wickedness on the earth had become, and that every inclination of the thoughts of his heart was only evil all the time. The LORD was grieved that he had made man on the earth, and his heart was filled with pain."[163] Chapter 9 of The *Bible · Genesis* writes: "every inclination of his heart is evil from childhood."[164] It can be seen from this that the Bible believes that one's sin is original. Moreover, the Bible argues that driven by original sin, the descendants of humankind have done more evil than their predecessors. *Bible* · Jeremiah Chapter 16 Jehovah said when the prophet Jeremiah commented on the descendants of Israel "But you have behaved more wickedly than your fathers. See how each of you is following the stubbornness of his evil heart instead of obeying me."[165] This intense self-blame of Christian culture is an extremely important feature. It always blamed the root causes of various disasters on its own original sin and fault. The 25th of the Bible · Psalms writes: "Remember O LORD! your great mercy and love, for they are from of old. Remember not the sins of my youth and my rebellions ways; according to your love remember me, for you are good; O LORD!" And wrote: "O LORD! For the sake of thy name, pardon my iniquity, for my sin is great". Such a strong sense of self-blame forms a stark contrast with the ancient Greek culture.

---

[162]*Bible* [22].

[163]Ibid., p. 9.

[164]Ibid., p. 12.

[165]Ibid., p. 1052.

It is well-known that ancient Greek culture attributes all the causes of disaster and tragedy to objective fate. There is very little deep remorse when compared to Christian culture. The famous tragedy "Oedipus the King", contains the protagonist Oedipus, whose widowed mother lived with him in sin. This was attributed to objective and irresistible fate. The effect it produced is also quite different. The fate of the tragedy gives rise to helpless sympathy, but the tragedy of the original sin can produce a strong soul shock. For if the root cause of sin lies in the original sin that would exist in everyone's heart, this causes not only remorse but also a strong reflection. At present, in the face of the increasingly serious ecological disasters in the process of modernization and industrialization, some people ignore or even become complacent and may not be able to correctly handle the notion that the ancient Greek tragedy attributes all disasters to objective destiny. Rather, we need to pay more attention to the spirit of the tragedy of original sin of Christian culture. At present, in the face of a series of serious problems brought by the ecological crisis to humankind, it is absolutely necessary that we conduct a self-examination of our past ideas and behaviors. Closely linked with sin is the theory of suffering. Because Christian culture recognizes one's original sin, in order to avoid the original sin, a very important human covenant with God has emerged. This is the famous "Ten Commandments", the ten prohibitions given by God to human beings to curb their original sin. However, human beings always disobey because of their sinfulness and difficulty in compliance. This has caused humankind to continue to be punished and plunged into misery. Therefore, the suffering of Christian culture, including the ecological catastrophes such as natural disasters, is caused by God to punish human beings, and belongs to the teleological scope of suffering. Of course, all these divine punishments are caused by humans by default. The Bible Leviticus chronicles God's warnings about humankind: "and if you reject my decrees and abhor my laws and fail to carry out all my commands and so violate my covenant, then I will do this to you: I will bring upon you sudden terror, wasting diseases and fever that will destroy your sight and drain away your life. You will plant seed in vain, because your enemies will eat it."[166] It is precisely because human beings have repeatedly defaulted to the promulgation of original sin that they suffer the disaster of God's punishment. The first was being driven out of the Garden of Eden and punished "with a life of hard work." Then they were overwhelmed by the flood. The Bible says that through the torrential floods, "Every living thing on the face of the earth was wiped out; men and animals and the creatures that move along the ground and the birds of the air were wiped from the earth."[167] At the same time, God also confronts mankind with other disasters. He turned their waters into blood, causing their fish to die. Their land teemed with frogs, which went up into the bedrooms of their rulers. He spoke, and there came swarms of flies, and gnats throughout their country. He turned their rain into hail, with lightning throughout their land; he struck down their vines and fig-trees and shattered the trees of their country. He spoke, and the locusts came, grasshoppers without number; they ate up every green thing in their land, ate up the

[166] *Bible* [22].

[167] Ibid., p. 11.

produce of their soil."[168] God also brought awesome droughts and earthquakes to humankind. The drought situation was such that "The ground is cracked because there is no rain in the land; the farmers are dismayed and cover their heads".[169] The earthquake is such that "the mountain shakes before him and the hills melt".[170] Most of these events listed in the Bible are natural disasters and are mostly natural disasters. However, today's disasters, such as nuclear radiation, AIDS, cancer, SARS, bird flu and so on, are mostly man-made disasters, and are the result of the human disruption to the environment. Is not this even more breathtaking? The Bible seems to have foreseen in general, speaking of the end times in the New Testament, 2Timothy: "But mark this: There will be terrible time in the last days. People will be lovers of themselves, lovers of money, boastful, proud, abusive, disobedient of their to their parents, ungrateful, unholy, without love, unforgiving, slanderous, without self-control, brutal, not lover of the good, treacherous, rash, conceited, lover of pleasures rather than lovers of God-having a form of godliness but denying its power." The above-mentioned selfish greed and the pursuit of pleasure are precisely the defects of human nature in which modern society is proliferating. The punishment caused by such ills should be even greater. Facts have shown that today's human living conditions beautify and the non-glorification of the two anti-serious facts does not exactly prove it?

Christian culture places salvation in a prominent position, the so-called salvation, the salvation of human suffering by God and Christ Jesus. Christian culture believes that this salvation is entirely determined by the loving nature of God and Christ Jesus. Psalm 30 and 31 wrote: "LORD my God, I called to you for help and you healed me. LORD! you brought me up from the grave; [Hebrew: Sheol] you spared me from going down into the pit. Sing to the LORD, you saints of his; praise his holy name. For his anger lasts only a moment, but his favor lasts a lifetime; weeping may remain for a night, but rejoicing comes in the morning." And he said, "Since you are my rock and my fortress, for the sake of your name lead and guide me. Free me from the trap that is set for me, for you are my refuge. Into your hands I commit my spirit; redeem me, O LORD, the God of truth."[171] From this we can see that the Bible believes that God's redemption of humankind as a refuge for humankind is entirely due to God's eternal grace, the holy name of the world, the character of faith, and the nature of love. In Christian culture, God's redemption of humankind is different from the general crisis of relief. The redemption is the ultimate concern for the future and destiny of mankind, and he is the universal hand to save humankind at the crucial moment of human existence. According to the Bible, at the very beginning of humankind, when it was engulfed in floods by sin, Noah, the righteous man of God, built the ark, and escaped this looting. Then, again, as humankind was in danger, God also gave birth to Jesus Christ, the only child, for a painful sin offering and resurrected the Gospel

---

[168] Ibid., p. 7976.

[169] Ibid., p. 1048.

[170] Ibid., p. 1284.

[171] *Bible* [22].

to "save his people from sin."[172] The Bible also predicts that in the future end of the world, Christ Jesus will come to earth to save mankind. Redemption of Christian culture is not only redemption of mankind, but redemption of all things, because various disasters are both human suffering and suffering of all things. Therefore, we must save all things while saving humanity. The Bible records that in the early days of humankind, with the arrival of the Great Flood, people and all things were inundated. God ordered Noah to build an ark that saved both humankind and all things. The Bible records that God said to Noah, "But I will establish my covenant with you, and you will enter the ark—you and your sons and your wife and your sons' wives with you. You are to bring into the ark two of all living creatures, male and female, to keep them alive with you."[173] Therefore, in Christian culture and in the Bible, human beings are redeemed by God as all things. It is from the perspective of human beings and everything being saved by God that there is also some equality between human beings and everything. Moreover, in Christian culture and in the Bible, God has not only redeemed humankind and all things, but also devoted his love to nature as a whole, with a strong love for nature and the earth. As we have already said, the Bible contains Sabbaths, and Sabbaths prescribe the rules governing the recuperation of man and nature. And God created man to make human form by the dust of the earth. God also said to mankind, "If you are dust, you will return to the dust."[174] More importantly, the Bible proposes the famous ethical thought of "blessing the earth", highlighting the worth of nature as a being. The 66th of the Psalms reads: "You care for the land and water it; you enrich it abundantly. The streams of God are filled with water to provide the people with grain, for so you have ordained it. You drench its furrows and level its ridges; you soften it with showers and bless its crops."[175] That is to say, the salvation of Christian culture includes God giving to mankind the beautiful nature of earth, rain, sunshine, grain and so on, so that people can survive beautifully. It also shows that in the Christian culture, the survival of human beings can't be separated from that of natural things.

In short, the theory of "Salvation through Word" in Christian culture is an extremely tragic ecological aesthetics of theological ontology. Not only does it give rise to fear and deterrence in the form of a huge and irresistible disaster, it also impresses itself with a strong self-condemnation and gives a deep revelation to the prophecy of a greater catastrophe in the future. With a vivid image and striking brush strokes, the Bible portrays a picture of disaster and redemption, infiltrating the rich tragedy. From the buoyant waves of Noah's ark to the tormenting picture of the crucifixion of Jesus Christ, even the horrific portrayal of the seven punishments in the future world remains in the minds of the world with its eternal and shocking image. This is indeed a kind of noble tragedy. As Kant said, this is a tremendous material object overwhelming the perceptual power of man, and finally by means of the rational spirit overwhelms the object of sensibility, arousing a lofty beauty. Christianity

[172]Ibid., p. 1388.

[173]Ibid., p. 10.

[174]*Bible* [22].

[175]Ibid., p. 752.

evokes a lofty beauty through the powerful rational spirit of redemption of Jesus Christ, triumph over nature and spiritual victory. The general "ecological beauty" mainly shows the harmony between human beings and nature, a beautiful picture, or attacked the evil deeds that destroy nature by means of art. However, only Christian culture, in its unique form of original sin—suffering—salvation displays the relationship between human beings and nature under the premise of "God's center" with rich tragedy, and highlights that in the face of natural disasters, human beings should have more responsibility and follow God's "care for the earth," the core theme. This give us a deep inspiration.

3. "Justification by Faith": the Inner Beauty

"Justification by faith" is the manifestation and emphasis of the belief in human beings and the intrinsic beauty of human beings' highly spiritual nature. This is a very important part of the Christian culture and the *Bible*. it is a very important part of the ecologically aesthetic view of theological ontology. It is also the only way to reach the truth of the theological beauty of theology. "Justification by faith" is a theological theory of Christianity that differs from determinism of the general epistemology. As the *Bible* · *Galatians* says, "know that a man is not justified by observing the law, but by faith in Jesus Christ. So we, too, have put our faith in Christ Jesus that we may be justified by faith in Christ."[176] The so-called "justification" is to get the word of Jesus. The *Bible* argues that it can't be achieved by the usual recourse to the law of moral reason, but by virtue of its belief in Christ Jesus, which is achieved by the pursuit of a spiritual inner spirit, and must abandon all kinds of extrinsic material temptation and inner desire, including property and even life. As the *Bible* · *Galatians* says, "Those who belong to Christ Jesus have crucified the sinful nature with its passions and desires. Since we live by the Spirit, let us keep in step with the Spirit. Let us not become conceited, provoking and envying each other."[177] And this "righteousness" is the pursuit of Jesus' "love", just as Jesus replied to the Levites: "To love him with all your heart, with all your understanding and with all your strength. This is the most important first commandment. The second is similar to it, that is, to love your neighbor as yourself. All the law and the prophets are based on these two commandments."[178] Those who have done all of the above are "new creatures"[179] who "remove all impurity from the body and soul, and to be one with Jesus." To do this, we must rely on the unique spiritual cultivation process in Christian culture, including baptism, prayer, repentance, and so on. Once we do, finally we can achieve the unity of God's Word and human beings—that is, "the Word became flesh." As Jesus said in the prayer of the disciples in the *Bible* · *John,* "My prayer is not for them alone. I pray also for those who will believe in me through their message, that all of them may be one, Father, just as you are in me and I am in you. May they also be in us so that the world may believe that you have sent me, I have

---

[176]*Bible* [22].

[177]Ibid., p. 1593.

[178]Ibid., p. 1368.

[179]Ibid., p. 1578.

given them the glory that you give me, that they may be one as we are one."[180] Here, the "justification by faith" of the Christian culture and the related spiritual cultivation process actually becomes the display process of theological phenomenology. This occurs through spiritual prayer that is justified by faith and incarnate, a repentance process, the "suspense" of all kinds of external material and the inner desires of people, and enter an aesthetic living state of an inner divine life. As Max Scheler, a German theological phenomenologist, put it, "This seems to be macroscopically reflected in the following doctrine: Christ's redemption not only redeemed the sin of Adam, but also brought man away from sin into a relationship with God that was deeper and more sacred than Adam's relationship with God, Although the rescuer among the faith and follow-up of Christ no longer has the immense perfection of Adam, there is always the unspeakable desires ('the lust of the flesh'). The alternation of the cycle of sinking and rising is again and again subtly shown in the *Gospels*: in paradise, the joy of a repentant sinner is more than the joy of a thousand righteous people."[181] This kind of "spiritual repair by virtue of Justification by faith" has many similarities to the "forgetfulness" and "carefreeness" of the minds of ancient Chinese Taoists are also ancient form of phenomenological aesthetic.

4. "New Heaven and Earth": the perfect beauty

Starting from theology ontology, Christian Culture and *the Bible* made a full discussion of the perfect beauty of ecological, aesthetic concepts. Of course, the Garden of Eden is an ideal place for the unity of God and human beings in the world. However, human beings were expelled from the Garden of Eden because of sin, and lost such a beautiful land. However, God in Christian culture and *the Bible* are still constantly creating new beautiful land for mankind. *The Bible· Deuteronomy* writes that Jehovah God is about to introduce humankind into the beautiful land of olive trees, oil, and honey, which does not lack food. The *Bible*, Isaiah specifically describes the new heaven and earth God will create.I It will be a beautiful material and spiritual homeland where people and things, people and people, and things and things get along well with each other. "For as the days of a tree, so will be the days of my people; my chosen ones will long enjoy the works of their hands. They will not toil in vain or bear children doomed to misfortune; for they will be a people blessed by the LORD, they and their descendants with them. Before they call I will answer; while they are still speaking I will hear. The wolf and the lamb will feed together, and the lion will eat straw like the ox, but dust will be the serpent's food. They will neither harm nor destroy on all my holy mountain," says the LORD.[182] The *Bible · Revelation* devotes itself exclusively to a new heaven and a new earth: "Then I saw a new heaven and a new earth, for the first heaven and the first earth had passed away, and there was no longer any sea. I saw the Holy City, the new Jerusalem, coming down out of heaven from God, prepared as a bride beautifully dressed for her husband."[183] This

---

180 *Bible* [22].

181 Scheler [23], p. 708.

182 *Bible* [22].

183 Ibid., p. 1711.

new heaven and new earth is wonderful: the city walls are made of jasper; the city is made of pure gold, and a bright, crystal river of life flows from the throne of God. There are trees of life on both sides of the river, which give birth to fruits 12 times….. In a word, this is a beautiful and prosperous homeland in harmony with each other. These narratives express the aesthetic ideal of the Christian culture and the ontology and ecological aestheticism of the Bible: the unification and harmony of God and human beings in heaven and earth, and the beautiful and harmonious material and spiritual homeland.

To sum up, we elaborate the basic connotation of ecological aesthetics of the ontology of Christian theology from four perspectives, which are "the Word was with God": Beyond the Beauty, "Salvation through Word": the beauty of the tragedy, "Justification by Faith" the inner beauty, "New Heaven and Earth": the perfect beauty. This shows that this is a special tragedy advocating that human benigs and all things are equally created, equally existing, equally valuable, and similarly redeemed. It also shows us of transgression, internality, ideality and sinful feelings of self condemnation, which has its own specific content and irreplaceable value.

References

1. [Italian] Vico. (1986). *New science*, translated by Zhu Guangqian. People's Literature Publishing House.
2. [U.S.] Santayona, G. (1982). *The sense of beauty*, translated by Yu Lingzhu. China Social Sciences Press.
3. Liyuan, Z. (Ed.). (1993). *History of modern western aesthetics*. Shanghai Literary Publishing House.
4. [U. S.] Dewey, J. (2010). *Art as experience*, translated by Gao Jianping. The Commercial Press.
5. [Russia] Chernyshevsky. (1979). *The aesthetic relationship between art and reality*, translated by Zhou Yang. People's Literature Publishing House.
6. [Germany] Hegel. (1979). *Aesthetics*, translated by Zhu Guangqian. Commercial Press.
7. [Germany] Kant. (1964). *Critical criticism*, translated by Zong Baihua. Commercial Press.
8. [Germany] Heidegger, M. (2006). *Being and time* (revised translation), translated by Chen Jiaying, Wang Qingjie. Sanlian Bookstore.
9. [Germany] Heidegger, M. (2009). *On the way to language*, translated by Sun Zhouxing. The Commercial Press.
10. [Germany] Heidegger, M. (2008). *Off the beaten track*, translated by Sun Zhouxing. Shanghai Translation Publishing House.
11. [Germany] Heidegger, M. (2005). *Speech and proceedings*, translated by Sun Zhouxing, Sanlian Bookstore.
12. [Canada] Carlson, A. (2006). *Nature and landscape*, translated by Chen Libo. Hunan Science and Technology Press.
13. [Finland] Sepanmaa, Y. (2006). *The beauty of environment*, translated by Wu Xiaoxi, Zhang Yi. Hunan Science and Technology Press.
14. [Canada] Carlson, A. (2006). *Aesthetics and the environment—The appreciation of nature, art and architecture*, translated by Yang Ping. Sichuan People's Publishing House.
15. [Canada] Carlson, A. (2005). *Aesthetics and the environment—The appreciation of nature, art and architecture*, translated by Yang Ping. Sichuan People's Publishing House.
16. [U.S.] Rolston, H. (2000). *Philophy gone wild*, translated by Liu Er et al. Jilin People's Publishing House.

17. [U.S.] Berleant, A. (2007). *Environment and the arts: A multidimensional perspective on environmental aesthetics*, translated by Liu Yuedi et al. Chongqing Publishing House.
18. [U.S.] Leopold, A. (1997). *A sand county almanac*, translated by Hou Wenhui. Jilin People's Publishing House.
19. [U.S.] Berleant, A. (2006) *The aesthetics of environment*, translated by Zhang Min, Zhou Yu. Hunan Science and Technology Press.
20. [U.S.] White, L. (1996). *The historical roots of our ecological crisis*, Containing in *The Ecocriticism Reader*. University of Georgia Press.
21. [Germany] Moltmann, J. (2002). *God in creation an ecological doctrine of creation*, translated by Kui Renlian et al. Life, Reading, and New Knowledge Sanlian Bookstore, 2002.
22. (1993). *Bible* (New Version). Hong Kong Tiandao Bookstore (《圣经》(新译本),香港天道书楼1993年版).
23. [Germany] Scheler, M. (1999). *Order of love*, translated by Linke, Liu Xiaofeng selected *Scheler Selected Works*. Shanghai Sanlian Bookstore.

Part IV
The Chinese Resources of Eco-aesthetics

Chapter 6
Book of Changes and Confucianism's Wisdom Pertaining to Ecological Aesthetics

Ancient Chinese wisdom may be divided up into three separate schools, that of Confucianism, Buddhism and Daoism, but Confucianism was the mainstream school, in which a fusion occurred linking together all three schools. No matter which school is under consideration, they all offered sufficiently rich ecological wisdom along with wisdom pertaining to ecological aesthetics. In examining the reason behind this, we need not look any further than Feng Youlan, who explains, "since China is a continental [rather than island] state, the Chinese people have lived solely off agriculture," "in such an economy, agriculture is not only important in times of peace, but is equally important in times of war."[1] Precisely because of this, as science and technology were underdeveloped, our ancient Chinese predecessors survived mainly on agriculture, which basically means that they ate what Nature provided—and that natural ecosystems were of paramount importance. For this reason, ecological wisdom and wisdom pertaining to ecological aesthetics became particularly plentiful in ancient writings. With respect to Confucianism and Daoism, even though there were disparities in their views, they "did both express an agricultural inspiration and aspiration." With respect to the ecological wisdom of Confucianism,[2] this chapter mainly introduces the ecological wisdom in the writings of Confucius and the other main figureheads of Confucianism as well as the Confucian classic *The Zhou Book of Changes*.

[1]Youlan [1], p. 21.

[2]Ibid., p. 31.

F. Zeng, *Introduction to Ecological Aesthetics*,
https://doi.org/10.1007/978-981-13-8984-9_6

6.1 The Ancient Ecological Wisdom and Wisdom Pertaining to Ecological Aesthetics in the Writings of Confucianism and the Main Figureheads of Confucianism

As the founder and representative figurehead of China's ancient school of Confucianism, Confucius ranks as China's most important thinker. His thought is collectively reflected in such classics as *The Analects* and *Record of Rites* recorded by his followers. Here, we analyze the ecological wisdom and wisdom pertaining to ecological aesthetics in the school of Confucianism as represented by Confucius himself with *The Analects* as our main support.

(1) **Ancient China's Wisdom of Ecological Ontology: "Nature and Humanity Integrate into One Whole"**

The most important thought in the doctrine of Confucianism represented by Confucius is "Nature and humanity integrate into one whole," which of course could be considered a thought equally shared by Daoism and Confucianism alike, namely "the relationship between Nature and humanity is integration and becoming one."[3] In the relationship between Nature and humanity, Daoism focuses more on Nature, while Confucianism focuses more on humanity. That said, Confucianism and Daoism both still see Nature (*tian* 天) as dominant in the relationship between Nature and humanity. There have been multiple interpretations of Heaven or Nature (*tian* 天). After Dong Zhongshu of the Eastern Han dynasty, interpretations tended to render "Nature as the mysterious Way," while in pre-Qin times most interpretations rendered "Nature as what is spontaneously so." So the idea of "Nature and humanity integrating into one whole" in those times already encompassed the thought of unifying "what is natural" and "what is human." "Nature and humanity integrating into one whole" actually expresses some relationship of humans being-in-the-world, a relationship between humanity and the world, including what is naturally occurring. This relationship is not one of opposition but rather of fusion, interdependence and singular substance, which is precisely the ecological wisdom of ancient China's Eastern ontology. In the *Analects*, Confucius states, "the function of ritual is to ennoble harmony." "Ritual" here does not refer to all of those rituals of everyday life, but rather the sacrificial rites, "the great rites synchronize the rhythms of Heaven and Earth" (*Record of Rites,* "Record of Music"); the "harmony" here could be understood as "the harmony of Nature and humanity," namely as a request for harmony between Nature and humanity. In the same vein, Confucius also said, "striking a medium (*zhong yong* 中庸) as a virtue is perfection itself," (*Analects*:), which is precisely to consider striking a medium as the highest virtue. Now, what is striking a medium? It means "excess is as bad as not enough," which is not entirely the everyday principle of practice that people generally understand, but rather the highest virtue. This highest virtue in those days should have been a morality upheld in relation to Nature. Here, "striking

[3] Zhongshu and Yui [2], p. 288.

a [good] medium" and "excess is as bad as not enough" are what the *Record of Rites* calls "the middle harmony," and what the *Zhou Book of Changes* calls "maintaining the blend of the supreme harmony" (I will elucidate this further in the next section). The middle harmony in the "Striking a Medium" chapter of *Record of Rites* is but a further elucidation of what Confucius spoke of as "harmony is ennobled" and "striking a [good] medium;" it is an ancient wisdom of ecological ontology." In "Striking a Medium" it states: "the moment at which joy and anger, grief and glee have still not been triggered is called equilibrium. Once the affects have been triggered, bringing them all in tune with equilibrium is called harmony. Equilibrium is the great root of the world under Nature's rule (*tian xia zhi ben* 天下之本); harmony is grasping the Way of the world under Nature's rule. When equilibrium is reached and harmony is struck, Heaven and Earth are situated and all creatures are nourished." It elevates "reaching equilibrium and striking harmony" to the height of "the great root of the world under Nature's rule" and "grasping the Way of the world under Nature's rule." Moreover, Heaven and Earth can only be respectively situated "when equilibrium is reached and harmony is struck," namely only with ecological balance do all creatures multiply, grow, and realize a state of harmony between humanity and all beings that is beneficial to the growth and survival of humanity and all beings. This is in fact an ancient wisdom of ecological aesthetics. Because of this, "situating nourishes equilibrium and harmony" became the famous classical saying of Confucianism, and were etched on the lintel of the Confucius temple.

(2) Reverence for Nature and the Appeal for the Objective Beauty of Nature

Confucius did not approve of the witchcraft flourishing in the common world at the time, and opposed the disruption of people's lives with such superstitious activity. "The Master did not speak of miracles, forces, chaos or gods," (*Analects*: *shu er*), but Confucius did however embrace the affect of reverence for Nature, stating that "someone of mastery is to have three reverences: reverence for the mandate of Heaven, reverence for great men and reverence for the sayings of wise men." (*Analects ji shi*), where he puts "reverence for the mandate of Heaven" first. When talking about Lord Yao of ancient times, Confucius believes Yao became great, because Yao emulated Heaven or Nature, "how great was Yao as someone of mastery. How towering! Heaven or Nature (*tian* 天) alone is grand, and alone Yao emulated it." (*Analects*, *tai bo*). Of course, aside from connoting Nature, *tian* 天 in Confucius was still tinted with a mysterious godliness. When he responds to a request for instruction from the senior official Wang Sunjia of the State of Wei on how to perform the sacrificial rite, Confucius answers: "this is not so, whoever is caught in the wrong by Heaven or Nature (*tian* 天) has nothing left to pray for." (*Analects*, 3.13). This mystical sense of reverence for Heaven or Nature was perfectly natural in a scientifically and technologically backward age, and the fatalism and mysticism in it was perfectly apparent as well, but enveloped in it was still an appropriate degree of reverence for the natural heavens; moreover, Confucius to a certain extent also saw Nature as an objective being unmoved by the will of man. People should have had a yearning and desire for this objective beauty of Nature. Once when talking about education with the disciple Zi Gong, Confucius spoke of the unspoken teachings of Nature. "Confucius said: 'I would much rather have nothing to say.' Zi Gong responded, 'If you do not speak, what would we your lesser ones have to transmit?' Confucius responded, 'what does Heaven have to say? The four seasons carry on and all creatures come to life, what does Heaven have to say?" (*Analects*: 17.19). This is to say, Nature teaches humanity with its objective workings that make all creatures prosper, and becomes humanity's perpetual yearning and expectation.

(3) **The Symbiotic Thought of Harmony without Homogeneity**

Confucius said: "Those of mastery coexist harmoniously but are not the same, those of inferior standing are all the same but do not coexist harmoniously." (*Analects*: 13.23) "Harmony without homogeneity" thus became an important thought in the Confucian doctrine. Confucius here opposes "harmony" to "sameness" or "homogeneity" (*tong* 同). What he calls *he* 和 (harmony) is but the harmonious compatibility of different things, and what he calls *tong* 同 (sameness) is but a homogeneous state of things that disallows the existence of different things. Just as the *Zuo Commentary* states, "harmony actually creates things, while homogeneity discontinues [the creation]." This reveals a classical thought of "symbiosis": nature can only prosper under the condition of many different kinds of living things mixing together; and vice versa, ecological flourishing cannot be sustained with only one kind of living thing. This is the conclusive thought of "no reproduction in the same family," which ancient people derived as the manifestation of an ecological law through the reproduction of humankind; at the same time, Confucius was also a thinker who fervently loved life and living. On one occasion, Zi Lu asked him for teaching on the problem of "death," and Confucius responded: "you do not yet know what life is, how could you know what death is?" (*Analects*: 11.12) In other words, Confucius believed that life, which is intimately bound to death, is more important than death. Moreover, Confucius also applied this thought of "life" to the Way of filial conduct in the famous discussion of "upholding filial conduct for 3 years [after the death of a parent]." Zai Wo asked Confucius why one ought to remain filial for 3 years, stating 3 years is too long of a period of time, and that 1 year is enough. Confucius responded to this question, stating: "You lived for 3 years and only afterward weaned off of your parents' nursing embrace. Three years of mourning is needed for the world to thoroughly mourn." (*Analects*: 17.21) This is to say, it generally takes 3 years for parents to nurse a life, enabling it to separate from the bosom and move and eat on its own. These 3 years are the beginning phase of a person's life, the critical period of life, and also the period where parents are the most attentive of and caring for their children. So, a person repays the affection of parental nurturing with no less than 3 years of remaining filial, which certainly shows great respect for life and the inheritance of life.

(4) **The Ancient Ecological Wisdom of Following Agricultural Timing**

In the era of Confucius, agriculture was the basis of economic livelihood, because of which, not violating the proper agricultural timing became the most important affair of government and people alike. "Not violating agricultural timing" refers to arranging agricultural production according to the natural ecological law of the four seasons climate. When speaking about ruling a state, Confucius put forward three principles, "solemnly tend to affairs but in trustworthy fashion, be economical in expenditure but care after the people, and work the people with seasonal timing." (*Analects*: 1.5) Here, "work the people with seasonal timing" refers to not messing up the agricultural timing, but instead undertaking agricultural production in accordance with the seasonal timing of the four seasons climate. This is classical ecological wisdom, and is extremely important. The "Lunar Ordinances" chapter of the *Record of Rites* documents in great detail the case of how government and people undertake agricultural activities in the four seasons of spring, summer, autumn and winter. The "Lunar Ordinances" considers the spring season for example to be the time when "the matter-energy (*qi* 气) of Heaven falls down, the matter-energy of Earth rises up, Heaven and Earth harmoniously unite, and flora sprouts forth with activity," in which case the government should not violate the agricultural timing, but instead lead the people to undertake timely agricultural production. First is "The Ritual of Greeting Spring," where "on the day of the inauguration of spring, the son of Heaven personally leads the Three Ducal Ministers, the Nine High Ministers, the Marquises and Grand Officers to greet the spring in the Eastern suburbs." (*Record of Rites*: 6.4) Second is "the ceremony of initiating the plough" with "the day of the first conjunction of the sun and moon having been chosen, with the handle and share of the plough in the carriage, placed between the man-at-arms who is its third occupant and the driver, he conducts his three Ducal Ministers, his Nine High Ministers, Marquises and Great officers, all with their own hands to plough the field of God. The son of Heaven turns up three furrows, each of the ducal ministers five, and the other ministers and feudal princes nine." (*Record of Rites:* 6.5) Next are the ecological protections needed to guarantee agricultural production, "[t]he canons of sacrifice are examined and set forth, and orders are given to sacrifice to the hills and forests, the streams and meres, sacrifices that do not use female animals. Prohibitions are issued against cutting down trees. Nests should not be overturned; unformed insects should not be killed, nor creatures in the Womb, nor very young creatures, nor birds just taking to the wing, nor fawns, nor should eggs be destroyed." (*Record of Rites*: 6.7) Confucius also of course expressed thoughts about corresponding "ecological protections," namely the well renown idea of "fishing but not casting nets," and "shooting but not aiming at nesting birds," which shows that ancient China was indeed conscious of ecological protection.

(5) **Forcefully Advocating a Prudent and Simple Way of Life in Harmony with Ecological Laws**

We previously mentioned that Confucius set "being economical in expenditure and caring for the people" as one of the most important principles of ruling the state, which is a manifestation of advocating a way of life in harmony with ecological laws. On yet another occasion, Lin Fang of the state of Lu asked Confucius for teaching on the foundation of "ritual," whereupon Confucius responded: "A great question indeed! In festive ceremonies, it is better to be sparing rather than extravagant. In the ceremonies of mourning, it is better that there be deep sorrow than a minute attention to observances." (*Analects*: 3.4) This is to say, for Confucius, the most basic intention behind "ritual" is avoiding extravagance and seeking frugality, while seeking and advocating an economically frugal way of life, which was of the utmost importance for the times in which Confucius lived. Since according to Confucius' contemporary, Mozi, political chaos repeatedly erupted at the time, the ruling class excessively overindulged and extravagantly overspent, while the broader mass of people could not bear the heavy burden, driving them into a life of vagrancy and extreme poverty. The extravagant and overindulgent way of life of the ruling class in one respect exacerbated social contradictions, and in another respect severely destroyed the natural environment and ecological balance. "Frugality" is a way of life in harmony with ecological laws, according to which "enough is enough."

(6) "The Wise Enjoy Waters, while the Humane Enjoy Mountains" and the Feeling of Affinity with All Natural Beings

A renowned saying attributed to Confucius goes: "the wise enjoy waters, while the humane enjoy mountains," (*Analects*: 6.23) which is a poetic thought of nature-friendly metaphors, where waters are used as an analogy for the wisdom of the wise, and where mountains are used as an analogy for the virtue of the humane. The fluidity of water and the liveliness of surging flows of water in rivers and creeks analogically show the wit of the wise in their clever responsiveness and the joys of acquiring wisdom; whereas the deep stillness and perpetuity of mountains figuratively illustrate the resoluteness of those with higher morality in holding firmly to the principle of being humane along with their value on par in radiance with the sun and moon. This shows Confucius' feeling of affinity with natural beings and his application of the method "analogical incitement" of affinity with nature in dialogues. There is of course another saying in Confucius: "those of mastery cherish virtue, while those of inferior standing cherish their soil; those of mastery cherish discipline, while those of inferior standing cherish rewards." (*Analects*: 4.11) Those of mastery here refer to those occupying the position of rulers, while those of inferior standing refer to common folk occupying the position of the ruled; neither those of mastery (*junzi* 君子) nor those of inferior standing (*xiaoren* 小人) here refer to what were ordinarily called *junzi* (gentleman, princes) and *xiaoren* (petty folk), and have nothing to do with evaluative judgments about good people and bad people. The meaning here is that those occupying the position of ruler seek refinement in ritual observances, while common folk seek security in living on their native soil; the rulers seek the lawful ruling of the state, while common folk expect the implementation of humane governance and improvements in daily life. "Cherishing their soil" here is actually an expression of common people's passion for their soil, because war and servitude in those times drove most of the people into lives of vagrancy, driving them to abandon home and well, so the affective connection to a homeland is one of the connotations of ecology.

(7) The Eastern Classical Ecological Humanism of "The Humane Love Humanity"

The core of Confucius' thought is that of learning how to be humane, which is also to say that Confucius' discourse revolves around "being humane" (*ren* 仁). So what is "being humane"? When Confucius responds to the student Fan Chi's request for teaching on the question of "being humane," he spoke of "loving humanity," which was precisely the thought circulating throughout the world that "the humane love humanity." Confucius gave many accounts of what "the humane love humanity" means, several of which encapsulate classical ecological humanism. This is to say that it is a caring love for humanity encapsulating ecological ideas, among which an important thought is "cultivating oneself to pacify humanity." When Confucius responds to Zi Lu's question about "how to become someone of mastery?" he states [someone of mastery] "cultivates himself to pacify humanity," then says "cultivates himself to pacify the people, which even Yao and Shun had pains in doing." (*Analects*: 14.42) This is to say, an important connotation of "being humane" is cultivating oneself to make the people enjoy a peaceful life and delight in work, which even the sage-kings Yao and Shun had difficulty accomplishing. "To be pacified" here has the meaning of "living securely," namely enabling the people to have their own secure living household, which is an important mark of ancient ecology on one hand and of the philosophy and aesthetics of life on the other. With respect to "being humane," Confucius elsewhere gives an explanation about "compassion" (*shu* 恕). Zi Gong asked, "'Is there a teaching which one could practice one's whole life?' The Master said, 'it is compassion! What you yourself do not want, do not do unto others.'" (*Analects*: 15.24) "What you yourself do not want, do not do unto others" is precisely "inferring about others from oneself," which in the philosophical sense could also mean seeing the object as another subject, that is mutual subjecthood. Previously we mentioned there are also similar sayings in the *Bible*, which ecological philosophers and ecological ethicists apply in ecological theory to explain the relationship between humanity and Nature. The human being could treat nature in the spirit of equality by inferring about others on the basis of oneself. If we expand the scope of "others" in "what you do not want, do not do unto others" to include all other natural beings, that would be the humane mentality of humanity toward all natural beings, which is a typical case of classical ecological humanism.

(8) **The Classical Ecological Perspectives of Mencius and Zhang Zai: "Be Humane to People and Love Other Beings;" "I am the Companion of the People and Share Life with Other Beings."**

Mencius was another Confucian Master of pre-Qin times. His main theoretical viewpoint is the famous "human nature is good hypothesis," which claims that everyone possesses "the affective mindset of intolerance to cruelty" and "the affective mindset of compassion for others," and advocates the "inferential thinking" of making inferences about others on the basis of oneself: "take how you treat your own elders to treat the elders of other families, take how you treat your own young to treat the young of other families." (*Mencius*: 1.7) From this we can infer how to treat natural beings. In Liang Hui Wang Book I, it is recorded that King Xuan of Qi saw someone butcher a cow for the "blood sacrifice" at the great hall, but upon seeing the cow tremble out of fear, he could not tolerate the cruelty, and finally substituted a lamb for the cow. Mencius thus deduced: "The prince's disposition toward animals is, upon seeing them alive, he cannot tolerate seeing them die; upon hearing the sounds they make, he cannot tolerate eating their flesh. So the prince stays far away from the butchery."

This is precisely an inferential thinking of the "affective mindset of intolerance to cruelty" and the "affective mindset of compassion for others," an expansion of "humane care." Furthermore, Mencius also put forward the thought of "being humane to the people and caring for other beings." In "Fully Actualizing the Affective Mind: Book I" Mencius states that the prince's disposition toward other beings is caring for them without being humane to them; his disposition toward the people is being humane to them without loving them as family. He loves his family but is humane to the people; he is humane to the people but cares for other beings." Here, regardless of what difference there is between other beings, the people and one's family, it is a matter of progressing from loving one's family to being humane toward people and then caring for other beings in a process of inferring about others on the basis of oneself. This humane thinking that progresses from what is near to what is further away is perfectly clear; it could be called a classical ecological humanism.

The Northern Song dynasty philosopher, Zhang Zai starts with human beings coming to life by sharing the same energy flow from Heaven and Earth as all other beings, and states that "there is not one other being which is not a part of me" (Correcting Folly: "Enlarging the Mind"), and furthermore states, "I am the companion of the people and share life with all other beings," that is, the thought that "I am the companion of the people" is to see the people as one's own siblings, and to see all other beings as one's own partners. This thought became the typical declaration of the thought that "human being is equal to other beings" in ancient Chinese Confucianism.

6.2 "The Reproduction of Life is Change" in *Book of Changes* and its Wisdom Pertaining to Ecological Aesthetics

The *Book of Changes* is a Confucian classic as well as one of the sources of ancient Chinese philosophy and aesthetics. The *Book of Changes*, and especially *Commentary on Book of Changes*, envelops the poetic thought of life philosophy unique to the ancient Chinese people; it implies "the reproduction of life is change" and represents an Eastern style of wisdom pertaining to ecological aesthetics. It influenced the aesthetic ideas and artistic forms throughout ancient China.

(1) The Ancient Ecological Wisdom Implied in The *Book of Changes*

The core content of The *Book of Changes* is the ecological wisdom of "the reproduction of life is change." The *Book of Changes* (Xici I) remarks: "The reproduction of life is change, forming phenomena is called Qian, and emulating standards is called Kun." Xici Part II also remarks: "The great virtue of Heaven and Earth is being-alive." "The reproduction of life" or "being-alive" is an explanation of the growth of all creatures, living beings and human existence. This is the core content of The *Book of Changes*; it is also the basic spirit of ancient Chinese philosophy and aesthetics. Meng Peiyuan points out, "the problem of 'life' is the core issue of Chinese philosophy, and manifests the basic spirit of Chinese philosophy. Whether it is Daoism or Confucianism, there is no exception. We could fully say that Chinese philosophy is the philosophy of 'life.'" He insists, "the philosophy of 'life' (*sheng* 生) is the philosophy of becoming and not the Western form of ontology."[4] The *Book of Changes* largely elucidates the ancient Chinese ecological and aesthetic wisdom of a theory of life and survival. We should say that in the genealogy of ancient Chinese philosophy, this philosophical thought of life and survival is largely an aesthetic one as well.

① The Ancient Eco-Ontological Thought of "The Reproduction of Life is Change"

"The reproduction of life is change" in the *Book of Changes* is actually a philosophical thought that explains the ancient eco-ontology in the briefest possible language. Here, "the reproduction of life" refers to the life and survival of individual life forms, and "change" refers to developmental change, which is to say "change means transformation." "The reproduction of life is change" refers to the principle growth and living development in living individuals. The *Book of Changes* "Commentary Explaining Hexagrams" states, "the sages of the past made The *Book of Changes* to follow the principle of life," namely ancient people wrote The *Book of Changes* mainly to explain the principle of life's creation, existence and development. The *Book of Changes* "Xici II" states: "Heaven and Earth envelop one another, all creatures transform enriching one another; male and female share energies, and all creatures develop into birth." This illustrates that humans are not diametrically opposed to Heaven, Earth and all natural creatures, but are rather one with Heaven, Earth and all natural creatures. Humans can only breed and develop, grow and survive among all natural creatures. The *Book of Changes* "Xici II" states: "Therefore, changes have the ultimate pole, which produces the two standards. The two standards produce the four phenomena, which, in turn, produce the trigrams. The trigrams stipulate fortune or misfortune, and fortune and misfortune generate great enterprises." The "ultimate pole" here is the *Dao* as the source of life and birth, while "the two standards" are "Heaven and Earth," "Yin and Yang," and it is precisely in this inseparable sympathy of affecting and being affected between Heaven and Earth, Yin and Yang that the production, existence and transformation of life becomes possible. Because of this, the Qian hexagram of The *Book of Changes* clearly champions "the integral oneness of Nature and humans," stating: "the greater man integrates his virtues with those of Heaven and Earth, blends his light with that of the sun and moon, integrates his order with that of the four seasons, and mixes his fortunes and misfortunes with those of ghosts and spirits; innately, Nature does not go against it; in learning, one follows the seasonal timing of Nature." In the philosophical system of The *Book of Changes*, "the reproduction of life is change" and "the integral oneness of Nature and humans" are interconnected with the "verge between what is natural and what is human,"

[4]Peiyuan [3], p. 4.

reaching the middle harmony, and the "doctrine of three natures," namely Heaven, Earth and Man, which altogether reflect an eco-ontological philosophy in ancient Chinese form that differs from the epistemological philosophy separating subject and object, man and world, in the ancient Western theory of Ideas or "imitation theory." In this eco-ontological philosophy, man and Nature fuse together and compose a whole. Meng Peiyuan insists, "objectively speaking, man is one part of Nature; subjectively speaking, Nature is a component part of human life. On a certain level, although there is a differentiation between inside and outside, subject and object, on the whole, inside and outside, subject and object integrate into one."[5] The wisdom of ecological aesthetics unique to Ancient China represented by "the reproduction of life is change" in The *Book of Changes* could only have been produced in this eco-ontological thought of the whole composed of Nature and man.

② Qian and Kun, Yin and Yang and "the ultimate pole" are theoretical ideas of the source of living creatures

The *Book of Changes* further elucidates the theory of the source of living things on the foundation of ancient eco-ontological thought. The *Zhou Book of Changes* is the Zhou dynasty's revelation of the secret of "change." The *Zhou Book of Changes* explains the rising Yang force of Qian in the workings of *yin* and *yang*; that rising force becomes the source of living things. The Qian Hexagram section in The *Book of Changes* "*Tuan Commentary*" states: "how great is the originating force of Qian! All creatures are seeded to begin covering all of Nature," which regards Qian and Yang as the "beginning" of the world and the origin of all creatures. As for Kun which corresponds with Qian, the *Book of Changes* insists, Kun is also the foundation of all creatures. The *Tuan Commentary* on the Kun Hexagram states: "How perfect is the originating force of Kun! All creatures are seeded into birth, following and abiding Heaven." What Kun symbolizes is Earth, the source of all things "seeding into birth." But, the ultimate source of the production of life and all creatures is the "ultimate pole" chaotically mixing *yin* and *yang*, Qian and Kun together. This is the "ultimate pole" of chaotically inseparable sympathetic affectivity between Qian and Kun, *yin* and *yang*, to which Eastern cultures, including China, reduced the origin of all creatures. This differs from the West, which took "matter" and "Ideas" as the source of all things. Life in The *Book of Changes* includes all entities on earth, whether organic or inorganic; namely what is produced out of Qian and Kun, *yin* and *yang*, Heaven and Earth is equally in possession of vitality. This differs from the modern Western philosophy of life, which limits life to organisms, plants, animals and especially humans. This philosophy and aesthetics of life in the West could be said to be clearly shaded by "anthropocentricism, while the theory of life in The *Book of Changes* is much more markedly ecological. Of course, The *Book of Changes* does not overlook the role and status of humans; it places humans in an important position above other entities in the famous "doctrine of three natures." Heaven, Earth and Man in The *Book of Changes* considers human being an important dimension next to Heaven and Earth; it argues for the impossibility of separating humans from the Heaven-Earth

[5]Peiyuan [3], p. 6.

dyad of Qian and Kun; it insists humans only come into being, reproduce and survive through the sympathetic affectivity between Heaven and Earth. Because of this, The *Book of Changes* envelops a simple and ancient eco-humanist spirit of symbiosis and coexistence between humans and natural creatures.

③ The Idea of the intermixing of *yin* and *yang*, Qian and Kun, Heaven and Earth out of which living creatures are produced

The *Book of Changes* makes a concrete description of the production process of creatures and life forms, which is a picture of mutual affection between Heaven and Earth, Qian and Kun, *yin* and *yang*. The *Book of Changes* "*Tuan Commentary*" on the Tai 泰 hexagram states: "Heaven and Earth cross the verge between them and all creatures emerge through the passage." This is to say, the *yin* and *yang* forces of Heaven and Earth affect and are affected by one another, generating all creatures, which is called "Tai." Conversely, "Heaven and Earth do not intercross and there is no passage for all things" (The *Book of Changes* "*Tuan Commentary*" on the Pi hexagram). This is the state of Pi 否, which is the hexagram sign blocking the growth of all things. The *Tuan Commentary* on the Xian 咸 or "conjoining" hexagram further points out: "The supple on top and the firm on bottom, the two forces (*qi* 气) sense and respond to one another to conjoin," "Heaven and Earth affect one another and all creatures are born." Its inner trigrams are "bound" (*gen* 艮) on top and "open" (*dui* 兑) on bottom, *gen* symbolizes firmness and Heaven, and *dui* symbolizes suppleness and Earth, so the supple rises up and the firm sinks to the bottom, the *qi* of the Earth rises and the *qi* of Heaven drops, firm and supple, Heaven and Earth, *yin* and *yang* intermix and all creatures are born. Here, The *Book of Changes* shows us a philosophical picture of ancient eco-ontology. Here, there is the state of *tai* 泰 conjoining Nature and Man, and there is the state of *pi* 否 preventing the conjunction of Nature and Man. *Tai* and *pi* symbolize two typical states of survival of Man and all natural creatures between Heaven and Earth. Man survives at the verge between what is natural and what is human, namely in the close interconnection between the natural and the human. The positive ecological environment characterized by the harmonious intermixing of Heaven and Earth will bring humans a better survival, while the degrading ecological environment characterized by failure of Heaven and Earth to intermix will make humans live in a poor state of survival. Natural ecology and human survival are intimately correlated.

④ All Creatures in the Universe is a Theory of Living Chains

The *Book of Changes* constructs a cosmic chain of repetitive cycles involving the poles of Nature and Man, Qian and Kun, *yin* and *yang*, firm and supple and humane and righteous; this cycle is an endless, living periodic repetition. The function nine line judgment of the Qian hexagram goes "seeing a flock of dragons without a leader, fortune." This image of the Qian hexagram symbolizes the organic connections and periodic repetitions of the natural world. The six lines of the Qian hexagram are all unbroken *yang* lines, symbolizing a group of dragons flying in circles, a periodic repetition, but no leader is seen. Only this accords with the power of Heaven, and accords with the affective state and regulating laws of interconnection in the natural world. But the "ultimate pole" that gives birth to all creatures is actually the Taiji schema of cyclical interconnection and mutual affection between *yin* and *yang*, Qian and Kun. "There is an ultimate pole to changes, which produces the two standards, and the two standard produce the four phenomena, the four phenomena produce the eight trigrams," and the eight trigrams couple and overlap, generating 64 hexagrams; *yin* and *yang* succeed one another, and cyclically repeat, thereby composing a chain of developmental changes in Heaven, Earth and Man as well as all other creatures in the universe. Actually, to a significant degree, this describes the cycle of all things in the universe and human life, which is a process of alternating exchange of matter-energy and the laws of movement; it is one of the characteristics of life.

⑤ Ancient Earth Ethics and "The Earth is Thick Supporting All Things"

The *Book of Changes* passionately praises the great, efficient virtue of the earth symbolized by the Kun hexagram, "the earth is thick supporting all things," "virtues integrate infinitely." The characteristic of the Kun hexagram in The *Book of Changes* reflects the noble motherly character of giving birth to and nourishing all creatures. First of all, the earth is the source of creation of all things, namely "all creatures are seeded into birth"; second, the earth rests in the subordinate position in relation to "Heaven" with the noble virtue of the fully respectful way of the wife and servant. The *Tuan Commentary* on Kun calls the power of Kun "following and embracing Heaven," the third SIX of the Kun hexagram in the *Tuan Commentary* states: "the way of the earth is the way of the wife, the way of the servant, the way of earth is incomplete and awaits completion"; third, the earth has the virtue of embracing widely and broadly nourishing; it also has the mastery of resourcefulness, "its comprehension is wide, and its brightness is great; the various things obtain by it their full development." The *Wenyan Commentary* lauds Kun "is most gentle and weak, but, when put in motion, is hard and strong; it is most still, but is able to give every definite form." Finally, the earth has the noble character of selfless contribution. The *Tuan Commentary* on the Kun hexagram hails, "the (capacity and sustaining) power of the earth is what is denoted by Kun. The superior man, in accordance with this, with his deep virtue supports things." The *Shuogua Commentary* states: "Kun is the earth, by which all things attain nourishment, so it is said attain service from Kun." In early human culture, being able to make such a complete description and praise of the motherly character of the earth—which is extremely rare in the world—the famous Western notion we are all familiar with, "the Gaia hypothesis," is already something of the 1960s. The Earth Ethics in The *Book of Changes* represented by "Kun is thick and supports all things" still has extremely high value even in the context of modern Earth Ethics; it should constitute a precious wealth and resource of ecological theory that includes ecological aesthetics.

6.3 The Ancient Wisdom of Ecological Aesthetics in The *Book of Changes*

The ancient ecological wisdom that "the reproduction of life is change" found in *Book of Changes* possesses rich and profound aesthetic implications that gave birth to and influenced ancient China's wisdom pertaining to ecological aesthetics. The facts show that this statement in The *Book of Changes* is an ancient "poetic thought" of ecological wisdom containing rich aesthetic implications.

① Describing Art and Aesthetics as One of the Ancient Chinese Ways of Survival

Book of Changes is a book of divination from ancient China that mainly lays out the divination practices and ideas of ancient people. In the primitive age of natural shamanism and the thick atmosphere of religion, divination was a spiritual way of life and basic mode of survival for ancient people. Moreover, primitive divination activities were tightly combined with witchcraft, ceremonies, song and dances, and could even be considered all one thing. Because of this, we could say that divination life existed in accompaniment with religion, art and aesthetics. The *Zhou Book of Changes* "Xici I" points out with Confucius' words: "the master said: 'the sages establish what the image-sign is to fully get the intention, sets up the hexagrams to fully determine reality from appearance, and connect the judgment to fully articulate it; they change and connect them to fully determine what is beneficial, drums to them and dances to them to fully embody the spirit.'" Here, the entire activity from establishing what the image-sign is and setting up the hexagrams to dancing and fulfilling the spirit has profound meaning with important value for our understanding of ancient aesthetics and art. "Drums to them and dances to them to fully embody the spirit" shows the ancient Chinese people's mode of survival, which tightly combined song, dance, music, witchcraft and ritual. The Chinese character *wu* 舞 (dance: Trans.) in bone inscription script is the image of a shaman holding two cow tails in hand shaking and dancing, which illustrates that art and aesthetic activities were a way of life tightly combined to the existence of life for ancient peoples. An important characteristic of ancient Chinese aesthetics and art was intimately associated with the most basic way of human survival; it was inseparably tied to music and ritual, and permeated every respect and facet of human life. People not only found entertainment in music, "music is joy," but at the same time found communication with the cosmos and harmony of social life in rituals and music as well. As The *Book of Rites* "Book of Music" puts it, "the greater music harmonizes with the heavens and the earth, the greater ritual is in synch with the heavens and the earth," "music plays in the ancestral temple, if ruler and servants listen to it together joining lower and higher status, none will not harmonize in respect; playing in the village of the clan chief, if elders and youth both listen to it together, none will not harmonize in obedience; playing in the houses of wealthy families, if father and son, older and younger brothers all listen to it together, none will not harmonize in affinity." Art and aesthetics permeated every aspect of life, which is a characteristic of the vital aesthetic activities of ancient Chinese ecological life.

② Classical China's Basic Aesthetic Ideas of "Holding Together the Greater Harmony" and "the Beauty of Supple *yin*"

The *Book of Changes* "*Tuan Commentary*" on the Qian hexagram makes the assertion, "holding together the greater harmony is beneficial and auspicious." "The greater harmony" was the most basic philosophical and aesthetic mode of being in ancient China, referring to a state of harmonious oneness between Nature and humans attained when Heaven and Earth, *yin* and *yang*, humanity and righteousness came into proper alignment. As The *Book of Rites* "Doctrine of the Mean" states, "the state prior to the arousal of love, hate, sadness and joy is called finding the middle, after the arousal [of any such affects], modulating them is called harmony. Finding the middle is the great origin of the realm; harmony is attaining *Dao* in the realm. Striking the middle harmony, Heaven and Earth come into proper alignment and all things grow." Here, the "middle harmony" was the form of beauty unique to ancient China; it was the beauty wholeness, the beauty of life, the beauty of compliance, the beauty of suppleness, whose core was harmony between Nature and humans. The third SIX line of the Kun hexagram, divided, "(shows the subject) keeping his excellence under restraint, but firmly maintaining it. If he should have the occasion to engage in the king's service, though he will not claim the success (for himself), he will bring affairs to a good end." The *Wenyan Commentary* on the Kun hexagram states: "Although (the subject of) this divided line has excellent qualities, he (does not display them, but) keeps them under restraint. 'If he engage with them in the service of the king, and be successful, he will not claim that success for himself.'—this is the way of the earth, of a wife, of a servant. The way of the earth is-'not to claim the merit of achievement,' but on behalf (of heaven) to bring things to their proper issue." Visibly, the beauty of the feminine suppleness of *yin* is the beauty of the way of the earth characterized by the hexagram Kun; it is the typical manifestation of the beauty of the earth. This is a beauty of inner implications, a beauty referring to resting in the subordinate position, of embracing social morals. The fifth SIX of the Kun hexagram states, "yellow skirt with blouse, the fortune of beginning"; the *Wenyan Commentary* on the Kun hexagram explains: "The superior man (emblemed here) by the yellow and correct (color), is possessed of comprehension and discrimination. He occupies the correct position (of supremacy), but (that emblem) is on (the lower part of) his person. His excellence is in the center (of his being), but it diffuses a complacency over his four limbs, and is manifested in his (conduct of) affairs:–this is the perfection of beauty." The fifth SIX line of the Kun hexagram is in the middle position of the top trigram, representing even more so the beauty of feminine suppleness. The *Wenyan Commentary* goes on, stating, the fifth SIX line states the judgment that the superior man inside wears a yellow skirt with a blouse on the outside, symbolizing that he is in the correct position of the middle, also implying the virtue of resourceful restraint, so he is capable of unleashing his virtue into the enterprise of ruling the state and pacifying the realm, thereby attaining the highest state of beauty. The trigram judgment of the Kun hexagram judges: "Kun (represents) what is great and originating, penetrating, advantageous, correct and having the firmness of a mare. When the superior man (here intended) has to make any movement, if he take the initiative, he will go astray; if he follow, he will find his (proper) lord. The advantageousness will be seen in his getting friends in the south-west, and losing friends in the north-east. If he rest in correctness and firmness, there will be good fortune." This is also the manifestation of the aesthetic consciousness of "the yellow and correct (color), is possessed of comprehension and discrimination. He occupies the correct position (of supremacy)." "Great and originating," "resting in correctness and firmness, good fortune," etc., illustrates that the *yin* and *yang* forces of Qian and Kun find proper alignment, and each fulfill their positions. In this way, even if there is a spell of becoming lost and losing the way, ultimately the way back will become clear, one will find the master's reception, and will be greatly of benefit; even if one loses a friend for some time, one will once again bounce back from lostness and regain the friendship in peace and fortune. In the Kun hexagram, "although *yin* is beautiful" manifests the proper positioning and alignment between the *yin* and *yang* forces of Qian and Kun as revealed by The *Book of Changes*, such that "Heaven and Earth interlock and all things find passage," "Heaven and Earth transform, and vegetation grows," fostering the vital and robust growth of humanity and all things, which is finally "the perfection of beauty." This idea is fully consistent with the notions of "striking the middle harmony, heaven and earth aligning, and all creatures growing" found in "Doctrine of the Mean." "He occupies the correct position of supremacy" here means the *yin* and *yang* forces of Qian and Kun find their proper alignment, thereby enabling the "beauty of middle harmony" that grows and multiplies all creatures in the world. This is the beauty of Nature and humans harmonizing; it is also the beauty of feminine suppleness and life characterized by the maternal quality of earth.

③ The Poetic Thought Unique to China that Expresses "Establishing What the Image-Sign Is to Fully Get the Intention"

The *Zhou Book of Changes* explains the principles of change through image-signs (*xiang* 象), numbers and judgments among other methods. So-called "image-signs" means the hexagram signs, diagrams representing the principles of change, which belong to a category of "poetic thought" unique to ancient China. "Xici II" states: "Therefore the changes are signs. Signs are images." The basics of the changes are the trigram signs, which show images of phenomena; the principle of changes are housed in them. For instance, if you see the Kun hexagram on top and the Xun 巽 hexagram on bottom, Kun is earth and compliance, Xun is wind and entrance, showing wind blowing all things on the earth, and in the blowing of the wind, you can see everything making it impossible for anything to hide. Its hexagram sign is two unbroken *yang* lines high up at the top being looked up to by four broken *yin* lines on bottom. The practice of viewing hexagrams in The *Book of Changes* uses such hexagram image-signs to house acutely observed principles of change. Then take the doubling of the Zhen 震 trigrams, which powerfully suggests the strength of thunderous movement. Cheng Yi of the Northern Song dynasty said in the *Yi Chuan Commentary on Book of Changes*: "Zhen as a trigram is one *yang* line generating below two *yin* lines, moving and rising, and hence it is thunder. Thunder is movement. …Thunder with movement means inciting shock." He also states: "Its image-sign is thus lightning; its meaning is thus movement. Lightning has the image-sign of inciting thunder and the meaning of moving and startling fear."[6] All of the trigram image-signs in The *Book of Changes* imply human patterns through patterns of the heavens and the earth; they imply human images beneath natural images, which is consistently connected to the techniques of "analogy and incitement" in ancient Chinese literary creations. As for "analogy," according to the explanation in *Explaining Graphs and Analyzing Characters*, "The character *bi* 比 (for analogy) is two people, close together."[7] This illustrates that the ancient Chinese practice of "analogy and incitement" implies considering Nature a partner and ecological friendship. This idea mainly comes from *Book of Changes*, whose *bi* 比 hexagram reveals the meaning of being so-closely akin and dear to one another that no distance is in between. The *Tuan Zhuan* commentary on the *bi* hexagram states: "*bi* 比 is auspicious. *Bi* 比 means to aid…," expressing *bi's* meaning of dear closeness and no distance in between. The *Xiang Zhuan* commentary on the *bi* hexagram states: "earth with water on top, close (*bi* 比)." The *bi* hexagram is the Kun trigram on bottom and the Kan 坎 trigram on top, Kun is earth, Kan is water, the earth gets watered and becomes supple, water finds earth and flows, earth and water are close without space between, which is the essence of the meaning of *bi* 比. This along with the practice of "analogy and incitement" in The *Book of Poetry* similarly express the eco-aesthetic relationship of close affinity between humans and other beings, other beings among themselves and humans with humankind. The relationship between "signs" and "meaning" in The *Book of Changes* is actually one of some "meaning" of cosmic movement and change in life being implied by "signs," which is fully consistent with ideas like "poetry is the expression of earnest thought" in The *Book of Documents* "Yao Dian," and "the mind uses signs to make connections" in Liu Xie's *wenxin diaolong* 文心雕龙, implying the meaning of some kind of life in the relationship between Nature and humans. This differs greatly from the Western doctrine of "art types" that unifies commonality and individuality.

④ Praising "Peace" (*tai* 泰) and the Beauty of Strong Healthy Life

The ancient Chinese eco-aesthetic vision represented by The *Book of Changes*, whose basic intension is life, also praises the beauty of healthy life. The Qian hexagram in the *Book of Changes* symbolizes "Heaven, in its motion, (giving the idea of) strength…," and reflects the aesthetic pursuit of "the superior man, in accordance with this, nerving himself to ceaseless activity." This is the praise of the beauty of the masculine strength of a health robust in vitality. The Tai hexagram makes an even more profound exposition of this beauty of masculine strength; it states: "In Tai (we see) the little gone and the great come. (It indicates that) there will be good fortune, with progress and success"; in the Tai hexagram, the Qian trigram is on top and the Kun trigram on bottom, the *yang* force of Heaven drops from high to low, and the *yin* force of Earth moves up from below, *yin* and *yang* forces interlock and combine, making everything between heaven and earth find fluid passage, smooth pursuits and

[6] Yi [4], p. 292.

[7] Shen [5], p. 386.

robust life. The *Tuan Zhuan* commentary on the Tai hexagram interprets it, stating: " 'In Tai (we see) the little gone and the great come. (It indicates that) there will be good fortune, with progress and success.' It is thus Heaven and Earth intercrossing, and all creatures interlinking; the higher and lower (in society) cross paths, and their wills converge. *Yang* on the inside and *yin* on the inside, assertive within and compliant without, superior master within and minor human being without." The forces of Heaven and Earth form an intercross through which all things find fluid passage, the higher and lower in society cross paths and their wills converge. *Yang* and the inside and *yin* on the outside, strong on the inside and compliant on the outside, healthy vitality overflows, and thus there is fluid passage. The Da Zhuang hexagram is another praise of healthy, strong vitality. The *Tuan Zhuan* commentary states: "In Da Zhuang we see that which is great becoming strong. We have the (trigram) denoting strength directing that which denotes movement, and hence (the whole) is expressive of vigor. 'Da Zhuang indicates that it will be advantageous to be firm and correct:'—that which is great (should be) correct. Given correctness and greatness (in their highest degree), and the character and tendencies of heaven and earth can be seen." The Da Zhuang hexagram is the Qian (heaven) trigram on bottom and the Zhen (thunder) trigram on bottom, Qian is strength, thunder is movement, and so denotes strength directing movement, and moreover, brimming strength. Of course, while the *Book of Changes* praises the beauty of masculine strength, it also lauds highly the beauty of feminine suppleness in "the earth being thick supporting all things," and especially emphasizes the virtue of the earth carrying the load of all things, nurturing their life. This shows that The *Book of Changes* proposed two aesthetic forms of "masculine strength" and "feminine suppleness," which are the two ontological forms of beauty in human life.

⑤ Explaining the Ancient Chinese People's Simple Expectation and Pursuit of a Better Life and Home

The content of the 64 hexagrams of the *Book of Changes* basically encapsulated all aspects of early people's most basic existence and life, including making things and producing, eating and drinking, making shelter, social interactions, marriage and family, advance and retreat, gain and loss, existential encounters and the joy and sadness of separation and union, etc. Summarily speaking, what early peoples sought in these basic aspects of survival and life was a better life and poetic dwelling, namely the four virtues lauded in the judgment of the Qian hexagram, "originating, penetrating, advantageousness and correctness." The *Wenyan* commentary on the Qian hexagram explains: "'the great and originating' is (in man) the first and chief quality of goodness; what is called 'the penetrating' is the assemblage of excellences; what is called 'the advantageous' is the harmony of all that is right; and what is called 'the correct and firm' is the faculty of action." Here, goodness, excellences, harmony and action are all statements about doing things successfully and better human survival, a pursuit of an eco-aesthetic state of harmonious coexistence between Nature and human society. Gao Heng's *Current Annotation on the Great Commentaries of the Zhou Book of Changes* insists, "penetrating means beauty," "penetrating development of the variety of things" hence "everything gets refined."[8] The *Book of Changes* considers Qian to be the source of all things, symbolizing Heaven, which brings the energy of infinite life to the earth and humankind and is capable of making "the clouds move and the rains ensue, with the variety of things circulating." In the case of normal conditions, all creatures grow, prosper and multiply, while people find the auspicious moment of peace and health. Therefore, the *Tuan Zhuan* commentary on the Qian hexagram states, "the way of Qian is transformation, each entity rectifying its fate, holding together the greater harmony, which is what is good and correct." This is also the result of "correctly assuming the proper position," and "holding together the greater harmony" is the beauty of "striking the middle harmony." The *Zhou Book of Changes* gives ample expectations for peaceful and harmonious family life. One could say, The *Book of Changes* is the more earlier ancient Chinese canons in which the sense of "home" appears, thus making sense of home become the earliest aesthetic idea in the world with the implication of the appreciation of beauty. Such a concept only appeared in Western ontological and environmental aesthetics by the 20th century. The *Book of Changes* first proposed the notion of safety and auspicious correctness in the trigram judgments of the Kun hexagram, insisting that man could only safely return home and put affairs in smooth order by walking the right path. The entirety of what is spoken of in the Kun hexagram involves "originating and penetrating," which reveals Earth as the safe place where humans find happiness. The Jia Ren hexagram in The *Book of Changes* pins hope on parental authority illuminating the way to govern the family. As for realizing the greater hopes of one's home and beauty, the Jia Ren hexagram's fifth NINE states, according to the *Xiang Zhuan* commentary that: "'The influence of the king extends to his family:' - the intercourse between them is that of mutual love." The *Wenyan* commentary on the Kun hexagram exposes the expectation of happiness and safety, stating: "The family that accumulates goodness is sure to have superabundant happiness, and the family that accumulates evil is sure to have superabundant misery." Meanwhile, the Fu 复 hexagram considers those that leave home and return as a happy affair with the second SIX line stating, "[it] shows the admirable return (of its subject). There will be good fortune." Here, return originally refers to the forces of *yin* and *yang* correctly realigning, namely returning to the right path of the cosmic movement. Meanwhile, "returning" which includes returning to family, means "rest," and this is to say a good affair. With respect to the second SIX, the *Xiang Zhuan* commentary states: "'The good fortune attendant on the admirable return (of the subject of the second line)' is due to his condescension to the virtuous (subject of the line) below." This means that being capable of enabling many people who go out for service in the name of the state to return home is the result of those occupying the higher position in society enacting humanity and righteousness. Because of this, the *Book of Changes* asserts that even in disadvantageous terrain, if those occupying the higher positions in society act in accordance with the way of Nature as a whole, they can still, in effect, "be a great fruit which has not been eaten…," while "the superior man finds (the people again) as a chariot carrying him." This is to say, with the great fruit not ready to drop the supe-

[8]Heng [6], pp. 53, 77.

rior man can find the proper chariot he duly needs. This is precisely what the *Xiang Zhuan* commentary on the Bo 剥 hexagram states, namely "Superiors, in accordance with this, seek to strengthen those below them, to secure the peace and stability of their own position." This is to say, those in higher position have the humane heart to make the people find safety in dwelling territory. This illustrates that such better states of survival as "returning" and "resting in safety" were the aesthetic survival goals pursued by early Chinese people. The exposition on the fortune of returning home in The *Book of Changes* was the earliest description of the "consciousness of home" related to ecological aesthetics, which has important theoretical value indeed.

The four virtues and notion of the fortunate nature of returning home found in The *Book of Changes* are accounts of people's aesthetic experiences of better states of survival. Considering contemporary ecological aesthetics' pursuit of humanity's "poetic dwelling," this is a true sense of beauty, a wisdom pertaining to ecological aesthetics that is intimately related to ecology and life.

(3) The Influence of Ecological Aesthetic Wisdom in The *Book of Changes* Upon Later Generations

The ancient ecological aesthetic wisdom of the statement that "the reproduction of life is change" found in *Book of Changes* has had profound and important impacts, and has largely determine China's difference from the West in terms of aesthetics and art in both basic form and appearance; this has made it present a fully robust beauty of vitality in terms of harmonious wholes and partnership with humans and nature.

First of all, with respect to direct influences, the *wenxin diaolong* largely carried on the eco-aesthetic wisdom found in the statement that "the reproduction of life is change." First look at its chapter *Principle of Origination* (*yuandao* 原道). As the lead chapter of the whole book this chapter, one of the origins opined by this chapter includes the basic thought of the *Book of Changes*. Just as Liu Xie states in the same work, "the patterns as virtues are grand indeed; how is it that they are co-produced with Heaven and Earth? Colors blend obscurely in the yellow light, and shapes, square and circular, are separated. Sun and moon overlap their spheres to shine down images from heaven; mountains and rivers shine like silk, laying out the shapes of the earth's patterns.......the mind generates and words are established, words are established and patterns become clear; this is the way of Nature." He also states: "the origin of human writing, commences with the ultimate pole (*taiji* 太极). Obscure and obvious, mysterious and clear, only the sign-images of *Book of Changes* stand premier." *Book of Changes* states: 'what incites the movement of the world is preserved in language.' The reason why language can incite the movement of Nature is because of the patterns of *Dao*." Clearly, the Liu Xie's chapter *yuandao* includes the "way of nature" and "way of civilization" expressed in *Book of Changes*. Zong Baihua mentioned the two hexagram signs brought up in the *wenxin diaolong* in his work *Initial Exploration of Important Problems in the History of Chinese Aesthetics*. First is the Bi 贲 hexagram mentioned in his work *qingcai* 情采, about which he states, "therefore, wearing silk clothing, fear of patterns being too ornate; the sign-image of Bi is fully white, ennobling the return to the original." This is to say, Liu Xie insists that wearing silk clothes with fine fiber blouse on the outside is for the sake of avoiding the appearance of being overly ornate. The upper NINE line's judgment for the Bi hexagram "the ornateness of white" reflects precisely a return to the simple beauty of colorlessness. He argues "the need for natural, simple beauty adorned in white is the highest state of mind"[9]; second, Zong Baihua asserted in the work *zhengsheng* "articles written clearly depend on emulating the Li hexagram." Liu Xie insists that lucidly written articles are modelled after the Li hexagram in The *Book of Changes*. The *Tuan Zhuan* commentary on the Li hexagram states: "the Li hexagram is adherence. The sun and moon adhere to the sky, and all grains and plants adhere to the soil, doubling the brilliance (of the two trigrams) requires adherence to doing what is right, and the result is transforming and completing the realm." Thus, the Li hexagram is the sign of adherence, but has the meaning of brightness and clearness. Liu Xie demands that articles should meet these demands of the Li hexagram. By extension, Zong Baihua deduces four aesthetic implications: adherence is beautiful, inside and outside are penetratingly consistent, opposites are symmetrical and thoroughly connected like a web.[10] In addition, the chapter "Analogy and Incitement" in *wenxin diaolong* was also clearly influenced by the *Book of Changes*. It states: "so analogy is adherence; incitement is arousal," and "seeing that incitement depends on metaphor, it is round-about and logical, the names called may be small, but the category called is simultaneously large." Such descriptions mostly extract from The *Book of Changes*. The *Tuan Zhuan* commentary states: "analogy is auspiciousness; it is an adjacent pair helping one another." This sheds light on analogy's meaning in complementary aiding and the auspiciousness of friendship. The relationship between "sign-images" and "meaning" in the *Book of Changes* principally used the technique of "analogy and incitement" like using Heaven and Earth as images of *yin* and *yang*, and using natural phenomena as metaphors for relationships in human affairs, etc. Next, *Book of Changes* "Xici II" discusses the aesthetic characteristics of The *Book of Changes*, stating: "the names that it calls up are small, but the category it extracts is big. Its reference is far-reaching, its diction is patterned, its language is round-about but hits the mark, the end it serves is all over the place and hides." Obviously, this aesthetic characteristic of *Book of Changes* mainly originates from the relationship of analogical incitement between "images" and "meaning." The influence of the *Book of Changes* can be seen all over the place in other chapters and relevant expositions in *wenxin diaolong*. One could say, the basic philosophical and aesthetic spirit of *Book of Changes* already permeates the *wenxin diaolong*. As China's first systematic literary theory and work on aesthetics, the *wenxin diaolong*'s influence upon Chinese aesthetics and literary thought was massive, but carries with it always the immortal merits of the aesthetic influence of the *Book of Changes*.

Second, even more important is that the eco-aesthetic wisdom of "the reproduction of life is change" and "the beauty of striking the middle harmony" in the *Book of Changes* already appears as a way of human survival in every different facet of the

[9]Baihua [7], p. 333.

[10]Ibid., pp. 334–335.

life and thought of Chinese people; in a deep way it determines the characteristic of Chinese people's way of appreciating beauty; it has also pervaded the entire process of development of Chinese aesthetics and art. Especially the "beauty of striking the middle of harmony" in "assuming the proper position" revealed by the *Book of Changes* has had a profound role in the formation of an aesthetics and form of art unique to China and different from those of the West. The literary thought of ancient China's theory of poetry in the notions of "analogical incitement," "images," "intentional states," "force of writing" and that "poetry expresses earnest thought" was fully influenced by such aesthetic thoughts in the *Book of Changes* as "the reproduction of life is change," "assuming the proper position" and the beauty of striking the middle harmony. For instance, the famous theories of images, intentional states and spiritual thought whose impact on later generations has been far and wide were clearly and profoundly influenced by the "image thinking" of the *Book of Changes* from "establishing the image-sign to fully get the meaning" to "drumming to them and dancing to them to fully embody the spirit." The image thinking in the *Book of Changes* enlightens us, the "image" in ancient Chinese literature is not stuck to the "image" in-itself, but furthermore reflects "celestial phenomena" and the "way of the heavens" through the "image of creatures" and reflects "the patterns of heaven and earth" through "the patterns of human society," which constitute the deeper meanings of "the image beyond the image," "the message between the lines" and the "implication between the sense" in ancient Chinese poetics and aesthetics.

The artistic Idea of "moving rhythm" in the Chinese theory of painting also bears traces of ancient theories of ecological and life aesthetics. The doctrine of rhythm in the aesthetics of painting holds that the principle of "life" and the "rhythm" of *qi* penetrates concrete images of things, which is even more so a typical manifestation of the aesthetics in the *Book of Changes*. The different styles in and poetic methods in the *Book of Poetry*, from ode and hymn and praise to analogy and incitement, all carry the shadow of the aesthetics of the *Book of Changes*. As to the sensible characteristics unique to Chinese aesthetics and art, from friendship with nature and harmonious wholes to the over-brimming fulness of the tensions and implications of life, they are even more closely related the ancient eco-aesthetic wisdom of the *Book of Changes*, from "the reproduction of life is change" to "the beauty of striking the middle harmony." Ancient Chinese arts, especially, poetry and painting, mostly adapt Nature, mountain-waters and plant life as their objects of depiction, and in such depictions, some kind of spiritual rhythm is pervasive. Such aesthetic and artistic characteristics differ widely from the ancient Western notions of sculpted beauty like still harmony and alignment with perspective. Differing from the ancient Western aesthetic ideas of "harmony, proportion and symmetry" along with the variety of epistemological forms of aesthetics, ancient Chinese aesthetics was a much more expansive beauty of middle harmony, a beauty of the earth and a beauty of life formed out of the interlocking of Nature and humans in the verge between them against a philosophical backdrop. The bone script version of the character for beauty, *mei* 美, is precisely the image of ornaments like feathers attached to a human head; the ancients found this to be beautiful, which is to say, what the ancients called "beauty" originated from wearing feathers to dance in celebration or using a goat head to sacrifice in a dance; all of this had the meaning of praying to the heavens to rain down fortune, health and luck, which does not have much of any relationship to specific harmonies, proportions and symmetries.

The facts show that we have an extremely rich Chinese form of ecological and biological aesthetics, whose basic ideas were "the reproduction of life is change" and "holding together the greater harmony," represented by the *Book of Changes*. Such a system of aesthetics requires us to better research the consummation of it on the grounds of our predecessors, in order to construct contemporary Chinese aesthetics and participate in a dialogue with other forms of aesthetics in the world.

References

1. Youlan, F. (1985). *A brief history of Chinese philosophy*, translated by Tu Youguang. Peking University Press (冯友兰:《中国哲学简史》,涂又光译,北京大学出版社1996年版).
2. [Han] Zhongshu, D., [Qing] Zheng, Y., & Yui, S. (1992). Collated by Zhong Zhe: *Chun Qiu Fan Lu Yi Zheng*. Zhonghua Book Company.
3. Peiyuan, M. (2004). *Man and nature*. People's Publishing House (蒙培元:《人与自然》,人民出版社2004年版).
4. [Song] Yi, C. (2011). *Zhou Yi Cheng Shi Chuan*, Collated by Wang Xiaoyu. Zhonghua Book Company.
5. Shen, X. (1988). Notes of *Shuo Wen Jie Zi zhu*. Shanghai Ancient Books Publishing House.
6. Heng, G. (1979). *Zhou Yi Da Chuan Jin Zhu*, Qi Lu Shushe.
7. Baihua, Z. (1987). *Yi Jing*. Peking University Press.

Chapter 7
Daoism's Wisdom Pertaining to Ecological Aesthetics

Traditional Chinese culture is fully set upright by three traditions, Confucianism, Buddhism and Daoism, which supplement one another and mesh together into one culture. Although Confucianism has been authorized over the dynasties as the orthodox tradition by the feudal rulers, the major roles played by both Buddhism and Daoism in traditional Chinese culture cannot be easily overlooked. An examination of the wisdom pertaining to ecological aesthetics contained in traditional Chinese culture would basically fail to produce any comprehensive and profound grasp without analyzing these two schools. Investigating the wisdom pertaining to ecological aesthetics in Daoism and Buddhism carries fully important enlightenment and significance for today's protection of the environment and construction of aesthetics.

In traditional Chinese culture's complementary pairing of Confucianism and Daoism, that Confucianism has acted the mainstream culture in feudal Chinese society is obviously an uncontested fact, but the culture of Daoism represented by Laozi and Zhuangzi has increasingly shown its own unique value and significance in contemporary society, thereby drawing broader attention across the academic world in the East. As an ancient Eastern form of ontology, aesthetic thought and life wisdom possessing a complete theoretical system and profound implications, the philosophy and aesthetic thought of Lao-Zhuang Daoism has already become an extremely important piece of heritage and medicine treating contemporary social and mental sicknesses among the treasure keep of human thought. This chapter tries to explore the profound implications of Lao-Zhuang Daoism's philosophy and aesthetic thought from the perspective of eco-ontological aesthetics.

F. Zeng, *Introduction to Ecological Aesthetics*,
https://doi.org/10.1007/978-981-13-8984-9_7

7.1 The Ecological Ontology of "Dao Emulating What is Naturally So"

What "Dao emulates what is naturally so" states is the basic law, origin and propensity of the movement of everything in the universe, but the law, origin and propensity spoken of here differs entirely from the epistemological model of thought in the ancient West characterized by the subject-object dichotomy like "the theory of Ideas" and "the theory of imitation"; it also differs from Pythagorean "numbers" as well. What "Dao emulates what is naturally so" reveals is a philosophical and aesthetic proposition in ontology, not a philosophical and aesthetic proposition in an epistemology presupposing the subject-object dichotomy. It is only possible to resolve the problem of symbiosis between everything in the universe and humankind from the ontological perspective. One could say, this proposition implies profound ecological wisdom. Thus, "Dao emulates what is naturally so" cannot be understood from the epistemological perspective.

First of all, the "Dao emulates" in "Dao emulates what is naturally so" speaks of the origin of the birth and development of everything in the universe, and this "Dao" is the most basic Being of everything in the universe and humankind. Laozi says: "the Dao that can be spoken is not the eternal Dao" (Daodejing, 1.1). Here, Laozi distinguishes the Dao that can be spoken from the eternally existing, unspeakable Dao. The Dao that can be spoken is the "beings" present in phenomena, while "the eternal Dao" or eternally existing, unspeakable Dao is the absent Being behind phenomena. The task of Daoism is to explore the absent Being behind phenomena through phenomena, to grasp the eternal Dao through the Dao that can be spoken, which is the ancient form of the problem of "Being."

Moreover, Laozi and Zhuangzi also consciously drew a clear boundary in their own theory between Daoism as ontology and epistemology as a system of knowledge. They divide human knowledge into two kind: one being common knowledge, what they call zhi 知; the other being the highest knowledge, what they call zhi zhi 至知. Daoism basically holds a negative attitude towards common knowledge; Daoism does not deny the existence of such knowledge, but denies its value. They advocate "doing away with sagehood and abandoning knowledge" (19.1). They affirm the highest knowledge, however. Zhuangzi says, knowing what is done naturally, and knowing what is done by human [agency] is the highest knowledge." (6.1) This is to say, Zhuangzi argues, the capacity to know the working laws of Nature and human society is the highest knowledge or zhi zhi 至知, which only "true human being" may attain through Dao, namely the true human being's "knowledge can ascend to the peak with help from Dao" (6.1). This illustrates that there is a strict boundary between "Dao" and common "knowledge." Since there is a distinction between "the Dao that can be spoken" and "the eternal Dao," there is also a distinction between forms of expression, namely that between "propositions" (yan 言) and "sense." "Propositions" are "the Dao that can be spoken" and "sense" is the "eternal Dao." Propositions and sense are different in terms of presence and absence, but they are also closely connected. You have to get the "sense" from the "proposition," but also have to "get the sense" and "forget the proposition." Zhuangzi states: "The bamboo trap in relation to the fish is such that once the fish is caught the trap is forgotten; snares are employed to catch hares, but once the hare is caught the snare is forgotten; propositions are like this with the sense, once the sense is caught, the proposition is forgotten." (26.13) Zhuangzi shows via this vivid analogy that the fisherman's aim is to catch fish, which, once caught, requires forgetting the trap used to catch them; the hare hunter's aim is to catch hares, which, once caught, requires forgetting the snare used to catch it; the aim of those speaking is to express a deeper sense, which, once expressed, requires forgetting the linguistic proposition used to express it. Here, Zhuangzi illuminates the relationship between propositions and sense with the relation between bamboo traps and fish, snares and hares, which actually touches on the relationship between language and Being. In his eyes, language as a combination of sounds belongs to what is present, the phenomenal world, but tries to express the absent Dao behind the phenomena (namely, "Being"). This expressing, in Zhuangzi's view, is almost impossible. He argues that Dao, as the root of birth and development for all things in the universe, can only be gathered by sense, and cannot be transmitted by words. Therefore, Laozi states: "the great voice spares sound, the great image lacks form, Dao hides unnamed." (Laozi, 41) Zhuangzi argues: "Heaven and Earth have great beauty beyond words, the four seasons have clear order beyond debate, all creatures have consummate reason beyond words." In the view of Laozi and Zhuangzi, the great voice, the great image, the great beauty cannot possibly be expressed with sounds, shapes and words. Because they do not exist on the epistemological level, but rather belong to the category of ontology, they can only be comprehended through aesthetic imagination and spiritual freedom. This thought in Laozi and Zhuangzi laid the foundation for the deeply influential doctrines of "intended image" and "intentional state" in ancient Chinese arts and aesthetics, so its implications are extremely profound, differing widely from the "typology" uniting commonality and individuality on the foundation of epistemology in the West.

"What is naturally so" in "Dao emulates what is naturally so" is the spontaneous being opposed to "human effort"—no need for external force, shapeless and speechless, effortless and unconscious. This is but the essence of Dao. Zhuangzi states, using Laozi's words, that: "the water is to the spring, effortless and naturally so." "Naturally so" is like flowing water, which when encountering obstacles, generates water waves, and this is effortless done, naturally so. The gushing of water requires no outside force or effort, it is done naturally so; it is effortless action (wuwei 无为); it is also an action without object of desire. Laozi states: "Therefore, be constantly without object of desire to observe its subtlety." Without object of desire, and without external force, one may finally observe the subtle aspect of Dao, the true affective state of everything in the universe. What is "observing?" It is to behold. To behold is not to possess; it is not to know, but rather to realize. To "observe" is to realize, but this realization must maintain some distance. "To observe its subtleties means to eternally maintain the original disposition of effortlessly acting naturally, solely by virtue of which one can realize the profound subtle height of Dao. Here, "being without object of desire" is already an unique aesthetic attitude that is neither one of possession, nor of cognition, that is both transcendent in relation to phenomena and yet physically observed. Laozi insists, "do nothing intentionally and yet nothing is left undone" (48), which is to say, one can only effect everything by doing nothing. "Doing nothing" and "leaving nothing undone" correspond to one another, one can only leave nothing undone by do nothing at all. Conversely, if one cannot accomplish doing nothing, one cannot but leave things undone. Laozi furthermore explains: "All creatures finish but do not initiate, produce but do not possess, act but do not hold onto it, consummate forces but do not dwell in their success. Only by not dwelling in success, nothing leaves." (2) To understand this through the relationship between humans and nature, that would require humans starting with the naturally spontaneity with everything that is happening, not reforming their initial growth, but producing without possessing, and with the aid of everything growing, having something to do, but not seeking to dwell at the center of things after successfully doing something to everything's growth. It is by virtue of not dwelling at the center of everything that the growth and position humans are due on the contrary become guaranteed.

Laozi also put forward an important idea of eco-ontological aesthetics, which is that of "not contending." He states: "not promoting those of superior character will prevent the common people from becoming contentious." (3.1) Here, "not promoting those of superior character" means not excessively praising superiors and the famous, and thereby making the common people not contend for merit and fame, but rather return to the natural state of being. Laozi and Zhuangzi expand the thought of not promoting those of superior character to human being's relationship with Nature, namely "water benefits all things but does not contend." Laozi argues, that similar to water, Dao nourishes all creatures, but does not contend with them. Precisely by not contending with the realm in following the non-coercive action of Dao, "none in the realm can contend with you."

Zhuangzi furthermore develops Laozi's "non-coercive action" into the thought of xiaoyao you 逍遥游, which according to its essential meaning refers to reaching the relaxed, carefree state of being through unconstrained maneuvering. For Zhuangzi, it specifically refers to a spiritual state of freedom. Free movement you 游 in Zhuangzi refers to two distinct states: one is freely moving with dependencies, the other is freely moving without dependencies. In everyday life, you 游 is always moving with dependencies. The massive bird, Peng, soaring up ten-thousand li must depend on the winds, the travelling spirit of a wild horse and kicked up dust all have something to depend on. Only the free movement of affective mind is without dependencies, needing and depending on nothing, and thereby attains the soaring movement of freedom. He states: "You freely move the affective mind in blandness, integrate forces into dispassion, follow how things naturally are and tolerate nothing from private selfishness." ("Ying Di Wang") "Freely moving the mind" puts one's spirit in the state of detached non-coercive action, following what happens naturally and forgetting about the existence of the self, which is precisely the mind rejecting the obstacle of partiality and moving toward natural non-coercive actions. This is also a process of moving from concealing to clearing, which is true spiritual freedom. This spiritual freedom may comprehend everything that is happening relying on the free movement of thought. The chapter Autumn Waters (qiu shui 秋水) covers Zhuangzi dialogue with Huizi: "Zhuangzi and Huizi were moving over a bridge. Zhuangzi said: 'those fish swimming at such ease, this is the joy of fish.' Huizi said: 'You are not a fish, how could you know the joy of fish?' Zhuangzi said: 'you are not me, how do you know that I don't know the joy of fish?' Huizi responds: 'I am not you, so I don't know what you know; since you are not a fish, it comes full circle to say you do not know the joy of fish.'" Zhuangzi says to Huizi, the fish in the river swim at such ease, they enjoy it so much. Huizi retorted: you are not a fish, how could you possibly know the fish's joy? Zhuangzi said: you are not me, so how could you possibly know that I don't know the fish's joy? This implies: that petty epistemological style wit of Huizi made it impossible for him to know the joy of fish. Only Zhuangzi's detached free movement allows him to comprehend the joy of fish. Because the fish's joy is not known, but comprehended or realized. Only the free movement of thought, carefree and non-coercively acting, can comprehend the joy of fish, which is "freely moving the affective mind." "Freely moving the affective mind" is Zhuangzi's further deepening and developing of the thought of non-coercive action in "Dao emulating what is naturally so," which is a process of moving from unconcealing to clearing.

7.2 The Ecological Theory of Genesis in "Dao is the Mother of the World"

Daoism's thought about the becoming of everything in the universe concentrates on the idea of "Dao being the mother of the world," which is a theory about the source of the genesis of everything in the universe. Everything in the world is born and nourished by dao. Dao in Daoism is neither material nor spiritual substance; it does not fall under the scope of epistemology, and does not have a dichotomy between subject and object. Dao falls under the scope of ontology; it is the ultimate origin of the genesis of everything in the universe and even "Being"; it is also a process or way of human life. Using the Western epistemological model of thought with the subject-object dichotomy makes it impossible to understand the Dao theorized by Laozi and Zhuangzi and their Daoist thought. Laozi and Zhuangzi clearly argue that the ultimate source of the genesis of everything in the universe and even human development is dao. This puts forward a thought of the common origin of humans and all creatures, and thereby makes the philosophical and aesthetic thought of Laozi and Zhuangzi become "non-anthropocentric." Laozi states: "There is something in mixed becoming, emerging prior to heaven and earth. Silent and empty, standing alone as all that is, it does not suffer alteration. All pervading, it does not pause. It can be thought of as the mother of the heavens and the earth. I do not yet know its name. If I were to style it, I would call it Dao, and if forced to give it a name, I would call it grand. Being grand, it is called passing. Passing, it is called distancing. Distancing it is called returning. Therefore, Dao is grand, the heavens are grand, the earth is grand, and the king is also grand. Within our territories there are four "grandees" and the king is one of them. Human beings emulate the earth, the earth emulates the heavens, the heavens emulate Dao, and Dao emulates what is naturally so." (Laozi 25) Laozi argues, Dao is an indivisibly mixed whole, which exists silently and independently, forever unaltered, transforming periodically without dwindling; it is the mother of everything that is happening. Zhaungzi furthermore puts forward the thought of "Dao being the root of everything that is happening." He states: "Even they, with their spirit-like and most exquisite intelligence, as well as all the tribes that undergo their transformations, the dead and the living, the square and the round, do not understand their root and origin, but nevertheless they all from the oldest time by it preserve their being. Vast as is the space included within the six cardinal points, it all (and all that it contains) lies within (this twofold root of Heaven and Earth); small as is an autumn hair, it is indebted to this for the completion of its form. All things beneath the sky, now rising, now descending, ever continue the same through this. The Yin and Yang, and the four seasons revolve and move by it, each in its proper order. Now it seems to be lost in obscurity, but it continues; now it seems to glide away, and have no form, but it is still spirit-like. All things are nourished by it, without their knowing it. This is what is called the Root and Origin; by it we may obtain a view of what we mean by Heaven," (22.2). Zhuangzi argues the "Root and Origin" that everything's growth, everything rising and falling under the sky, and the workings of the four seasons depend upon is actually Dao, which is the origin of genetic movement of everything in the universe.

So, how is Dao the mother or root that creates everything in the universe? In other words, how does Dao create everything in the universe? This requires the involvement of an intermediary link, which is qi 气 or the transformative genesis of everything that is happening through the process of the forces of yin and yang mutually affecting and being-affected by one another finding middle harmony. Dao is a bundle of qi, qi splits into yin and yang, which reciprocally affect one another to generate everything that is happening. Laozi states: "Dao generates unity, unity generates polarity, polarity generates plurality and plurality generates everything that is happening. Everything that is happening carries yin and embraces yang, qi charges to discharge into harmony." (42) Zhuangzi further explains this viewpoint. Using the words of Confucius, he points out: "When the state of Yin was perfect, all was cold and severe; when the state of Yang was perfect, all was turbulent and agitated. The coldness and severity came forth from Heaven; the turbulence and agitation issued from Earth. The two states communicating together, a harmony ensued and things were produced. Some one regulated and controlled this, but no one has seen his form." (21.4) Zhuangzi argues, complete yin is cold and severe qi; complete yang is turbulent and agitated qi. Cold and severe qi affecting and being-affected by turbulent and agitated qi makes the fuse together, which is responsible for creating everything that is happening between heaven and earth. This is the effect of invisible and formless Dao. This shows that Laozi and Zhuangzi both believe the birth and genesis of everything that is happening in the universe is the result of the forces of yin and yang affecting and being affected by one another and finding harmony in the middle. Laozi even more concretely uses the birth and growth of human being as a metaphore for the birth and development of all creatures. Laozi states: "the life0force of the valley never dies—this is called the dark female. The gateway of the dark female—this is called the root of the world. Wispy and delicate, it only seems to be there, yet its productivity is bottomless." (85) Laozi argues, the vast but empty original spirit will not die, because this is subtle femininity. This subtle female gateway is the root of birth and development of everything that is happening between heaven and earth. Its unceasing duration is like existing forever, and its effect is inexhaustible. This is ancient Chinese Daoism's thought of everything in the universe transforming and growth through the charge and discharge generated by yin and yang interacting. It is a cosmogenesis based on ontology, which differs entirely from the theory of genesis in the West from the epistemology-based theories of material substance or mental substance to the Christian theory of God creating man and creatures. This is a "middle harmony theory" of philosophy and aesthetics; it is a theoretical viewpoint in traditional Chinese philosophy and aesthetics carrying a fundamental character; it has extremely rich implications from the communication and fusion of yin and yang to the genesis of everything that is happening; it differs completely from the Western philosophical and aesthetic theory of harmony that began in ancient Greece and matured in classical German philosophy. "Harmony theory" is entirely based on epistemology; its model of thought is the dichotomized opposition of subject-object and sensibility-rationality; its theoretical implications include proportion, symmetry and extreme and mean ratio in external material form and even relationships of commonality and individuality. But ancient China's theory of middle harmony is the mixed becoming of middle harmony between yin and yang, the genesis and development of everything that is happening in the universe is the combined production of a macroscopic whole. This is to say, this is an ontological theory of philosophical thought that differs widely from the epistemological model of philosophical thought characterized by harmony theory. Zhuangzi tells a story in the chapter "Ying Di Wang," which could figuratively illustrate the grave consequences brought about by applying the subject-object model of thought to the ontological thought of middle harmony. He states: "The Ruler of the Southern Ocean was Shu, the Ruler of the Northern Ocean was Hu, and the Ruler of the Centre was Chaos. Shu and Hu were continually meeting in the land of Chaos, who treated them very well. They consulted together how they might repay his kindness, and said, 'Men all have seven orifices for the purpose of seeing, hearing, eating, and breathing, while this (poor) Ruler alone has not one. Let us try and make them for him.' Accordingly they dug one orifice in him every day; and at the end of seven days Chaos died." "Chaos" here symbolizes the forces of yin and yang harmonizing in the middle of affecting and being-affected by one another as the origin of the genesis of everything that is happening in the universe. It is naturally and spontaneously so, a Being that follows what is natural. But according to common knowledge, namely the epistemological viewpoint or Shu and Hu's viewpoint, "Men all have seven orifices for the purpose of seeing, hearing, eating, and breathing, while this (poor) Ruler alone has not one. Let us try and make them for him.' Accordingly they dug one orifice in him every day; and at the end of seven days Chaos died." This illustrates that according to common knowledge, using the epistemological model of thought to determine the ontological philosophy and aesthetics of middle harmony does not work, the two necessarily find themselves in sharp contradiction to one another.

Laozi and Zhuangzi argue that the role of Dao characterized by middle harmony is exceptionally grand; not only does the charge and discharge between yin and yang give birth to and nourish everything that is happening in the universe, the Dao of middle harmony could also be used to rule the state. Zhuangzi borrows the words of the Yellow Emperor to argue, the "perfect music" of middle harmony between yin and yang could "blend and pattern the four seasons, ecstatically harmonizing everything that is happening—he succession of the four seasons one after another, the production of things in their proper order. Now waxing and now waning, civil and military relations managed well. Now clear and now turbid, yin and yang blend in harmony, those notes now flow away in waves of light" (14.3) The Dao of "middle harmony" can also nourish oneself. Zhuangzi borrows the words of Guang Chengzi to say: "There heaven and earth have their controllers; there the Yin and Yang have their Repositories. Watch over and maintain your body, and all things will of themselves give it vigor. I maintain the (original) unity (of these elements), and dwell in the harmony of them. In this way I have cultivated myself for 1000 and 200 years, and my bodily form has undergone no decay." (11.3) This Dao of "maintaining the original unity of these elements and dwelling in the harmony of them" is precisely the source of China's theories of medical self-cultivation.

In summary, Laozi and Zhuangzi's thought of "Dao being the mother of the world" enables humans and other natural creatures to possess a common source in terms of genesis. This is one of the implications of eco-ontological aesthetics that Laozi and Zhuangzi fully focused on.

7.3 Ecological Equality in the Theory of "All Creatures Being Equal"

"All creatures are equal" means everything is equal to the point that there is no distinction of noble and base, high and low between them, and there is no problem of humans being higher than Nature. This Daoist thought obviously falls under ontology and not epistemology. From the epistemological perspective, natural beings objectively speaking have differences in terms of long and short, superior and inferior such that we cannot treat them equally; only from the ontological perspective can we argue that everything possesses "intrinsic value." Because of this, whatever exists is rational, everything has its own inherent status and value and should be treated equally.

A famous statement is attributed to Laozi: "So, Dao is grand, Heaven is grand, Earth is grand and Man is grand. There are four grandees in the realm, and the king stands as one among them. Man emulates earth, the earth emulates heaven, heaven emulates Dao and Dao emulates what is naturally so." Dao along with heaven, earth and man are the four greats in the realm, and man is one among them, which determines humans and everything else in the universe to have equal status. Laozi states: "Heaven and Earth are not humane, they treat all creatures as straw dogs." Laozi follows the Dao of Nature and opposes the theory of humanity and righteousness praised since the Spring and Autumn period. He argues, Heaven and Earth follow the Dao of Nature, not the ethical maxims of humanity. The Dao of Nature is "Dao emulates what is naturally so," "acting non-coercively without object of selfish desire," and rejecting partial love for things, equally considering everything as "straw dogs." "Straw dogs" are lowly things, which illustrates that Laozi opposed dividing humans and other creatures into noble and base, and opposed giving unequal love to humans and other creatures, but rather advocated considering humans and other creatures equally without distinction between high and low, noble and base. In qiwulun 齐物论, Zhuangzi states: "I emerge in the conjunction of heaven and earth, all creatures are one with me." In Autumn Waters, Zhuangzi gives the allegorical dialogue between He Bo and Bei Hainuo, declaring the principle that there is no disparity of value between things and that humans are one with all creatures. When we would set forth the number of things (in existence), we speak of them as myriads; and man is only one of them. Men occupy all the nine provinces; but of all whose life is maintained by grain-food, wherever boats and carriages reach, men form only one portion." In the mixture of the universe, why stop counting the types of things at millions, while man is only one among them; within the nine provinces, among things eating grain to live and the places reached by boat and cart, humans are only one 10,000th of the overall figure. Zhuangzi borrows the words of the Earl of the He to ask: "Whether outside of things or inside of things, how could we differentiate noble and base? How could we differentiate big and small?" Bei Hainuo replies: "From the perspective of Dao, there is nothing noble or base in things; from the perspective of things, oneself is noble and everything else is base by comparison; from the perspective of common sense, nobility and baseness does not depend on oneself. From the perspective of difference, if something is considered big because of what it is bigger than, then there is nothing that is not big; if something is considered small because of what it is smaller than, then there is nothing that is not small; If you know that heaven and earth are as small as grains of rice, and you know the tip of a hair is as big as a mountain, then different measurements become clear." (17.5) Why is there no distinction of value between things? Why is there no difference of size between things? Zhuangzi argues, from the ontological perspective of Dao, things do not possess noble or base value; considering the problem from one's own perspective, it can only lead to belittling things because one prizes oneself; from the perspective of common sense, nobility and baseness does not depend on the self, it is in external things; from the perspective of comparing quantities, the size of things is relative. The largeness of heaven and earth could be seen as a large mountain. Proceeding from different perspectives, different conceptions of value will hold. Here, Zhuangzi raises different perspectives of observing the problem, and raises the problem of proceeding from the perspective of Dao. From the perspective of Dao, it is determined by the natural non-coercive nature of Dao. Dao's essence is what is naturally so without object of desire or narrow purpose to act on, and thus, proceeding from the perspective of Dao, there is no such distinction between noble and base, high and low among things originally; there is no difference between high and low among humans and other natural creatures, nor is there any distinction between "center" and "non-center."

Zhuangzi argues, Dao is everywhere and manifests in every single thing, which raises the issue of "all things being equal." Zhuangzi 22.6 records a dialogue between Zhuanngzi and Dong Guozi: "Dong Guozi asked Zhuangzi, saying, 'Where is what you call the Dao to be found?' Zhuangzi replied, 'Everywhere.' The other said, 'Specify an instance of it. That will be more satisfactory.' 'It is here in this ant.' 'Give a lower instance.' 'It is in this panic grass.' 'Give me a still lower instance.' 'It is in this earthenware tile.' 'Surely that is the lowest instance?' 'It is in that excrement.' To this Dong Guozi gave no reply." This dialogue illustrates that ants, grass, and excrement, which are ordinarily considered base and lowly things, are no less homes to Dao than anything else, or in other words, all things are equal. The famous Hegelian statement—"What is real is rational" has a certain truth from the ontological perspective. Each and everything between heaven and earth becomes an indispensable member of the ecological world in its own distinct way, each possesses a value due to its own being.

Zhuangzi's viewpoint that Dao is everywhere is very important; it shows an ancient idea that all things possess intrinsic value. Zhuangzi 8.1 states: to cut off a portion of them, it produces grief." This is to insist that each and everything in the world possesses its own natural self-sufficient essence, alterations to which produce suffering. Zhuangzi 5.1 describes several people with disfigured bodies, each of whom although suffers from bodily impairment still offers noble moral deeds. One hunchback's appearance could be considered lowly, but because of his highly noble virtue, men wish to be his servant, and women wish to be his wife. Zhuangzi also raised the issue of the "use of uselessness," which is to say although something may not have value in the ordinary understanding of usefulness, precisely because of this it can maintain its own original nature and exist for a longer period of time, and hence "is useful." The use of uselessness also means that innate value in something, including natural things, which do not show to the outside, hence the saying "virtue does not take shape," and virtue "is maintained in-itself and is not paraded outside," (Zhuangzi 5.4).

Zhuangzi 4.5 tells a fable:

> A (master) mechanic, called Shi, on his way to Qi, came to Qu-yuan, where he saw an oak-tree, which was used as the altar for the spirits of the land. It was so large that an ox standing behind it could not be seen. It measured a hundred spans round, and rose up eighty cubits on the hill before it threw out any branches, after which there were ten or so, from each of which a boat could be hollowed out. People came to see it in crowds as in a market place, but the mechanic did not look round at it, but held on his way without stopping. One of his workmen, however, looked long and admiringly at it, and then ran on to his master, and said to him, 'Since I followed you with my axe and bill, I have never seen such a beautiful mass of timber as this. Why would you, Sir, not look round at it, but went on without stopping?' 'Have done,' said Mr. Shi, 'and do not speak about it. It is quite useless. A boat made from its wood would sink; a coffin or shell would quickly rot; an article of furniture would soon go to pieces; a door would be covered with the exuding sap; a pillar would be riddled by insects; the material of it is good for nothing, and hence it is that it has attained to so great an age.' When Mr. Shi was returning, the altar-oak appeared to him in a dream, and said, 'What other tree will you compare with me? Will you compare me to one of your ornamental trees? There are hawthorns, pear-trees, orange-trees, pummelo-trees, gourds and other low fruit-bearing plants. When their fruits are ripe, they are knocked down from them, and thrown among

> the dirt. The large branches are broken, and the smaller are torn away. So it is that their productive ability makes their lives bitter to them; they do not complete their natural term of existence, but come to a premature end in the middle of their time, bringing on themselves the destructive treatment which they ordinarily receive. It is so with all things. I have sought to discover how it was that I was so useless; I had long done so, till (the effort) nearly caused my death; and now I have learned it—it has been of the greatest use to me. Suppose that I had possessed useful properties, should I have become of the great size that I am? And moreover you and I are both things—how should one thing thus pass its judgment on another? how is it that you a useless man near death know all this about me a useless tree?'

The trunk of the oak tree grew a hundred spans around and as high as a hill; the tree's shade could cover 1,000 heads of cattle, and the number of people who came to see it were as many as at the market. But its wood is neither good for boat building nor for coffin making or any other appliance, much less for doors and bridge posts, so the mechanic did not think of it as material, but as a useless tree. The spirit of the oak tree believes the mechanic views himself from the perspective of how he sees those "ornament-trees," by which he means "hawthorns, pear trees, orange trees, pummelo trees, gourds and other low fruit-bearing plants" whose wood is useful or whose fruits are edible, but once the fruits ripen, they are picked, and when picked off the trees, the trees are injured; the tree trunks and tree branches, by becoming useful materials, saw down and hack apart the trees which they were. These ornament trees cut their own lives short because they are useful to people, namely "do not complete their natural years but die halfway." The oak tree however can grow to flourishing heights precisely because of its total uselessness, which for that reason, becomes the tree's "greater usefulness." What the oak tree tells us is "all things are so." Humans and trees are equally "things," and "comparing things" with the "useful" and "useless" like the mechanic, one could say, makes you a "useless man near death." This fable raises several such questions: first, it raises the idea of things, specifically natural things, having a greater usefulness in being totally useless. What is said to be totally useless is said with respect to common sense utilitarian criteria and the cognitive perspective of judging the value of things, and "greater usefulness" refers to judging the value of things from the perspective of eco-ontology, which argues that anything, and specifically any natural thing has its own "intrinsic value." This is fully consistent with Laozi and Zhuangzi's viewpoints concerning the omnipresence of Dao pervading all things; second, it proposes Daoism's eco-ontological philosophy of "doing nothing and leaving nothing undone" in "being useful for nothing in particular but for everything in general." Only by being useful for nothing in particular can something be useful for everything in general. If something seeks intentionally to be useful, it will on the contrary become useless. This cautions humanity, that if we overly pursue the economic and utilitarian aims of "usefulness," overfell forests, abuse resources, and pollute the environment, we will ultimately reach the point of exhausting all resources and degrading the environment into a totally useless one; third, it raises the issue of equality between humans and other things. Human beings and other natural beings are equally beings, and are equal in this sense, so we cannot use the criteria of useful and useless, or how much or how less useful to demand something else or select something else. In actuality, this is to deepen the notion that all things are equal; it is also a powerful critique of the anthropocentrism that rules humankind. Zhuangzi asserts that everyone knows of the use of usefulness, but no one knows of the use of uselessness. This passage is touching, but in Zhuangzi's age, whether it was the common sense ideas prevailing at the time or the Confucian doctrines so extremely influential in that era, all of them held that humans were the most precious beings among others, presupposing the conception of "the mandate of Heaven." Laozi and Zhuangzi advocated respecting Nature and treating all things equally. Xunzi criticized Zhuangzi, stating "Zhuangzi, obstructed by Nature, does not know human being" (jiebi 解蔽), meaning Zhuangzi only knows respect for Nature, and does not understand the principle of humanity. Actually, from the eco-ontological perspective, this is precisely where the noble element of Daoist thought is found. In contrast to the Daoists, the Confucians praised the value of human being over all other beings. Xunzi argues "Water and fire have energy (qi 气) but no life; plants and trees have life but no awareness (zhi 知); beasts and birds have awareness but no sense of what is righteous (yi 义); humans possess energy, life, awareness, and also a sense of what is righteous, and therefore rank of the highest nobility among beings in nature." Book of Filial Piety borrows Confucius' words to assert: "among the natural tendencies between heaven and earth, human being is the noble one." From this it becomes clear that Laozi and Zhuangzi's viewpoint of all things being equal could be considered rare and valuable at the time.

7.4 The Thought of the Ecological Chain in the Concepts of "Natural Differentiation" and "Natural Equalization"

Zhuangzi's Fables states: "All beings come in kinds, and succeed to one another in different forms. From beginning to end they link together, but none finds its match, which is natural equalization. Natural equalization is natural differentiation." Zhuangzi argues, each being has its own species and genus, and links to the others and transforms the others with its different shape. From beginning to end, they bind together and link together like a chain, and the transformational principle between them is unknowable. This is a naturally formed state of equality of cyclical transformation, which is "natural equalization." This illustrates that all things are equal. In book 2, qiwulun 齐物论, Zhuangzi further expounds on the concept of natural differentiation from the perspective that all things are equal: "What is call harmonizing things with the natural differentiation? Response: Whether right or not right, whether so or not so. If what is right is right, then the affirmation of it differs from the negation of it, but there is also no contradiction between them." Here, the "natural differentiation" Zhuangzi is talking about means no matter whether something is so or not so, both standpoints are the same. Because of this, Zhuangzi advocates not arguing right and not right, and rejecting the ideas of life and death and right and wrong. Because everything unifies in Dao. From the perspective of eco-ontological aesthetics, we can see the deeper meaning implied in it: first, clearly raising the thought of things composing a cyclical chain, namely all things succeed to one another in different forms, from beginning to end like a chain"; second, the important cause forming the chain of things is the connectivity and unity between things, "whether right or not right, whether so or not so"; third, the basic cause forming the chain of things is they all "inhabit the boundless," which is to say they have the same source of life—Dao; fourth, this thought of natural differentiation is a supplement to the theory of all things being equal, which is that the equality that all things enjoy is the equality of the position in the chain that they inhabit; they cannot overstep this position, which is finally "non-coercive action" and "what is naturally so." From this it becomes clear that Zhuangzi's doctrine of natural differentiation implies the basic idea of the chain of life in eco-ontology, and states the universal law of common symbiosis, inseparability and oneness between all things in the universe.

Laozi and Zhuangzi's Daoism also implies the extremely important ecological ideas that humanity should respect Nature, should do things in line with the original nature of things, and should not change the essence of nature out of convenience. Laozi put forward the thought of "aiding things do what they do naturally and not hoisting one's will on things (chapter 64), which also refers to respecting and assisting the natural essence of things rather than altering them at will. Zhuangzi 18.5 has the following fable:

> Formerly a sea-bird alighted in the suburban country of Lu. The marquis went out to meet it, (brought it) to the ancestral temple, and prepared to banquet it there. The Jiu-shao was performed to afford it music; an ox, a sheep, and a pig were killed to supply the food. The bird, however, looked at everything with dim eyes, and was very sad. It did not venture to eat a single bit of flesh, nor to drink a single cupful; and in three days it died. The marquis was trying to nourish the bird with what he used for himself, and not with the nourishment proper for a bird. They who would nourish birds as they ought to be nourished should let them perch in the deep forests, or roam over sandy plains; float on the rivers and lakes; feed on the eels and small fish; wing their flight in regular order and then stop; and be free and at ease in their resting-places.

The marquis of Lu "nourished the bird with one's own nourishment," using the music and foods he himself enjoyed to nourish the bird, resulting in the sea bird dying in three days. This actually uses an extremely typical example to illustrate the wrong attitude with which one should treat natural things—hoisting one's own ideas, desires and way of doing thing upon natural beings in line with the maxim of one's own will to control natural things. Zhuangzi advocates "nourishing the bird with a bird's nourishment," namely fully respecting the natural essence of the bird so that it can survive in the environment it is adapted to according to its own natural essence. Laozi had an important ecological thought, which was the famous doctrine of "three treasures." "I have three treasures, I uphold and maintain them: one is affection, two is thrift and benevolence, but rather what "Dao emulates what is naturally so" indicates, including non-coercive action, objectless desire, and not invasively approaching Nature; thrift refers to blandness and non-decadence; not willing to have priority over the world means not contending. These three treasures are linked to Daoism's conception of what is natural. Affection originally means respect for Nature; thrift means not plundering nature out of convenience and leading a life in tune with what is natural; not having priority over the world, namely affirming that all beings have their own space of survival, means not working in anthropocentricism, but rather adopting an attitude of equal coexistence with Nature.

7.5 The Ancient Eco-Phenomenology of "Maintaining Focus," "Purifying the Mind" and "Sitting in Forgetfulness"

Laozi and Zhuangzi's eco-ontological aesthetics possesses profound expositions and explorations of how to "embody Dao" and follow "Dao emulating what is naturally so." Generally speaking, Laozi and Zhuangzi argue Dao is not graspable by relying on learning knowledge, and requires reaching a transcendent state of mind that breaks free of material desires through the comprehension of the mind and the cultivation of spirit. This is also precisely an aesthetic attitude and state of mind, thereby making Laozi and Zhuangzi's eco-ontology of "Dao emulating what is naturally so" consistent with aesthetics, and become an eco-ontological aesthetics.

Laozi directly lays out the implications of Dao, and gives the necessary exposition of the grasp of Dao. Laozi states, "Maintaining focus is better than more words and exhaustive enumeration" (chapter 5). What "maintaining focus" refers to is a spiritual training and cultivation with specific meaning. Zhaungzi further develops Laozi's notion of "maintaining focus" into "sitting in forgetfulness" and "purifying the mind." Zhuangzi 6.9 uses the dialogue between Confucius and Yan Hui to bring up and expound upon the way to sit in forgetfulness: "Confucius changed countenance and asked: 'what is sitting in forgetfulness?' Yan Hui responded: 'laxing the limbs, discarding the intelligence, leaving behind set shapes and abandoning knowledge, and converging with the greater consistency; this is called sitting in forgetfulness.' Confucius said: 'thus converging [with the greater consistency], there is no preferences, thus transforming, there is no constancy. You have indeed become my superior! I must ask to follow in your footsteps.'" There are three essential points to sitting in forgetfulness: first, laxing the limbs and body, which is fully abandoning one's own physiological needs and every external material desire; second is discarding the intelligence, which is thoroughly abandoning every burden challenging one's mind, including the all-popular Confucian doctrine of being humane and righteous as well as the other forms of common knowledge and knowledge of everyday life; third, leaving behind set shapes and abandoning knowledge, converging with the greater consistency, which refers to losing these loads carried on one's mind and body, and after transcending every material desire and so-called form of knowledge, comprehend the Dao of empty and still Nature, connecting with it. What Zhuangzi means by "purifying the mind" is explained in 4.2 using a dialogue between Confucius and Yan Hui, "Hui asked: May I dare ask what purifying the mind is?' Confucius said: 'your will is one whole! Do not listen to things with the ear, but listen to them with the affective mind; do not listen to them with the affective mind, listen to them with force (qi 气)! Listening stops at the ears, the mind stops at signs. Only force (qi 气) is capaciously empty and receiving things. Only Dao is concentrated emptiness. Emptiness is the purification of the mind.'" Purifying the mind is a step further than sitting in forgetfulness, it tells of the training process of throwing out all the random ideas, nourishing forces and comprehending Dao; the first step, "not listening to things with the ears," namely abandoning the sense-organ's feeling of external things; second step, not listening with the mind, namely abandoning the rational cognition of external things; third step, "your will is one and whole, only force is capaciously empty and receiving things," which is concentrating the mind, and consolidating the empty and still forces of spirit, then comprehending the Dao of the workings of everything in Nature. Obviously, "purifying the mind" is a process of breaking free of the purely physiological pleasures of the sense-organs and theoretical knowledge, then reaching the mind's "independence" and the aesthetic state of free movement. Through this "sitting in forgetfulness" by laxing limbs and body, discarding intelligence and leaving set shapes and knowledge, and through purifying the mind by not listening with ears, not listening with the mind, but capaciously empty and receiving things, Laozi and Zhaungzi's Daoism rejects the pursuit of material desires, the strife of worldly affairs and utilitarian knowledge, thereby transcending beings in the non-beautiful state and directly reaching the Dao nourishing all beings in the universe while attaining an aesthetic existence. The free movement of the mind's "independence" is precisely the aesthetic existence that Zhaungzi sought. "Sitting in forgetfulness" and "purifying the mind" is a process of moving from concealment to unconcealment toward a state of clearing. Zhuangzi insists only the human being who transcends the ordinary and breaks free of common sense, the complete human being, the man of spirit, the true human being, the sage could attain the aesthetic state of mind of "perfect enjoyment" through leaving set shapes and discarding knowledge, going beyond material things, merge with Dao and reach true virtue (de 德). "The complete human being" is but he, "who wields the handle of power over the whole world, and yet it is nothing to him. His discrimination detects everything false, and no consideration of gain moves him. He penetrates to the truth of things, and can guard that which is fundamental. So it is that heaven and earth are external to him, and he views all things with indifference, and his spirit is never straitened by them. He has comprehended the Dao, and is in harmony with its characteristics; he pushes back benevolence and righteousness (into their proper place), and deals with ceremonies and music as (simply) guests: yes, the mind of the complete human being determines all things aright." (Zhuangzi 13.8) Zhuangzi argues "the complete human being" does not contend with people in the world over power and gain and mix-in with the filthy corruption that entails; the complete human being exists in a state of independent detachment, is not seduced by profit, fully penetrates the true nature of things, and firmly holds to their origin. For that reason, the complete human being can break free of concrete phenomena, shake off the burdens things bring to the self and with unperturbed spirit, the complete human being merges with Dao and virtue (de 德), says farewell the Confucian morality, disregards rituals and music, and attains complete peace of mind.

Laozi's "maintaining focus" and Zhuangzi's "sitting in forgetfulness" and "purifying the mind" is very much similar to the method of "eidetic reduction" in contemporary phenomenology, which suspends in brackets the day-to-day routine of one's body, the factual things in scientific research, the transcendent world in religion, and even the objects of mathematics and logic, and this finally leaves over just the pure flow of self-consciousness, that creates new images through its appearing and constituting. This is also a process of moving from epistemology to ontology, from beings to Being, from concealment to clearing, from not-beautiful to aesthetic. Laozi and Zhuangzi's sitting in forgetfulness and mind-purification is a process of "suspending" knowledge and external things in the phenomenal world, comprehending the Dao of everything in the universe through concentrating the mind and focusing the breath, and finally reaches the state of "perfect enjoyment" detached in free movement. This is precisely a process of transcending beings towards aesthetic and poetic Being, which is impossible to understand using the ordinary categories of the knowable and unknowable in epistemology.

7.6 The Eco-aesthetic Ideal of a "World of Complete Virtue"

Laozi and Zhuangzi also had some predictions of the "completely virtuous world" that agrees with the demands of eco-ontology, to which they pinned their perfect ideal and with which they profoundly inspired later generations. First of all, they expounded on the relationship between Dao and the world, which is to say, they pushed for the relationship between Daoism and social development. Their overall idea was: "When Dao declines the world declines with it, when Dao flourishes, the world flourishes with it." Zhuangzi said: "Looking at it from this, the world loses Dao and the Dao loses the world, the world and Dao lose one another. How could someone of Dao rise from the world, how could the world rise from Dao! With no way for Dao to rise from the world and the world no way to rise from Dao, even if the sage is not concealed in the mountain forest, his virtue hides." (Zhuangzi 13.3) "The world declines" means society fails, which includes the aspect of political economy as well as the aspect of natural ecology. Because in their cosmology, "Heaven," "Earth" and "Man" are all in Dao. From this it becomes clear that the problem of ecology is ultimately the problem of worldview, namely the problem of what Zhuangzi called "the rise and fall of Dao." If Dao flourishes, ecology and society flourish, and people can establish a way of life "aiding what things do naturally of themselves without hoisting their will upon them" in line with the idea that "Dao emulates what is naturally so of itself" and the idea of "pure and still non-coercive action," thereby establishing a society of universal symbiosis between man and nature, coordinated in the middle harmony of co-flourishing with ecology. Vice versa, when Dao declines, ecology and society decline. If people betray the idea of "Dao emulating what is naturally so," act according to the principle of anthropocentrism, and plunder and abuse Nature, the harmonious relationship between man and Nature will necessarily be destroyed, and social decay inevitably arrives. In light of the turmoil of the Warring States period, Laozi and Zhuangzi undoubtedly had some prescience with regard to this case of the world failing when Dao goes into decline.

Laozi also ruthlessly exposed the great disparity between the rich and the poor in society at the time, as well as the extreme decadence of extravagant luxury and the ecological destruction that came with it. He stated: "Their court is impeccably clean, yet the fields are overgrown with weeds and the granaries stand empty. Their clothing is embroidered and colorful and sharp swords hang at their sides; They are stuffed with food and have wealth and property to throw away. This is called highway robbery, which out not to be confused with Dao." (chapter 53) The facts at the time were the court was immaculately refined, but the fields were barren wastelands, the granaries were totally empty, but the aristocrats were decked out in elegant clothing, brandishing sharp precious swords, and eating delicious foods, opulently overflowing with wealth. Laozi argued, this was an act of forced robbery against Dao! Laozi further predicted from the anti-Dao actions of these officials and aristocrats, the coming of ecological and social crisis. He said: "Following this line of thinking, we could say that if the heavens had not become clear they may well have fallen to pieces; We could say that if the earth had not become stable it may well have collapsed; We could say that if the numinous had not become animated it may well have faded away; We could say that if the river valleys had not become full they may well have dried up; We could say that if all creatures had not come back to life they may well have gone extinct; We could say that if the lords and kings had not brought proper order to the world they may well have stumbled and fallen from power." (chapter 39) In Laozi's view, Nature and society can only find stability and peace by attaining Dao. If Heaven were to lose Dao, it would have no way to clear up and would fall apart; if the earth were to lose Dao, it would have no way to find stability and would collapse; if making things were to lose Dao, it would have no way to replenish forces and would exhaust reserves; if natural creatures were to lose Dao, they would have no way to regrow and would go extinct; if lords and kings were to lose Dao, they would have no way to correct the court discipline and it would collapse. This is a profound understanding of the relationship between Dao and the world, and a prediction and description of the ecological and social crises that result from going against the Dao. Zhuangzi 10.3 points out: "When those superiors do really love knowledge, but do not follow Dao, the whole world is thrown into great confusion. How do I know that the case is so? The knowledge shown in the (making of) bows, cross-bows, band-nets, stringed arrows, and contrivances with springs is great, but the birds are troubled by them above; the knowledge shown in the hooks, baits, various kinds of nets, and bamboo traps is great, but the fish are disturbed by them in the waters; the knowledge shown in the arrangements for setting nets, and the nets and snares themselves, is great, but the animals are disturbed by them in the marshy grounds. (So), the versatility shown in artful deceptions becoming more and more pernicious, in ingenious discussions as to what is hard and what is white, and in attempts to disperse the dust and reconcile different views, is great, but the common people are perplexed by all the sophistry." This should be a description of social reality at the time, and furthermore explicates the relationship between Dao and the world, illustrating that different ideas in thought determine different ways of life and attitudes, thereby having different impacts on society. Zhuangzi 11.4 further describes this natural crisis: "What disturbs the regular method of Heaven, comes into collision with the nature of things, prevents the accomplishment of the mysterious (operation of) Heaven, scatters the herds of animals, makes the birds all sing at night, is calamitous to vegetation, and disastrous to all insects—all this is owing, I conceive, to the error of governing men." Zhuangzi here writes of even more relations between man and Nature, pointing out the inevitability of wrong actions under the guidance of the wrong theoretical ideas, thereby destroying humanity's relationship with Nature, and resulting in ecological crisis. Moreover, Zhuangzi was took the case of going against Dao in full seriousness, and also predicted ecological and social crisis after 1,000 years. Zhuangzi 23.2 borrows the words of Geng Sangzi, who "had a greater grasp of Laozi's doctrines than others," stating: "I tell you that the root of the greatest disorder was planted in the times of Yao and Shun. The branches of it will remain for a thousand ages; and after a thousand ages men will be found eating one another." After one thousand ages, humans will be eating each other, which could be understood to mean serious ecological crisis of exhausted resources and degraded environment brought about by the destruction of ecology, resulting in humanity losing the most basic conditions of life. Humans destroying ecology makes humans lose the conditions of survival, making it impossible for them to survive.

Laozi and Zhuangzi also put forward their own social ideal of an ecological society established in line with Dao emulating what is naturally so. Laozi said: "You want a small state with a minimal population. Have ready to hand weaponry for a sufficient number of military units, yet have no recourse to use them. Make sure that the common people take dying seriously so that they have no taste for venturing far from home. Though you have ships and chariots enough, you have no reason to man them; Though you have armor and weapons enough, you have no reason to parade them. Bring the common people back to keeping their records with knotted string, to relishing their food, to finding beauty in their garments, to enjoying their customs, and to finding security in their homes. Although your neighboring states are within eyesight and the sounds of their dogs and cocks are within earshot, your people will grow old and die without having anything to do with them." (chapter 80) Laozi's passage here about a small state and minimal population is so famous, many people are fully familiar with it in detail, but in the past it was criticized much, mainly in the belief that it represented the social ideal of a regressive, small farming economy, an passive and backward utopian state. If we restricted ourselves to the scope of epistemology and socio-political aspects, such criticism would be made duly without err, but from the ontological perspective, these ideas are still filled with important theoretical value:

First, this is a powerful exposition and criticism of the dark reality of the Warring States period at the time with frequent military campaigns, difficulty with people finding subsistence, homelessness and great instability; it was a powerful rejection of the non-beautiful state of survival people found themselves in;

Second, it pursues the ideal of an ecological type of society mixing thrift, simplicity, peace, balance and harmony between human society and nature together in harmonious coexistence. This social ideal is still full of positive significance with respect to its implications in ecological ontology. For a rather long period of time, people weighed whether or not society should progress with the sole criteria of productivity; it seemed that if productivity would advance, society would certainly progress, which completely disregards ecological criteria. At present, the universally used system for evaluating quality of life in the international context already largely brings ecological criteria in to bear on the formulas. Laozi measured society at the time with the criteria of eco-ontology, and believed people at the time lived in a non-beautiful state of survival. If we were to evaluate Laozi's social ideal from this perspective, we would have to admit that it contains extremely valuable ingredients. He advocated returning to a primitive society prior to the era of Yao and Shun, when the people still kept records with string. Considering historical development, this is of course impossible, but it does ultimately hold the ideal of ecological society that Laozi tried to establish, one that follows the idea of Dao emulating what is natural. In all, this is no different from Confucius pinning his ideal on the Zhou dynasty and Westerners on the flourishing and undying pursuits of ancient Greece.

Zhuangzi thus housed his social ideals of a perfectly virtuous world and state of instituted virtue in many of the books attributed to him like books 9, 10, 16 and 20. For instance, in book 9, he professes of the "world of complete virtue": "Yes, in the world of complete virtue, humans lived in common with birds and beasts, and were on terms of equality with all creatures, as forming one family—how could they know among themselves the distinctions of superior and inferior men? Equally without knowledge, they did not leave (the path of) their natural virtue; equally free from desires, they were in the state of pure simplicity. In that state of pure simplicity, the nature of the people was what it ought to be." (9.2) The world of complete virtue in his understanding is one of friendship and partnership with animals, it is one of simultaneous symbiosis between humans and other natural creatures, so there is no need to distinguish between the superiority of men or lords over other creatures and other humans, who on the contrary live in a state of non-coercive action without narrow objects of desire, while the natural essence of the people is thus maintained and developed. Zhuangzi depicts a beautiful picture of symbiosis and friendly coexistence between man and Nature and among humans themselves through the Dao of non-coercive action without narrow objects of selfish desire; it is an ideal blueprint for ecological society. In book 16, Zhuangzi, like Laozi, places his world of complete virtue in the ancient age in chaos and confusion. He states: "The humans of old, while the chaotic condition was yet undeveloped, shared the placid tranquility which belonged to the whole world. At that time Yin and Yang were harmonious and still; their resting and movement proceeded without any disturbance; the four seasons had their definite times; not a single thing received any injury, and no living being came to a premature end. Humans might be possessed of (the faculty of) knowledge, but they had no occasion for its use. This was what is called the state of Perfect Unity. At this time, there was no action on the part of any one, but a constant manifestation of what is naturally so." This ancient world of complete virtue in Zhuangzi's ideal conception similarly reflects his criticism of current reality. The pursuit of an ecological society "perfectly one" with Dao emulating what is naturally so, embodies the idea of peaceful coexistence between humans and other creatures.

Chapter 8
Buddhism's Wisdom Pertaining to Ecological Aesthetics

As an organic component of traditional Chinese culture, Buddhism, whether with regard to its basic dogma or its religious practice, contains extremely rich ecological thought. This chapter summarizes Buddhism's wisdom pertaining to ecological aesthetics and Chan Buddhism's wisdom pertaining to ecological aesthetics, while uncovering the contemporary significance and inspiration within them.

8.1 Buddhism's Wisdom Pertaining to Ecological Aesthetics

The core of Buddhism's ecological wisdom is the equality of all living creatures, which encompasses six specific viewpoints related to the wisdom of ecological aesthetics.

(1) "The Conditioned Arising of Buddha Nature"

The theory of the conditioned arising of Buddha nature is the philosophical foundation of Buddhist ecological wisdom; it is Buddhism's fundamental idea with respect to the origin of the universe and human life; it falls under an Eastern style of cosmology or conception of the world and differs widely from the modern Western cosmology or worldview of anthropocentrism and the subject-object dichotomy. On the issue of the origin of the world, Buddhism strongly insists that Buddha nature is the root. "Buddha nature" refers to the Buddhist notions of "Buddha mind," "pure mind" and "suchness," which is the origin of the conditioned arising of all phenomena. This Buddha nature is neither material substance nor mental substance, but is rather "the state of Nirvana which is neither nothing nor something, i.e. "knowing there is no self, there is no human being, there is no life, the nature of the self is empty, there is no actor, there is no patient," which is to say, "the gate of attaining emptiness and detachment is now ahead."[1] In other words, Buddha nature is a "state

[1] [Tang] [1], (vol. 2), p. 650.

F. Zeng, *Introduction to Ecological Aesthetics*,
https://doi.org/10.1007/978-981-13-8984-9_8

of mind" and "process." This theory of Buddha nature and origination transcends ontology of matter and mind; it is a non-material and non-mental, neither subjective nor objective "following rightness without divisions." Without divisions "means every phenomenon should be 'undifferentiated' or transcend all distinctions. Volume of *Compendium of the Great Purport of the Mahayana*: 'speak monism means there is no difference, i.e. there is one real purport in the sutra. The principle of one reality, the state of subtle, silent truth, that all phenomena are equal, disappears in another dharma, and hence we say 'speak monism.'"[2] The monism of the conditioned arising of Buddha nature differs fundamentally from the anthropocentrism so opposed to natural ecology. "Conditioned arising" has the "basic meaning of everything in the universe causing and conditioning one another, with an infinite number of causes generating one effect, and one cause generating an infinite number of effects, each enveloping one another, and not obstructing one another, one another mutually contains human being, folding one another to infinity."[3] The famous Buddhist *Ye Dharma Hetu*: "Whatever from a cause proceeds, thereof the Tathāgatha has explained the cause, its cessation too he has explained. This is the teaching of the Supreme Sage," which means basically that "everything is one and one is everything." The *Hua Yan Sutra* uses "Indra's net" to illustrate the mutual reflecting of all causal relations in the conditioned arising of Buddha nature. As the *Hua Yan Sutra* puts it, "The next fragrant sea is called Pure Reflections of the Sun, and it has a world seed called Entirely Enclosed in Indra's Net."[4] According to the description, this net is co-linked together by an infinite number of jewels in the Heavenly palace of Buddhism, in it, every jewel will reflect the other jewels, so all jewels are infinitely interconnecting and overlapping infinitely. With this, the *The Flower Garland Sutra* metaphorically illustrates the mutual reflecting of everything in the world, each enveloping and penetrating the other. Buddhism insists that the dependent causation of Buddha nature is the basic cause of the movement of the universe. "Every action develops by unifying and does not develop without unifying. Conditions develop if they group together, and do not develop when they do not group together,"[5] this theory of dependent causation has important influence on world history. [The Chinese use the Buddhist term for "cause" *yinyuan* 因缘 to translate the word] "involvement" which the German philosopher Heidegger used describe the concept of "world" in which humans are interconnected and fused together with other things, and the American ecologist Rolleston used the idea of "Indra's net" to explicate his own land ethic. He argues, to develop the land ethic, we must understand the wholeness of the ecological web composed of different species. This tells us that the theory of dependent causation fully illustrates the causal interconnections of inseparability between humans and other natural creatures.

---

[2] *Religious Dictionary*, [2].

[3] Kesu [3], p. 236.

[4] [Tang] [1] (vol. 1), p. 170.

[5] [Tang] [1] (vol. 1), p. 651.

(2) "The Karmic Retribution of Good and Evil"

The karmic retribution of good and evil shows Buddhism's extremely important conception of human life and ethics, and holds important value for constructing contemporary ecological civilization. This is a theoretical idea held in common by all schools of Buddhism. As Feng Youlan put it, "although Buddhism has many schools, and every school proposed some different things, all school unanimously agree in believing the doctrine of 'karma.'"[6] Karmic retribution is the combination of the terms "karma" and "retribution," meaning the karma's retribution or the maturation of karma, which is used to explain the Buddhist principle of life and social distinctions. With the good and evil of the three karmas of body, mouth and intention, there is necessarily corresponding retribution.[7] Here, karma is human action, and retribution is the result of this action. Buddhism forcefully advocates that "good has good reward, and evil has bad retribution," and the criteria differentiating good and evil is the several commandments that Buddhism prescribes, mainly the "five prohibitions," namely do not kill, do not steal, do not engage in sexual misconduct, do not lie, and do not drink alcohol.[8] Do not kill life is the first of the five prohibitions, illustrating the important status ethically treating natural ecology holds in Buddhist ethics. The five prohibitions is the criteria with which Buddhism divides good and evil. Those respecting them are good, those disobeying them are evil; those respecting them can attain transcendence, those disobeying them will go down the path of evil. As the *The Flower Garland Sutra* puts it, "all sentient life enters the perception of dense forests, dwells in evil ways, and in states of mind arise distinctions of evil, in everyday actions bad karma in body, speech and intentions, misconducting all kinds of evil miserable deeds, producing correct perceptions in misperceptions, and negating correct perceptions in correct perceptions, ensnared by bad knowledge, producing bad visions, and falling into evil ways."[9] Buddhism forcefully advocates the doctrine of samsara with the doctrine of "three worlds," namely the world of desire, the world of forms and the world of formlessness; there is also the cyclical doctrine of "five ways," namely "hell, hungry ghosts, animals, humans and divinities. Hells, hungry ghosts and animals are evil ways, and humans and divinities are good ways. Those respecting the five prohibitions and doing good enter the good ways after death, and those disobeying the five prohibitions go down the evil ways. The *The Flower Garland Sutra* states: "They also realize this: bad actions at their worst, when developed and done a lot, are the cause of hell; at a middling degree, they are the cause of animality; at their least they are the cause of the underworld. Taking life leads to hell, to animality, to the underworld…"[10] Taking life will lead to evil ways, and even an eternity of failure at transcending life, which is Buddhism's important ecological ethics of clear distinction between love and hate, reward and punishment.

[6] Youlan [4], p. 208.

[7] *Religious Dictionary,* [2].

[8] Ibid., p. 167.

[9] [Tang] [1] (vol. 3), p. 1213.

[10] [Tang] [1] (vol. 2), p. 620.

(3) The Theory of the Equality of All Sentient Beings

This is Buddhism's basic conception of ecological values and wisdom of ecological aesthetics; it is Buddhism's representative ecological thought. The equality of all sentient beings is first of all determined by Buddhism's most basic philosophical conception of dependent causation. Because Buddhists insist that the source of all beings in the world is Buddha nature as "the formless," all beings are equal. The *The Flower Garland Sutra* "the sixth of the ten stages of the bodhisattva" specifically describes Buddha undergoing training to enter the sixth stage of presence, where the Buddha observes "ten equalities of things." Its content is: "the equality of signlessness of all things; the equality of nonorgination of all things; the equality of absence of characteristic marks of all things; the equality of nonbirth of all things; the equality of detachment of all things; the equality of primordial purity of all things; the equality of nonconceptuality of all things; the equality of all things in neither coming nor going; the equality of all things being like illusions, dreams, reflections, echoes, the moon's image in the water, or apparitions; the equality of nonduality of existence and nonexistence of all things."[11] Non-origination, non-birth, non-becoming, and neither coming nor going, are all the basic characteristics of the formlessness of the Buddha nature, illustrating that the root of all beings is formless Buddha nature, and hence are all equal. This is the theory of the fundamental distinctionlessness of all things determined by Buddhism's "middle path" of non-duality. At the same time, Buddhism furthermore argues that all sentient and living things are equal, which is the famous conception of the equality of all sentient life. The equality of all sentient life here encompasses extremely rich implications, firstly that all sentient life possesses Buddha nature. Just as the Mahayana Mahaparinirvana Sutra puts it: "all beings possess the Buddha-nature"[12]; moreover, all beings face samsara—all sentient creatures including humans face the fate of the good meeting the reward of good and the bad meeting the retribution of bad, without exception. On this point, humans are equal to all other sentient creatures. Another important thought of Buddhism is fully declaring that all sentient creatures are equal with respect to Buddha's compassion for all beings, and as only as one strives to do good one has the opportunity to reach Buddha enlightenment. Precisely because of this and solely because of this do people see a glimmer of hope in a life of suffering, and spot a bright future in terrible disasters. Buddhists believe, great mercy and great pity, transcending all life is the beginning of the Buddha-nature and the Buddha-dharma. The *The Flower Garland Sutra* states, "Thus [the Bodhisattvas] keep in mind such thoughts, by the inspiration of which they collectively fulfill the ten transcendent ways (paramita) of enlightening beings. How so? In that way do the enlightening beings dedicate each thought, led by great compassion, t the fulfillment of Buddhahood and enlightened knowledge." (*Flower Ornament Scripture*, p. 156)[13] This ecological equality of great kindness and compassion for all life is the extremely valuable ecological ethics unique to Buddhism; it illustrates that no matter what attitude a human had toward natural

[11][Tang] [1] (vol. 2), p. 648.

[12]Quoted from Kesu [5], p. 55.

[13][Tang] [1] (vol. 2), p. 657.

ecology in the past, as long as he or she is whole-heartedly devoted to good, changes his or her ideas and ethically treats nature, he or she immediately gains the possibility of becoming Buddha. This has positive significance with respect to encouraging people to change their attitude and establish the right ecological ideas.

(4) The Well-Rooted are Well-Rewarded

The well-rooted are well-rewarded is to say that as long as one does good and accumulates virtue, the earth will reward you with infinite treasures, which is a worldview unique to Buddhism. Although it is superstitious to be sure, it is not totally unreasonable with respect to humans ethically treating the earth and being well-rewarded for it. So, it still has points worth reflecting on. *The Flower Garden Sutra* says in "entering the world of creation" chapter, "Sudhana went to the site of enlightenment in Magadha, ten hundred thousand earth goddesses were all there." This passage describes Buddhism's conception of the earth in the well-rooted being well-rewarded, telling people as long as they pursue good, the earth will pay humans back with rich rewards. *The Flower Garland Sutra* tells us the earth "must be the universal place where all sentient life depends,"[14] which is to say, the earth is what all life depends on for life, it is the survival home of all sentient life. As long as all sentient life in general and humans in particular do good, accumulate virtue, and treat the earth ethically, the earth will treat all life well. *The Flower Garland Sutra* also tells us that the earth goddess Sthavara tells Sadhana: "Welcome. This is a spot of ground where you have planted roots of goodness, which I have witnessed. Do you want to see the results in one place?" (p. 1,283) Sadhana indicates that he does, and then, "the goddess, touching the earth with the sole of her foot, caused it to appear adorned with hundreds of billions of deposits of countless jewels, and said to Sudhana, 'These billions of treasuries of jewels accompany you and are at your service, to be used as you wish; they have been produced as a result of your good works and are preserved by the power of your good works. Take of them and do whatever should be done.'"[15] This illustrates that in Buddhist theory, if humans do good and accumulate virtue, and especially treat the earth well, the earth will return the gesture in kind, and build a better home for humanity.

(5) The Buddhist Practice of Liberation

Buddhism is a religion, whose followers have a practical process of cultivation, through which they transcend material desire and purify the mind, thereby liberating themselves from endless greed toward nature, and truly treating Nature with the attitude of ecological aesthetics. This is a little like Christian worship, Islamic "fasting," Daoist purification of the mind and phenomenological "suspension." Buddhism's cultivation practice tries to reach a state of mind that transcends mundane material desires through a process of prohibition, determination and enlightenment. Mahayana Buddhism insists that one can only truly realize liberation through "ten stages" of practice. *The Flower Garland Sutra* concretely describes a sincere monk,

[14] [Tang] [1] (vol. 3), p. 1208.

[15] Ibid.

Sadhana, who, under the guidance of the bodhisattva, Manjusri, underwent the ten stages of bitter ascetic cultivation practices. At the beginning Manjusri wanted Sadhana to seek knowledge of goodness without getting exhausted, to witness knowledge of goodness without becoming satiated, to follow all the teachings in knowledge of goodness and not to witness any mistake in knowledge of goodness and good skills.[16] After undergoing the ninth stage of aesthetic training, he ultimately reached the final tenth stage of attaining the fertilizing powers of the law-cloud, the state of Zen meditation. The Vajragarbha bodhisattva evaluated this thus: "Those enlightening beings who, having wisely reflected on the possibilities of knowledge up to the ninth stage, have made a thorough and discerning investigation, have thoroughly fulfilled pure practices, have gathered inexhaustible provisions, have acquired great stores of virtue and knowledge, have attained great, far-reaching compassion, are familiar with the distinctions and differentiations of worldly realms, have gone into the thickets of the realms of sentient beings, focus their perception and attention on approaching entry into the sphere of buddhas, and are intent on the powers, expertises, and other qualities of Buddhahood, are said to have reached the stage of coronation with omniscience in all its aspects."[17] This gathering of inexhaustible provisions, acquiring of great stores of virtue and knowledge, attaining great far-reaching compassion, becoming familiar with the distinctions and differentiations of worldly realms, and truly reaching the highest Buddhist state of "enlightenment" or "undeflected concentration." One can only truly liberate oneself, transcend mundane desires and truly treat natural ecology with the aesthetic attitude by reaching such a state.

(6) "The Pure Land of the West"

The conception of the pure land of the West is the ecological ideal of Buddhism, also called pure realm, buddhakṣetra, etc. It is the abode of Buddha, and the ideal state pursued by Mahayana Buddhism; it is also called The Western World of Extreme Joy as opposed to the dirty land where mundane creatures live. What condition is this pure land? Buddhism offers a description: perfectly ordered with plentiful and sweet waters, flowering and flourishing trees, splendid music, plentiful species of curious birds, reserves of rain water good for bodily health and clean, fresh air; Also look at the *Flower Garland Sutra* on the pure land of the West "the world adorned by treasuries of flowers": "The surrounding mountains of this Flower Bank Array ocean of worlds rests on lotus blossoms of sun-pearls. Their mass is made of sandalwood crystals, their peaks of majestic diamonds. Their curves are made of precious stones of exquisite fragrance. They are composed of fiery diamonds. Rivers of all perfumes flow between them. Their forests are made of all kinds of jewels. They are covered with beautiful flowers and fragrant herbs, and studded with bright pearls. Nets of precious stones are draped all around. There are as many such beautiful adornments as there are atoms in an ocean of worlds."[18] (p. 204) What a beautiful world indeed! It is truly a realm of purity without any pollution to defile it, it is somewhere human yearn to go.

[16] [Tang] [1] (vol. 3), p. 1100

[17] [Tang] [1] (vol. 2), p. 687.

[18] [Tang] [1] (vol. 1), p. 135.

8.2 Chan Buddhism's Wisdom Pertaining to Ecological Aesthetics

Chan Buddhism emerged after the middle of the Tang dynasty; it is a completely sinosized school of Buddhism, which absorbed the philosophical principles of the Confucianism, Daoism and metaphysics that preceded it in traditional Chinese culture. What took shape was a religious culture with Chinese characteristics. It is a religion based on meditation, rooted in the liberation of one's true nature, and encompasses rich ecological wisdom with brilliant characteristics.

(1) A State of Harmony Between Humans and Nature

The Platform Sutra states: "Good friends, what is seated meditation (*zuochan*)? In this teaching, there is no impediment and no hindrance. Externally, for the mind to refrain from activating thoughts with regard to all the good and bad realms is called 'seated' (*zuo*). Internally, to see the motionlessness of the self-nature is called 'meditation' (*chan*). Good friends,[19] what is it that is called meditative concentration (*chan-ding*; samādhi)? Externally, to transcend characteristics is 'meditation' (*chan*). Internally, to be undisturbed is 'concentration' (ding). If one concentrates on characteristics externally, internally the mind is disturbed." (Translated by John R. McRae) From this, one reaches the state of no impediments. This is a process of eliminating oppositions between human and natural, between material and spiritual. The Japanese Zen master, D.T. Suzuki, points out, the state in "state of mind" originates from the Sanskrit word, Gacara, which originally referred to the fields where cattle eat grass and roam about. Cattle have fields for eating grass, humans have fields for mental activity, which is a process of comprehending "fields"; it is also a process of harmoniously unifying humans and Nature. Nature becomes the manifestation of the state of Zen, the Chan master Hongzhi Zhengjue: "humans coming and going in the mountains, realize the green mountain is the body. The green mountain is the body and the body is I, where else would show roots and dirt." Here, humans and nature are one, and only by fusing into the green mountains and waters can one attain the *Dao* and become Buddha. After the Tang dynasty, large quantities of mountain water poems and mountain water paintings were greatly related to the growth of Chan Buddhism, and expressed the aesthetic state of human being's fusion with Nature, which manifested an ancient wisdom of ecological aesthetics closely tied to meditative concentration. For instance, Wang Wei's *Painting of Yuan An's House in the Snow*, places the plantains of the south and the white snow of the north together to allude to Buddhism's state of transcendence.

[19][Tang] Neng [6], p. 37.

(2) In Human Being's Fusion with Nature, the State of Nirvana Infinitely Pursued by Human Life

There is a distinction between Chan Buddhism and traditional Buddhism. Traditional Buddhism neglected Being-in-the-world, and practiced asceticism in this world to become Buddha in the next; Chan Buddhism concentrates on both the next world and this world, and seeks the peaceful equilibrium of real life. What is meditation? Chan Buddhism argues, all actions could be meditation. Chan Buddhism's outstanding feature is human being's preoccupation with life. Traditional Buddhism neglects life, while Chan Buddhism concentrates on life. Although human life is finite, Nature is infinite, and in the fusion of finite life and infinite nature, Chan Buddhism seeks the infinity of life. This is the famous conception of Nirvana. *The Platform Sutra* states: Huineng recited the fifth patriarch Hong Ren's *chuan yi fu fasong* 传衣付法颂, whose recitation goes: "Sentient beings cast their seeds; Because of the earth the flowers bloom; Insentient objects have no seeds, No natures, and no birth."[20] In Hong Ren's view, sentient human beings sow seeds, but only by relying on the insentient earth's infinity, do flowers bloom. Because of this, sentient human beings transform into insentient soil, and human life thereby makes the synthesis of the limitless, the lifeless, sentience and infinity, through which alone no birth is achieved. This is rarely attained in Buddhism; it is a theory that concentrates on human being's life in this world.

(3) The Theory of the Natural Tendency of the Mind and the Equal Value Possessed by All Things

Chan Buddhism is a theory of meditation that insists on realizing the Truth of the Buddha through gradual or sudden enlightenment. The Northern school practiced gradual enlightenment, while the Southern school lauded sudden enlightenment. The theory of the natural tendency of the mind is its important theoretical viewpoint. *The Platform Sutra* insists "The Dharma transmits from mind to mind, acting to lead to self-enlightenment."[21] When selecting successors Chan Buddhism masters generally had two avenues: one was everyone writing a verse; the second was orally transmitting a verse, manifesting enlightenment and wisdom. Here, the mind is human nature and human nature is the mind. The mind is the original constitution of the world, and human nature is the essence of the world, the mind and human nature are immediately one, which is a humanized theory of Buddha nature. Mahasattva is great, namely grand mental capacity, the Buddha mind is infinitely broad; it can contain Heaven and Earth, mountains and waters, grasses and trees, good and evil. In this envelopment, all existences, all beings have their own value in the face of Buddha nature, including evil, without which there would be no goodness. The theory of the natural tendency of the mind is the comprehension and embracing of wise nature, therefore, "Buddha does of his own self-nature, none seek it in themselves." With his self-nature illuminated, the Buddha is the natural tendency of the mind. Buddha is all sentient life. Buddha is the earth; With the enlightenment of the nature of wisdom, all living beings are

[20] [Tang] Neng [6], p. 104.

[21] Ibid., p. 19.

Buddha. *The Platform Sutra* states, Huineng came from a minority ethnic group in Guangdong, seeking grand master Hong Ren's way of practicing the Dharma. Grand master Hong Ren said: "If you're from Lingnan, then you must be a hunter. How could you ever achieve Buddhahood?" Huineng responded, saying: "Although people may be from north or south (*nan*), there is fundamentally no north and south in the buddha-nature. Although this hunter's body is different from Your Reverence's, how can there by any difference in the buddha-natures [within]?"[22] People are from the north or south, but Buddha-nature is neither from the north nor south. In the horizon of Chan Buddhism, there is no difference between southerners and northerners, races of people or humans and other creatures. In terms of equally embracing Buddha-nature, all are equal. This necessarily results in a Buddhist view of ecological equality.

(4) Suspending Material Desire, the Dharma of Meditative Concentration and Treating Nature Ethically

Chan Buddhism advocates a Dharma of meditative concentration and treating nature ethically; it splits up into two schools, that of gradual enlightenment and that of sudden enlightenment, but both choose the path of eliminating material desires and scattered thoughts. Thus, master Huineng's famous verse in the *Platform Sutra* goes: "The body is the *bodhi* tree; The mind is like a bright mirror's stand. Be always diligent in rubbing it—Do not let it attract any dust." Comprehending this verse, one can then see one's own Buddha-nature; According to this practice, one can eliminate material desire, but there are many paths and methods to attaining complete enlightenment in Chan Buddhism. For instance, there is meditative concentration, problems set by Chan masters, sharp witted and incisive remarks, blows of the stick, scolding Buddha and trashing patriarchs, normal mentality, etc. Meditators break through the boundary between forms and formlessness instantaneously through these methods. Forms refers to creatures, and the formlessness is Buddha; form is the formless, and the formless is Buddha. Only by destroying the forms can one enter the formless; only by breaking the boundary between what is human and what is natural can one make all creatures become one. *The Platform Sutra* records the story of master Huineng striking Shenhui with a stick: "When I hit you, does it hurt or not?' [Shenhui] answered, 'It both hurts and does not hurt.' The master said, 'I also see and do not see.' Shenhui asked, 'what is this seeing and not seeing?' The master said, 'My seeing is to see constantly my own mind's errors. I do not see other people's right and wrong or good and evil. This is to see and also not to see. You said it hurts and does not hurt. How about this? If you do not hurt, then you're the same as a tree or rock. If you hurt, then you're the same as an ordinary [unenlightened] person, who would become resentful. When you just said 'seeing and also not seeing' [you thought] they were two extremes, and your 'hurts and does not hurt' were [your misconception of] birth and death. But you don't see your self-nature, so you're just playing around.' Shenhui bowed in gratitude."[23] This illustrates, that enlightened

[22] Ibid., p. 8.

[23] [Tang] Neng [6], p. 90.

meditators eliminate the boundary between hurting and not hurting, seeing and not seeing, enlightened and ordinary, and reach the unity of human being and self-nature.

(5) The Ecological Practice of Pure Being-in-the-World

The Chan Buddhism conception of pure earth is very special. Previously we mentioned, Buddhism's conception of pure earth is an ideal, an ideal state. Chan Buddhism thus holds that the pure earth is in the mundane world, there is no need to seek it in the Western world, that is we may realize the pure world through reforming the actual world. This is another conception of a realistic ecological practice.

References

1. [Tang]. (2001). Translated by Shi Cha Nan Tuo, Collated by Lin Shitian et al.: *Hua Yan Jing* (vol. 2), Religious Culture Press (A[唐]实叉难陀译:《华严经》,林世田等点校,宗教文化出版社2001年版).
2. (1981). *Religious Dictionary*. Shanghai Dictionary Publishing House (《宗教辞典》,上海辞书出版社1981.年版).
3. Kesu, L. (1999). *Historical tales of Chinese buddhism*. Hebei University Press (刘克苏:《中国佛教史话》,河北大学出版社1999年版).
4. Youlan, F. (1996). *A brief history of Chinese philosophy*. Peking University Press.
5. Kesu, L. (2010). *The history of Chinese Buddhism*. Hebei University Press.
6. [Tang] Neng, H. (1983). Collated and Annotated by Guo Peng: *Collated and Annotated The Tan Jing*. Zhonghua Book Company.

Chapter 9
The Wisdom of Ecological Aesthetics in the Traditional Chinese Art of Painting

China is an ancient civilized state, whose culture, arts and aesthetic ideas have continuously aimed at "probing the verge between Nature and humanity." This not only implied rich classical wisdom of ecological aesthetics, but also developed into forms of aesthetics and art that differ from Western ones. This point is clearly reflected in traditional Chinese painting.

9.1 The National Painting of "Natural Ecological Art"

Originally, art is said in contrast to Nature, a civilized formation that is clearly distinct from natural formations. The West gave development to painting and matured in the renaissance and enlightenment periods, and such developments were closely related to industrial revolutions, which is fully illustrated in everything from tools and paints to the creative principle of "the mirror theory." But Chinese painting is an artistic field in closest proximity to Nature due to having emerged, having developed and having matured under natural economic conditions.

Let us first consider the tools employed in Chinese painting, the so-called "four treasures of the cultural room," namely the hair brush, ink, paper and inkstone, which are all natural products that differ from the artificial products used in Western painting from chemical paints to the paintbrush. Just as the famous contemporary Chinese painter Zhang Daqian put it, "the hair brush, ink and paper are all special materials, they are the elements composing the special style of Chinese painting. These are the features unique to Chinese painting that distinguishes it so greatly from the painting of other countries."[1] Hair brushes were made from sheep wool, rabbit and wolf hair, etc., ink was made from pine soot and oil soot, paper refers to Xuan paper made from plant fibers, inkstone was also made from natural cliff rocks or scorched mud and the paint used were also natural minerals or extracted from plants. With respect to types of painting, Western painting is centered on painting human

[1] Zhidong [1], p. 95.

F. Zeng, *Introduction to Ecological Aesthetics*,
https://doi.org/10.1007/978-981-13-8984-9_9

figures, while Chinese painting after the Wei-Jin era gave extremely privileged status to landscape painting. Tang Dai of the Qing dynasty said of "natural" in *Painting Things, Developing Subtleties*: "in using the naturalness of hair and ink to match the naturalness of sky and earth, one's painting is therefore absolutely singular." In the chapter "*Natural*," Tang Dai specifically theorizes: "all things take shape and phenomenally emerge from the cyclic opening and closing of sky and earth, none of which does not come about from the sway and natural becoming of Yin and Yang. The work of painting is no different. In painting, the ancients used the movement of the brush to enact Yang and the stillness of the ink to enact Yin. Collecting energy with the brush to enact Yang, generating color with the ink to enact Yin. By embodying Yin and Yang with the use of brush and ink, every painting composes the grandness of hills and valleys, the subtlety of trees and rocks, the minutia of sand and stream without any imprecision of brush, without even a point of lifelessness." This informs us that "natural becoming" is the natural law of everything of sky and earth coming about from the swaying reciprocal affectivity of Ying and Yang in ancient Chinese thought. Just as Laozi said, "*Dao* produces one, from one derives two, from two derives three, and from three derives all things. All things carry Yin and embrace yang, generating intensity with the goal of harmonious resolution." (*Laozi* Chapter 42). The artistic principle of "natural becoming" is exhibited in full clarity through Chinese painting, which basically relies on the reciprocal unity of such paired oppositions as movement and rest, brush and ink, concentration and dilution, ink and color along with paint and blankness to exhibit the power of art. For instance, the Song dynasty painter Su Shi's *Painting of Wood and Stone* is an extremely elegant withered stump and rough rock, most of the area in the painting is blank space, but through this natural contrast of paint and blankness, stone and tree, brush and ink, what is expressed is civilized humanity's spiritual temperament of unyielding firmness. Western painting on the contrary is a painting method informed by science. Just as the European renaissance painter Da Vinci said, "Truly this is science, the legitimate daughter of Nature," "noting that the sense of beauty is entirely established in the relationship of divine proportion between various parts, all characteristics must act simultaneously to generate the harmonious proportion that enraptures the viewer".[2] Da Vinci's famous work *The Last Supper* is the canonical example of this harmonious proportion: the whole painting is centered around the calm and tranquil Jesus with the two columns of disciples dividing respectively left and right of the central concentration of perspective in symmetrical proportion, facial expressions all different, fully exhibiting the humanist spirit unique to the renaissance period, that of punishing the wicked and raising up the good while saving the people.

[2]Quoted from Xingchen [2], p. 137.

9.2 "The Dispersed Perspective" of Chinese Painting

"Perspectives" refer to the angle of paintings and reflect different artistic ideas. Western painting generally adopts the method of focal point perspective also called the "far near method" [in Chinese], which is the method of starting from the fixed perspective of the painter and proceeding to paint the phenomenon according to how far or near objects are from the retina by making the nearer objects more full and solid and the more distant objects more empty and atmospheric. This focal point perspective is actually a creative method of painting guided by the scientific theory of optics and geometry, which Da Vinci so highly regarded. In his famous work *A Treatise on Painting*, Da Vinci states, "Practice must always be founded on good theory; to this, Perspective is the guide and entrance, without which nothing can be well done."[3] Clearly, this is a scientistic theory and method of painting, which of course has its own value and also made glorious achievements in the long term practice of Western painting. But this method only permits one central viewpoint in the painting surface, which is of course unproblematic if considered solely from geometric principles and the optics of invisibility increasing with distance, size decreasing with distance, light illumination and backside concealment, etc., but its defects are perfectly clear if considered from the principle of the equality of all beings in Nature. This convergent perspective painting method is unfair to those dark concealed objects, which is still the reflection of a scientism and anthropocentrism. Just as Wolfgang Welsch understood it, "the presentation of the entire scene depends on the eyes and standpoint of the viewer, the human scale is at the center of the entire painting surface, in such a way that anthropocentrism in perspective painting is deeply rooted, where nothing naturally appears of itself, but is rather based on our one-sided perception of it, in which case every detail in the painting is related to us and is determined by our perspective and standpoint; the object painted is intimately tied to our gaze at the world."[4]

The "scatter point perspective" adopted by Chinese painting differs from the focal point perspective of Western painting; it is a painting method according to which the scene follows human movement and the human being follows the movement of the scene, each step being visible, the painting surface presents multiple perspectives allowing near and far areas, light and dark areas and even internal and external things to gain the opportunity of appearing. Zhang Daqian once said, "Chinese painting is often criticized by people who don't understand it, assuming that it has no perspective. Actually how is it that Chinese painting has no perspective? The perspective of our nation's painting is captured from all directions, while the perspective of modern abstract painting only gets one aspect of it." He also says, "When painting a tree, one can only see the top of the tree if the method of overlooking is adopted, and one can only see the tree's branches and trunk if the method of looking up is adopted.

---

[3]Quoted from Yu [3], p. 254.

[4]Welsch [4], p. 475.

If both perspectives are combined, one can see the top of the tree as well as the tree trunk, giving people the view of the whole tree, what is bad about that?"[5]

One of the formulations of "the method of dispersed point perspective" in the traditional Chinese theory of painting is the method of "three distances." As the famous Song dynasty painter Guo Xi said in *The Heights of the Forest and Spring* that, "a mountain has three distances: looking up at the mountain peak from the foot of the mountain is called the distance of elevation; looking toward the back of the mountain from the front of the mountain is called the distance of depth; looking at a mountain far away from a mountain nearby is called level distance. The color of elevated distance is clear and bright, the color of the distance of depth is heavy and dark, level distance involves both bright and dark colors; The power of elevated distance is towering, the interest of depth distance is overlapping, and that of even distance is melding, and dimly discerning. Among human beings in these three distances, those of heights see clearly from far, those of depths distinguish minutia, and those of flat distance dilute conflicts. Those who see clearly from afar do not shorten, those who distinguish minutia do not lengthen, and those who dilute conflicts do not enlarge. These are the three distances." In using the method of "three distances" to paint pictures, may perspectives emerge on the picture surface, so all sides near and far, high and low, *yin* and *yang*, toward and away, inside and outside are given the opportunity to present themselves. This is largely at odds with the scienticism and anthropocentrism in Western painting, but it does strengthen the expressive forces of the art of painting. Thus, there emerged long scrolls expressively describing all of city life and entire rivers rarely seen in the human history painting. For instance, the Song dynasty painter Zhang Zeduan's famous *Along the River in the Qing Ming Festival*, 20.8 mm vertical and 528.7 mm horizontal, reflected the scenery on both sides of the river and the background of life during the Qing Ming festival in the Song dynasty capital, it touches on local habits, popular customs, architecture, land and river transport, porters, the practice of medicine and fortune telling, monks and priests, bandits, sedan and rickshaw drivers, boat merchants, the sexes and ages of people, the religious sects, etc., altogether over 550 people, around 50–60 heads of livestock, over 20 horse drawn carts, over 20 boats, and 30 houses, many different human figures in one macroscopic scene. It is only possible to artistically reflect such a magnificent scene with the method of diffuse perspective or moving perspective, where every human figure and piece of scenery on both sides of the river get the right of equal expression through diffuse perspective. Western painting's difference in this regard is extremely obvious. For instance, the well-known famous Dutch painter Hobbema's *A Road Winding Past Cottages* is the most typical work created with the method of concentrated perspective, presenting to us the cottage scenery of 17th century Holland. The said painting is composed according to the law of size decreasing tenuousness increasing with distance, the picture certainly has a sense of depth, but the true Dutch village is only a hazy shadow for us. This is perhaps the expression scientism and anthropocentrism in painting, whose limits led the later volume schools to break through this concentrated perspective.

---

[5]Zhidong [1], pp. 4, 52.

9.3 Chinese Painting's "lively rhythm"

Ancient Chinese philosophy believes, "I live in conjunction with the heavens and the earth, I am one with all beings." (Zhuangzi 2.9). This is to say, in the eyes of the ancient Chinese, humans and every natural being are alive and one. In the eyes of painters, the mountains and waters in the natural world share commonality with humans, and when they observe the seasonal changes of all beings, they are always compared with humans. Just as the northern Song painter Guo Xi says in *The Lofty Message of the Forest and Rivers*: "Hills as if smiling coquettishly in the spring, hills as if crying in the dark green of summer, hills as if solemn in the clear autumn, hills as if sleeping in the desolate winter." He uses the laughing, the tears, stern solemnness and sleeping stillness of human beings to describe the different figurative expressions of the four seasons, and of course, when painting hills he had to express the living form of each expression in different space-times.

In this respect, the ancient Chinese theory of painting raised the artistic demand of lively rhythm. Xie He of the southern Qi proposed the "Six Methods of Chinese Painting" the earliest in *The Record of the Classification of Chinese Painters*: "There are six methods of Chinese painting. One is spirit resonance, two is the bone method of using the brush, three is correspondence with the object, four is suitability to type, five is division and planning, and sixth is transmission by copying." "lively rhythm" is listed as the first of the "six methods." The "six methods" of which Xie He speaks is first of all a demand on the painting of human figures, but later it gradually became the basic precept of all Chinese painting. Guo Si of the northern Song insisted in *Experiences in Painting*, "The refined theory of six methods, is forever immutable. However, the five methods from 'the bone method of using the brush" on could be studied and learned, but its 'rhythm" is necessarily living knowledge, which cannot be attained through clever wit, nor reached in years time, it is tacitly grasped with spirit as being so without knowing how." He pushes "spirit resonance" to the highest state of mind of the art of painting. Zong Baihua gives an extremely important explanation of "spirit resonance": "the topic of lively rhythm in Chinese painting means 'the rhythm of life' or 'rhythmic life.'"[6] This is to say, "lively rhythm" is actually expressing the inspirited vitality of Nature. Because of this, Chinese painting does not seek the physical similarity of art with Nature, but the spiritual similarity of art with Nature, which is producing lively spirit. Just as Zhang Yanyuan of the Tang dynasty said in *Famous Paintings in History*, "with respect to spirits and humans, there is the depictable nature of lively movement, but it requires inspired rhythm to be complete. If the rhythm is not periodic, it is physically similar to empty display; without the powerful force of brush, the emptiness may well invite physical shapes, but not subtle movement." Lively rhythm is chiefly rhythm just as Gu Ningyuan of the Ming dynasty said, "the first among the six methods, 'lively rhythm' means that if there is rhythm there is lively activity. Sometimes the rhythm is in the scene, sometimes outside of the scene, depending on the coldness or heat, clear or rainy skies, darkness or brightness of the four seasons, not on how much ink is used."

---

[6]Baihua [5], p. 118.

As the rhythm "in the scene," Chinese painting proposes many means of expressing the vitality of all natural things, as Guo Xi said in *The Lofty Message of Forests and Streams*: "water makes up the arteries of the mountains, grasses and trees are the hair and fur of mountains, vapor and clouds their spirit and expression. So mountains come alive when watered, thrive when planted with vegetation, and are encharmed with vapor and clouds; water pours down the mountain like a face, pavilions and terraces make up the expressive eyebrows and eyes, and the fishing line is the vital spirit. So when the mountain is added to water, it becomes charming, when a pavilion is added it becomes lucid, when the fishing line is added, it becomes a wider settlement. This is the arrangement mountain water scenery." Of course, what is most important is to express the foundation of Nature's vitality—"the true intensive spirit of the heavens and the earth," which is the charm of all natural things. Tang Dai of the Qing dynasty said in *Uncovering Subtlety in Painting Things* that "Painting mountains and water is enriched by the intensive difference of *qi*, the intensive difference of *qi* is not clouds, smoke, fog or mist, but the genuine intensive difference between heaven and earth. No being is alive without the intensive difference of *qi*, the intensive difference of the mountain is released from within the stone. When looking at the mountain on the occasion of bright clear skies, the intensive difference of the mountain's vast moistness steams out, desiring activity, thus in painting mountain water scenery, the intensive difference of *qi* precedes all." "The genuine intensive spirit of *qi*" is the spirit of all things, and requires the painter's long term observation of everything around and the continuous elevation of one's state of mind to comprehensively grasp it. The modern painter Qi Baishi painted shrimp, after a long period of observation and realization, he painted the famous *Picture of Shrimp* with his spirit of "drawing portraits for all worms and expanding spirits for all birds"—each one as lifelike as the next, showing forth onto the paper full of vitality. In Western painting, there is the method of still life painting. The post-impressionist painter everyone is familiar with, Cezanne, made the famous still life painting *Still life with Cup, Jar and Apples*, where he painted apples in a ceramic cup. Even though as a post-impressionist painter, Cezanne already pinned much of his own subjective tones into this still life, this painting still expresses "the pursuit of the eternal image and solid structure." Qi Baishi's *Picture of Shrimp* has a different aim in seeking the thriving forces of life.

9.4 Chinese Painting "Learning to Create Externally and Grasping Mental Sources Internally"

The most basic creative principle of Chinese painting is the idea of "learning to create externally and grasping mental sources internally" proposed by the Tang dynasty painter, Zhang Zao. This is an extremely important theory of artistic creation possessing Chinese characteristics, which is fully in line with the ancient Chinese thought of "Nature and man integrating into one." The character for "Nature" in this phrase (*tian* 天) is extremely rich in content, it includes every natural thing as well as the physical appearance and affective spirit of natural events. "Man" here includes the mental grasping, experiential realization inner spiritual intensity in man's observation of external things, which are the so-called "mental sources." Learning to create externally and grasping mental sources internally are two unified and inseparable phases. Luo Dajing of the Song dynasty recounted the stories of Li Gonglin painting horses and Ceng Yunchao painting grass and insects in *Jade Dew from Crane Forest*. In order to paint good horses, Li Gonglin would, "make a general survey of people driving horses all day until he was busy with chatting guests. Accumulating in vitality and storing up in vigor, the steed in his spirit appreciated in value for so long that in his breast was full of horses. Sending his intention into his brush, himself transcendently subtle." So, Huang Tingjian wrote a poem appreciating him: "The noble Li painted horses and also painted flesh, dropping the brush to birth a horse as if splitting the bamboo handle." Luo Dajing believed "the character for 'birth' (*sheng* 生) in Huang Tingjian's poem was used most subtly. His breast was all horse, so from the end of the brush he gave birth." Ceng Yunchao's work in painting grass and insects was no different. Luo Dajing recounted the autobiography of Mr. Ceng: "From some point in youth, I would grab grass and insects, and observe them in a cage all day and night without getting tired of it. If I was afraid their spirit was lacking, I would again move them in between the grass and observe them, and so they began to get their Nature. At the verge of dropping the brush, would I know if I were becoming grass and insect or grass and insect were becoming me? This is being so close to creating living things it is no different." When Ceng Yunchao painted grass and insects, man, grass and insect already transformed and became one, which in actuality is the spirit of grass and insects and the spirit of man transforming and becoming one. This is also what Zheng Xie of the Qing dynasty said as well, the unity of "the bamboo in the eye," "the bamboo in the breast" and "the bamboo in the hand." Through this creative process, the work created is the unification of Nature and man, the unity of spiritual likeness and physical likeness, out of which permeates a rare style. Such artistic works are very stylistically different from the works created under the guidance of "the mirror theory" in Western painting. For instance, the famous impressionist Monet did the painting *An Impression of a Sunrise*, which although is distinct from traditional realistic works, did not take leave of the concrete image of the object itself, but rather sought originality in the coloring and light rays of the objective image. The Tang dynasty painter Wang Weiceng did the painting entitled *Snowy Picture of Yuan An's Bedroom*, in the snowy scene appears a plantain, he uses the empty core of the plantain to reflect the purity of the snowy skin, implying the Buddhist notion of purifying desire. This painting currently no longer exists, but Xu Wei of the Ming dynasty did *Pictures of Various Flowers*, in which all kinds of flowers and plants from peony, pomegranate, Chinese parasol, chrysanthemum, pumpkin, haricot bean, grapes and plantains to plum blossoms, narcissus and bamboo come together in one fluid breath, reaching the effect of "not seeking physical similarity but living rhythm.

9.5 "Workable, Viewable, Explorable, Inhabitable" Chinese Painting

Chinese painting does not just transform natural scenery into objects of people's appreciation, but goes further by pulling Nature and man closer together, making Nature become a loveable thing intimately close to human being, and even makes Nature enter into human being's living world. This is the famous "observable, inhabitable, explorable" theory. Guo Xi said in *The Lofty Message of Forests and Streams* that: "The sincere theory of the world says among mountain waters, there are the workable, the viewable, the explorable and the inhabitable. Any painting that gets these is outstanding to everyone. But workable and viewable isn't as good as what the inhabitable and explorable grasp. Why? Of many 100's of *li* of earth, the explorable and inhabitable areas do not even reach 3 or 4 tenths. But it is necessary to get inhabitable and explorable works, and the reason why princes love works of forests and streams is precisely because they are excellent places to live. So, the painter creates with this intention, and the appreciator also exhaustively explores with this intention. This is what is called not losing its original intention." Guo Xi speaks very clearly here, namely that one of the original intentions of the creation is not in the simple artistic appreciation, but rather in creating a natural scenery intimately bound to one's living world. This is a Chinese idea of nature painting, namely that external things in nature are not extrinsic to human being, but rather exist in a relationship of involvement with human being, and become a component part of human life. For instance, the famous Song dynasty painter Wang Ximeng's work, *Picture of 1000 li of Rivers and Mountains*, 51.3 mm vertical, 1191.5 horizontal, is a long scroll, colored mainly in the theme of blue-green, paints the glorious scenery of beautiful elegant rivers and pristine mountains. Even though it is natural mountain waters, it is human being's living world. The painting is dotted with fishing villages, adorned with paths, small bridges and people, interspersed with disparate forest trees with a humanly observable, habitable and explorable air; it has become a precious work of Chinese painting. Western painting usually focuses on expressing the beautiful vigor of the natural scenery itself, while not caring for its relationship with human being. For instance, Rousseau of France made the landscape painting *The Oak in the Rocks*, which although outstandingly depicts the grassy ground under the sunlight and the dark shadow of the tree, does not make any point to express the tree's relationship with human being.

9.6 "Intention Precedes the Brush" and "Invest Interest in the Ink" in Chinese Painting

The Tang dynasty painter Wang Wei pointed out in *A Theory of Mountain Water Painting* that "In any painting of mountain water, intention precedes the brush," emphasizing that one must process the relationship between "intention" and the "brush" well in the creation of mountain water paintings. By "intention" he means the painter's interest, and by "brush," he means "ink." The former is emotional interest while the latter is the ink figure, and the two of them make up a relationship between "interest and investing it." Chen Zi'Ang's *Preface to My Poem "The Bamboo" Composed with Inspirations from the Scribe to the Left Dongfang Qiu* proposes the theory of "interest and investing," pointing out an artistic method of "entrusting things to spark interest" and "us ing things to house aspiration." The rise of Chinese mountain water painting was related to the political chaos of the Wei-Jin era. The political scenario at the time was unstable, war frequently broke out, and the plight of literate people was difficult, so by investing emotion in mountain water, the mountain water painting began to thrive. Literate painters painted mountain waters not so much to copy the image of mountain water, but rather to invest it with emotional interest, which in turn found expression in the ink image, the relationship between "interest" and "brush" was one of friendly analogy. Early in the pre-Qin era, Confucius brought up the issue of "the wise enjoy waters, the humane enjoy mountains," using the mountain to metaphorically convey the deep and thick generosity of the humane person's moral deeds and waters to metaphorically convey the unsettling fluidity of the wise person's wisdom. The friendly co-habitation of humans and Nature in art was indeed an excellent tradition of the ancient Chinese people in treating Nature as friendly. Li Bai's poem *At Mt. Jingting Seated by Myself* goes, "Soaring high, now all birds have flown—A slow cloud drifts away alone. With fondness we watch one another—Mount *Jingting*, I'll seek no other!" (Trans. Frank Cyue). What he writes about here is the mutual respect and mutual fondness between the poet and Mount *Jingting*, the beautiful relationship of harmony between self and things. This was even more clearly expressed in mountain water paintings with flowers and birds. The famous painter at the beginning of the Qing dynasty, Shi Tao pointed out in *Record of Words in the Bitter Mellon Monk's Paintings* that "the ancients invested interest in the brush's ink, and hid the *Dao* in mountain rivers. Unchanging and yet coping with all changes, effortless and yet intentional action, the bodily being does not dazzle the world and yet the name stands up on its own." In Shi Tao's view, the painter invests interest in the brush's ink and imparts the *Dao* to mountain water through painting and in this way is capable of coping with all changes without changing and realizing intentional action through effortless action. In actual fact, he himself was well adept at applying the force of investing interest in painting. He was a famous representative of the Huangshan Mountain School, who lived in Huangshan mountain for a long time and came up with such thoughts as "treating Huangshan mountain as one's teacher," "treating Huangshan mountain as one's friend," and "grasping the natural tendency of Huangshan mountain." At the same time, through is own painting of Huangshan mountain, through dancing brush strokes and drenching ink rain and the towering power of the mountain, he expressed his own national thought as an old adherent of the Ming dynasty tradition after the rise of the Qing, or as he said "Golden branch, jade leaves, kinsman noble to the bygone age—the inkstone, outstanding, far removed from native soil." We may look at his distinctive feature of investing interest through is representative work *Bold Ink Mountain Water Scroll.* Of course, there are also the famous three friends in Chinese painting, the plum tree, pine tree and bamboo tree, which everyone is well familiar with and which the ancients used metaphorically to convey the noble integrity of the superior man in his capacity to survive frost and snow. This is naturally the manifestation of the theory of "analogizing virtues" in art ever since the pre-Qin era. Bian Jingzhao of the Ming dynasty made the famous *Picture of the Three Friends and Hundred Birds*, drawn up in the dead of the winter season with one hundred birds dwelling betwixt the pine, bamboo and plum trees, some flying, some singing, some resting, looking around echoing one another, each displaying to their fullest, expressing the noble moral integrity of the painter enduring the cold, an extraordinary use of intention. Zhang Daqian once pointed out: "Chinese painting cares about where spirit is invested. For instance, Chinese painters have historically loved painting the 'four princes' (plum tree, orchid, bamboo and chrysanthemum), some people believe this amounts to an ossified mindset, but actually this is not the case, it is where the spirit of Chinese painting lives. If a painter paints a plum tree and chrysanthemum, and gifts it to someone, in one respect it metaphorically conveys the frost-enduring integrity of these two plants along with their unique strengths, and in another respect it also draws the receiving party into an equal state of mind. This is expecting from oneself and respecting in another. This spirit of "investing" in Chinese painting is really quite a noble tradition, and is also the greatest characteristic differentiating it from Western painting."[7]

In summary, extremely plentiful wisdom pertaining to ecological aesthetics fills the traditional Chinese art of painting, which is deeply inspiring for the development of contemporary aesthetics. Of course, us affirming that traditional Chinese painting possesses the superior strength of natural ecological art does not mean that we deny the superior aspects of Western painting. Both have their unique strengths, and are fully capable of complementing each other in the new era. In 1956, Zhang Daqian hosted an art exhibit in Europe, where he met Picasso, the two gifted one another paintings they did and chatted with great joy. Picasso highly appreciated Eastern art including Chinese painting, and Zhang Daqian sighed with deep emotion after the affair that: "deeply feeling art is the common language of humankind, and although the forms of artistic expression vary in particular, they are all one with respect to seeking the poetic state of mind, efficacy and skill."[8]

References

1. Zhidong, C. (Ed.). (1998). *Records of Zhang Daqian Talks about Ar*. Henan Fine Arts Publishing House.
2. Xingchen, L. (1994). *Western aesthetics course*. Peking University Press.
3. Yu, L. (1980). *The outline of western art history*. Liaoning Fine Arts Publishing House.
4. [Germany] Welsch, W. (2006). *How to transcend anthropocentrism*? Gao Jianping and Wang Keping, Editor-in-Chief, *Aesthetics and culture, east and west*. Anhui Education Press.
5. Baihua, Z. (1987). *Yi Jing*. Peking University Press.

[7]Zhidong [1], p. 3.

[8]Ibid., p. 129.

Part V
The Implications of Ecological Aesthetics

Chapter 10
The Implications of Ecological Aesthetics (Part I): Ecological Ontology's Conception of Aesthetics

The most basic characteristic of ecological aesthetics is that it is a conception of aesthetics that envelops the ecological dimension, and is thus distinct from a form of aesthetics characterized by "anthropocentrism." The "ecological dimension" here is distinct from "ecocentrism"; it is a new ecological humanism of "the ecological whole" that fuses human being and Nature into one; it is the philosophy and aesthetics of ecological ontology. This eco-ontological conception of aesthetics is the most basic category of ecological aesthetics, which differentiates ecological ontology from both the traditional conception of aesthetics guided by "anthropocentrism" and the Darwinian conception of aesthetics indicated by "ecocentrism." Lastly, it is both connected to and yet distinct from contemporary environmental aesthetics.

10.1 The Guidance of Marx's Practical Ontology

Ecological ontology was first brought up by the contemporary ecological theorist David Ray Griffin, who is a retired professor of theology and philosophy at the Claremont School of Theology and Claremont Graduate University in the United States; he was the first director of the Center for a Postmodern World, directed the Center for Process Studies, and is a famous contemporary philosopher of religion and ecological theorist. In the article *Peace and the Postmodern Paradigm*, he criticized the negative consequences appearing in the modern paradigm of research, pointing out that "[a] fourth feature of the modern paradigm, which has had all sorts of negative consequences for world peace, is its nonecological view of existence."[1] The ecological conception of existence spoken of here refers to the philosophy of ecological ontology, whose central interpreter in the West was the famous contemporary philosopher Heidegger. Heidegger's positing of ecological ontology marked contemporary Western philosophy's realization of the turn from traditional epistemology to contemporary ontology as well as the turn from "anthropocentrism" to "the ecological whole."

[1]Ray Griffin [1], p. 224.

F. Zeng, *Introduction to Ecological Aesthetics*,
https://doi.org/10.1007/978-981-13-8984-9_10

What we need to think about further now is what role Marx played in this most vital shift of contemporary philosophy. Our answer is: Marxism not only reflected this shift in concentrated fashion, but also represented the correct direction forward for this shift. The positing of Marxist practical ontology and its development was the revealing of this shift. In the Spring of 1845, Marx remarked in his famous *Theses on Feuerbach* that "the chief defect of all hitherto existing materialism – that of Feuerbach included – is that the thing, reality, sensuousness, is conceived only in the form of the *object or of contemplation*, but not as *sensuous human activity, practice*, not subjectively."[2] Marxism criticizes the old materialism represented by Feuerbach for only understanding things from the epistemology of passive mechanical materialism, and for only seeing matter as an object of knowledge opposed to the subject, instead of viewing matter from the perspective of the subject's activity, where it first forms a connection to the human being and hence first becomes the object of human being's understanding through human being's sensuous practical activity. This shift from grasping things through objective intuition to grasping things through subjective practice was precisely the shift from epistemology to ontology. In the philosophy of Marxism, practice is equal in character to being. The practice of material production is the first precondition of the human's being, it is human being's most basic way of being. Marx states: "In creating a *world of objects* by his personal activity, in his *work upon* inorganic nature, man proves himself a conscious species-being."[3] (*Estranged Labor*) He also states: "the first premise of all human existence and, therefore, of all history, the premise, namely, that men must be in a position to live in order to be able to "make history." But life involves before everything else eating and drinking, a habitation, clothing and many other things. The first historical act is thus the production of the means to satisfy these needs, the production of material life itself."[4] Because of this, we believe that Marx stood at the forefront of the philosophical revolution of breaking through the model of thought brought about by the industrial revolution known as instrumental reason and the subject-object dichotomy early in the middle of the 19th century. Afterward, Marx and Engels would go on to expand the three dimensions of philosophy, political economy and scientific socialism on the basepoint of unifying the proletariat, human liberation and the free development of human beings, thereby constructing a profound system of materialist practical ontology. This theoretical system shares commonalities with contemporary Western Ontology in transcending the traditional epistemological model of thought characterized by the subject-object dichotomy, but it also has extremely important and fundamental differences with the latter.

First of all, in terms of philosophical foundation, Marxism's materialist practical ontology is materialist; its premise is the practice of material production, and its foundation is "the practical world"; contemporary Western ontology is idealist; it is entirely founded on the basis of subjective "intentionality"; its foundation is the rather abstract "life world."

---

[2]Marx and Engels [2] (Vol. 1), p. 16.

[3]Marx and Engels [3] (Vol. 42), p. 96.

[4]Marx and Engels [2] (Vol. 1), p. 32.

Second, in terms of implications, Marxism's materialist practical ontology has the deeply profound foundation of actual society, and unifies the all-around development of everyone's freedom with the proletariat class and human liberation; contemporary Western ontology is mainly rooted in the concern for the state of survival of persons represented by middle class intellectuals; its path is the cultural and mental atonement of a society relatively divorced from the real economy, which is unavoidably abstract and divorced from actuality.

Third, in terms of its keynote, the aim of Marxism's materialist practical ontology is the realization of human emancipation and the ideals of communism; it is glowing with a vigorous and actively rising keynote; meanwhile, contemporary Western ontological philosophy reflects the "dejected" and "fearful" mood of the middle class intellectuals in the fierce competition capitalism; it harbors the negative mindset of "the other is hell."

Finally, in terms of human being's relation to Nature, Marxism's materialist practical ontology not only forcefully champions "building according to the laws of beauty" and also insists that the harmonization of humans with Nature and the unification of humanism with naturalism must be accomplished through reforming the relations of production and social institutions; meanwhile, contemporary Western ontology pins excessive hope on the cultural dimension, and is thus unavoidably constrained by historical idealism.

What specifically requires explanation is Marxism's materialist practical ontology is actually Marxist anthropological theory, because, the important theoretical implication of contemporary ontology is the elevated concern with human being's state of survival, and the distinguishing feature of Marxism's practical ontology is the concern with human being's state of survival in relations of production with its goal of emancipating human beings and the unification of human emancipation with the proletariat class. Marxism's materialist practical ontology is actually ecological humanism with a revolutionary spirit; it contains the important guiding principles of contemporary ecological theory, which includes ecological aesthetics.

10.2 The Implications of Ecological Ontology's Conception of Aesthetics

"The ecological conception of existence" is the most basic philosophical support and cultural standpoint of contemporary ecological aesthetics; it was brought up by the American constructive postmodern theorist David Ray Griffin, who, from the critical perspective, proposed this extremely important philosophical notion of the ecological conception of existence, which is the continuation and development of contemporary ontological philosophy represented by Heidegger; it contains fully rich implications, and marks contemporary philosophy's and contemporary aesthetics' transition from epistemology to ontology, from anthropocentrism to the ecological whole, from the complete "disenchantment" of Nature to the partial "re-enchantment" of Nature.

The transition from epistemology to ontology was Heidegger's chief creation, which provided the harmonious union of human being and Nature with a theoretical basis.

As everyone knows, epistemology is a bipartite relationship of human being and the world marked by the "subject-object dichotomy." In this relationship of being-in-the-world, human being is fundamentally opposed to Nature and cannot possibly unify in harmony with Nature. Meanwhile, contemporary ontological philosophy is a relationship of "Dasein and the world," which finally provides human being and Nature with the possibility and precondition of unifying harmoniously. Just as Heidegger said, "But subject and object do not coincide with Dasein and the world."[5] The reason why "being-in-the-world" relationship of Dasein and the world can provide the precondition for the unification of human being with Nature is because "Dasein" exists and unfolds in this relational state, namely human being's here and now relational state of existence composed with surrounding things. Here, there is only "relation" and "involvement," no "separation" and "opposition." "The concept of "facticity'.' implies that an entity 'within-the- world' has Being-in-the-world in such a way that it can understand itself as bound up in its 'destiny' with the Being of those entities which it encounters within its own world."[6] Heidegger furthermore understands those entities with which Dasein is bound up and encounters in the world "things ready-to-hand" like the innumerable things people face in life, but only those things people truly use and are concerned with are things "ready-to-hand," while others are things "present-at-hand," namely although such things are at hand, they are not used or focused on, and hence have not formed true relations with human being. Heidegger calls things "ready-to-hand" an "involvement," and states: "The character of Being which belongs to the ready-to-hand manifests as just such an involvement. If something has an involvement, this implies letting it be involved *in something*."[7] This is to say, human being becomes involved with Nature in human being's factual existence, and Nature is an indispensable component of human being's factual existence. Nature is enveloped in "Dasein"; Nature is not outside of Dasein. This is the philosophical ground that contemporary ontology proposes for the unification of human being and Nature; it marks the transition from the "subject-object dichotomy" to "Dasein and the world" and the transition from epistemology to contemporary ontology. Just as the contemporary eco-critic Harold Fromm put it, "[t]hus, 'the problem of the environment,' which many people persist in viewing as a peripheral arabesque drawn around the 'important' concerns of human life, must ultimately be seen as a central philosophic and ontological question about the self-definition of contemporary humanity."[8]

"The ecological conception of existence" also includes the important content of the transition from anthropocentrism to the ecological whole. "Anthropocentrism" became the dominant thought in the philosophical realm since the industrial revolution. For a time, such thoughts as "man legislates for Nature, "man is the center

[5]Heidegger [4], p. 70.

[6]Ibid., pp. 65–66.

[7]Ibid., p. 98.

[8]Fromm [5], p. 38.

of the universe" and "man is the noblest" became the theoretical ideas suppressing all else, which is one of the important causes behind humanity's endless demands on Nature and the increasing seriousness of ecological problems. "The ecological conception of existence" is the sublimation of this "anthropocentrism," and simultaneously, the advocation of the contemporary "ecological holism." The contemporary ecocritic William Ruekert pointed out, "[i]n ecology, man's tragic flaw is his anthropocentric as opposed to biocentric vision, and his compulsion to conquer, harmonise, domesticate, violate and exploit every natural thing." (*Ecocriticism: A Study of Environmental Issues in Literature*) He calls this human "compulsion" "the ecological nightmare."[9] The most powerful ground for breaking out of this ecological nightmare of anthropocentrism and moving into "ecological holism" was the positing of "biosphere" thought, which tells us that species on earth constitute a completely integrated system; between species and species, between species and the earth and air are inseparable connections making up a balanced organic whole of circulating energy, and the destruction of this whole implies the birth of the ecological crisis threatening human survival. From the famous Rachel Carson to Tang Yinbi all the way to Barry Commoner, profound expositions of this "biosphere" thought have been conducted. Commoner pointed out in *The Closing Circle* that "Any living thing that hopes to live on the earth must fit into the ecosphere or perish. The environmental crisis is a sign that the finely sculptured fit between life and its surroundings has begun to corrode. As the links between one living thing and another, and between all of them and their surroundings, begin to break down, the dynamic interactions that sustain the whole have begun to falter and, in some places, stop."[10] It is knowable from this that the connections between one organism and another as well as the connections between all organisms and the things surrounding them are the basic contents of ecological wholeness. The destruction of this ecological whole is the cause generating the ecological crisis, which will necessarily threaten human survival.

As Griffin understands it, the ecological conception of existence also necessarily involves the partial "reenchantment" of Nature, which reflects the transition in contemporary philosophy and aesthetics from the total disenchantment of Nature to the partial reenchantment of Nature. By "enchantment" we refer to the sense of mystery surrounding Nature and the human feeling of respect for and fear of Nature in distantly ancient times which came about due to the underdevelopment of technology. Since the industrial revolution, the development of technology greatly strengthened the human capacity to know Nature and reform Nature, so humankind felt they could know everything about Nature, which is precisely what Max Weber meant by the human "disenchantment" with Nature due to instrumental reason. This disenchantment became one of the important causes behind humanity's reckless plundering of Nature that has brought about the ever serious ecological crisis. As Griffin put it, "The disenchantment of nature has thereby resulted in a more acquisitive type of human being. Human beings for whom the very meaning of life is tied to acquisitions are more prone to want more than they need, and to resort to violence to get

[9]Ruckert [6], p. 113.

[10]Commoner [7], p. 7.

it." (*Spirituality and Society*, p. 148) "[t]he self-destructive path on which the world has been set, as it has been increasingly dominated by the paradigm of the modern West, can only be altered by developing a new worldview and thereby a new ethic. This will require 'the reenchantment of the world.' A postmodern paradigm will help bring about just that."[11] Of course, this reenchantment of the world will by no means restore the pre-enlightenment period of humankind nor is it the all-around negation of the industrial revolution; it is rather the contemporary age's partial "reenchantment" of Nature after the great achievements of the industrial revolution, namely it partially restores the sanctity, mysteriousness and potential aesthetic quality of Nature. We only gain the possibility of establishing an ecological humanism that unifies humans and Nature as well as humanism and ecologism on the aforementioned theoretical foundation of "the ecological conception of existence," which then becomes the philosophical ground and cultural standpoint of contemporary ecological aesthetics. Precisely because of this, we call contemporary ecological aesthetics the contemporary eco-ontological conception of aesthetics.

Previously we mentioned that the eco-ontological relationship of being in the world between "Dasein and the world" resolved the problem of the unity of ecology and humanism, but how is the unity between ecology and aesthetics achieved? Why do we say that eco-ontological philosophy is simultaneously an aesthetics? In ontological philosophy, the meaning of beauty already differs greatly from the theory of beauty as "the perfection of sensible knowledge" in traditional epistemological aesthetics—in it, the meaning of beauty is no longer fundamentally distinguished from the meaning of truth and being, but is intimately tied to them. So-called beauty is the opening up of being and the unconcealing of Truth. Heidegger remarked that: "[b]eauty is one way in which truth occurs as unconcealedness." He further cites an example to explain: "Truth happens in the temple's standing where it is … what is as a whole is brought into unconcealedness and held therein…."[12] Here, what Heidegger is speaking of is another type of beauty differing from the ordinary one of "proportion, symmetry and harmony." This beauty is not the beauty of knowing, it is not the beauty of accurately describing and expressing things, but is rather the beauty of an "ecological existence;" it is the opening up of Truth, the manifestation of Being. Heidegger explains this unique beauty of ecological being with the example of the ancient Greek temple. He states that "standing there, the building rests on rocky ground. This resting of the work draws up out of the rock the mystery [*das Dunkle*] of the rock's clumsy yet spontaneous support. Standing there, the building holds its ground against the storm raging above it and so first makes the storm itself manifest in its violence. The luster and gleam of the stone, though itself apparently glowing only by the grace of the sun, yet first brings to light the light of the day, the breadth of the sky, the darkness of the night. The temple's firm towering makes visible the invisible space of air. The steadfastness of the work contrasts with the surge of the surf, and its own repose brings out the raging of the sea. Tree and grass, eagle and bull, snake and cricket first enter into their distinctive shapes and thus

[11] Ray Griffin [8], pp. 221, 222.

[12] Heidegger [9], p. 276.

come to appear as what they are... We call this ground earth... [it] is that whence the arising brings back and shelters everything that arises without violation. In the things that arise, earth is present as the sheltering agent." It is on the "earth" where the temple enfolds everything including human beings into a unity, from which unfolds act after act of the living drama of humankind, out of which Truth opens up and being is manifested. "It is the temple-work that first fits together and at the same time gathers around itself the unity of those paths and relations in which birth and death, disaster and blessing, victory and disgrace, endurance and decline, acquire the shape of destiny for human being. The all governing expanse of this open relational context is the world of this historical people."[13] Because of this, what the unique "world" of sky, earth, humans, all things and one thousand years of history the earth composes by erecting the temple reveals in its opening up is the ups and downs of one thousand years of Greek life, the rising and falling history of an entire people along with their extraordinary fate. This is precisely the manifestation of Truth, the beauty of the opening up of being; it is the beauty of "ecological being." But if you divorce the temple from the rock it had been erected on for one thousand years, if you divorce it from the world it breathed for so long, and place it in a museum or exhibition hall, the beauty of this ecological being will cease to exist. Here, in this structure of being-in-the-world known as "Dasien and the world," wholeness, humanity and aesthetics may find unity. Clearly, this structure of being-in-the-world becomes the key and secret of eco-ontological aesthetics.

Actually, the law of beauty that Marx expounded in his famous *Economic and Philosophic Manuscripts of 1844* is precisely the unity of man and Nature as well as the unity of humanism, ecologism and aestheticism, because the laws of beauty involve the problem of the intrinsic unity of three levels. First is the "intrinsic measure" which mainly concerns human needs and falls under the scope of humanism; the second is the "the measure of things" which mainly concerns the needs of species and falls under the scope of ecology, and the unity of the two is "the law of beauty" and falls under the scope of aesthetics. This is actually The unity of humanism, ecologism and aestheticism, which envelopes strong implications of ecological aesthetics.

10.3 The Ancient Chinese Wisdom of Eco-ontological Aesthetics

There is some compatibility between the ecological ontology upon which the establishment of ecological aesthetics depends and the ancient Chinese classical wisdom of "Nature and humanity integrating into one." The ancient Chinese theory of "Nature and humanity integrating into one" has complex implications. Here, we are mainly referring to the explanation of "Nature and humanity integrating into one" found in pre-Qin classics. In the view of the ancient Chinese, human being's relationship to the world is not an epistemological relation of knowing and reflecting,

[13] Heidegger [9], p. 262.

but is generally an axiological relation whereby humans gain auspicious health in the world at "the verge between Nature and humanity." "The verge between Nature and humanity" is the human world, and "Nature and humanity integrating into one" is the pursuit of human beings, while auspicious health is the goal of life. This is what is meant in *Doctrine of the Mean* by "striking the mean harmony" (*zhi zhonghe* 致中和) namely "reaching the mean harmony, heaven and earth are properly positioned, and all things grow." Here, humans, the heavens, the earth and all beings constitute a unified whole, become harmoniously integrated and inseparable, on the basis of which alone beings grow prosperously, humans gain auspicious health, and survive happily. It is precisely in this relationship of being-in-the-world "at the verge between Nature and humanity" in the harmony of Nature and human beings where ancient China's classical wisdom of ecological aesthetics emerged, "production and reproduction is called change," "follow the heavens and await the seasons." It is also precisely in this relationship of being-in-the-world "at the verge between Nature and humanity" in the harmony of Nature and human beings where the beauty of the four virtues of "beginning, expanding, benefiting and rectifying" took shape out of what the *Commentary on the Book of Changes* called "holding together the great harmony," and "interconnecting patterns in the Yellow Center with bodies occupying their proper positions." In *About the Character Kun*, it is remarked: "the *junzi* interconnects the patterns in the Yellow Center with bodies occupying their proper positions, beauty is in the center, it flows out into his four limbs and grows out into his occupations, which is the perfection of beauty." The crucial point here is "bodies occupying their proper positions," which refers to Yin and Yang, Qian and Kun, each being in their proper positions and fulfilling their respective roles such that "heaven and earth cross into one another and all beings intercross, thereby reaching the state of "the perfection of beauty." What is this state of beauty? The *Zhou Book of Changes*, differing from the West, uses the "four virtues" of "beginning, expanding, benefiting and rectifying" to summarize it. *About the Character Qian* states: "beginning is the growing of good; expanding is the converging of good; benefiting is the harmonizing of what is right; rectifying is the working of affairs." Here, goodness, harmony and working are all descriptions of human being's better state of survival; they describe an Eastern beauty of ecological being in classical form.

10.4 Environmental Aesthetics Through the Perspective of Ecological Ontology

To study ecological aesthetics it is necessary to bring up the problem of the relationship and distinction between ecological aesthetics and environmental aesthetics. At present, considering the international sphere, what draws widespread attention is not so much ecological aesthetics as much as it is environmental aesthetics. Why do we still insist on "ecological aesthetics" instead of calling it "environmental aesthetics"? This is determined by the most basic category of ecological aesthetics, "the eco-ontological conception of aesthetics," because "the eco-ontological conception of aesthetics" defines human being's relationship to all natural beings through the ontological perspective of being-in-the-world, where humans and all natural beings interjoin into one whole, forming "the world" in relation to which the being of humans gradually expands and reaches the state of clearing through the long river of time.

The environment in "environmental aesthetics" has different connotations than the "ecology" in ecological aesthetics. According to the dictionary, "the environment refers to the space surrounding human groups and the sum of all the natural and social factors influencing human production and human life".[14] The contemporary environment aesthetics theorist Yrjö Sepänmaa thinks that "the environment surrounds us (we as the observers are positioned in its center), and we use all of our senses to perceive it, act in it and exist inside of it".[15] From this we can see that the "environment" in environmental aesthetics is the space surrounding humans, humans exist inside of the environment, but are still dualistically opposed to the environment. Some environmental aesthetics theorists regard the "environment" in environmental aesthetics as "humanized Nature," which illustrates the inseparable relationship between environment and human being. To a certain degree, this is to ignore the intrinsic value possessed by Nature itself along with the fact of the world being co-constituted by humans and all natural beings, and unconsciously exhibits the tendencies of anthropocentrism. The famous contemporary American environmental aesthetics philosopher Arnold Berleant more or less sees the anthropocentric tendency implied by the "environment" in environmental aesthetics, and tries to correct it with a new interpretation. He asks, "'The' environment circumscribes environment and turns it into a bounded object like any other, including art objects, and that we can think of and deal with as if it were outside and independent of ourselves… It would beg the aesthetic question to assume that environment is separate from human involvement. Where, moreover, can we locate "the" environment? Where is "outside" in this case? Is it the landscape that surrounds me where I stand? Is it the world beyond my window? Outside the walls of my room and house? On the other side of the clothes I wear? Is environment the air I breathe? The food I eat?"[16] He insists on defining the environment from the perspective that "humans and the environment are one." We should say, Berleant's definition of the environment gets closer to "ecology." In terms of content, environmental aesthetics puts forth gardens, cities, farms and human dwelling spaces as the important objects of research, which in some sense has greater operability, but it still carries traces of "man-made scenery" and "anthropocentrism," while ecological aesthetics more thoroughly breaks through the dichotomized subject-object model of thought, anthropocentrism and epistemological aesthetics, and in the philosophical ontology of developing from epistemology to ontology it is more complete. Of course, it is rather complex inside of environmental aesthetics, there are environmental aesthetics that fail to complete break free of anthropocentrism, there are environmental aes-

[14] *Dictionary of Environmental Studies*, Science Press, 2003 edition.

[15] Sepanmaa [10], p. 237.

[16] Berleant [11], p. 6.

thetics that advocate the "ecocentrism" of the "total beauty of Nature," and there are environmental aesthetics philosophers like Arnold Berleant who move toward "ecological holism." Berleant's many expositions have become important resources for ecological aesthetics, but all environmental aesthetics regard Nature as their aesthetic object and seek human being's "happy dwelling" and "suitable dwelling." Because of this, environmental aesthetics is largely the allied force of ecological aesthetics. In the broader sense, our ecological aesthetics should also bring environmental aesthetics into itself.

References

1. [U.S.] Ray Griffin, D. (1998). *Postmodern spirituality and society,* translated by Wang Chengbing. Central Compiler Publishing House.
2. Marx and Engels. (1972). *Selected works*. People's Publishing House (《马克思恩格斯选集》,人民出版社1972年版).
3. Marx and Engels. (1979). *Collected works*. People's Publishing House.
4. [Germany] Heidegger, M. (2006). *Being and time* (revised translation), translated by Chen Jiaying, Wang Qingjie. Sanlian Bookstore.
5. [U.S.] Fromm, H. (1996). *From transcendence to degeneration: A road map*. Containing in *Ecological Criticism Reader*. University of Georgia Press.
6. [U.S.] Ruckert, W. (1996). *Literature and ecology: An experiment of ecological criticism,* Containing in *The Ecocriticism Reader*. University of Georgia Press.
7. [U.S.] Commoner, B. (1997). *The closing circle nature, man, and technology*, translated by Hou Wenhui. Jilin People's Publishing House.
8. [U.S.] Ray Griffin, D. (1998). *Peace and postmodern paradigm, postmodern spirituality and society*, translated by Wang Chengbing. Central Compiler Publishing House.
9. [Germany] Heidegger, M. (1996). *Selected works of Heidegger,* selected by Sun Zhouxing. Shanghai Sanlian Bookstore (孙周兴选编:《海德格尔选集》,上海三联书店1996年版).
10. [Finland] Sepanmaa, Y. (2006). *The beauty of environment,* translated by Wu Xiaoxi, Zhang Yi. Hunan Science and Technology Press.
11. [U.S.] Berleant, A. (2006). *The aesthetics of environment*, translated by Zhang Min, Zhou Yu. Hunan Science and Technology Press.

Chapter 11
The Implications of Ecological Aesthetics (Part II): The Object and Method of Ecological Aesthetics

11.1 Ecological Aesthetics' Object of Research—The Aesthetics of Ecosystems

The object of research in traditional aesthetics is art, which of course is due mainly to the influence of Hegel. Hegel called aesthetics "philosophy of art" in his *Aesthetics*, and his definition of beauty as the sensuous manifestation of the Idea was said with reference to art. Hegel believed that Nature, and specifically the earth and the inorganic world could not have any "Idea," so natural beauty is imperfect beauty. "The living beauty of nature is produced neither *for* nor *out* of *itself* as *beautiful* and for the sake of a beautiful appearance. The beauty of nature is beautiful only for another, i.e. for *us*, for the mind which apprehends beauty."[1] This is to say, Hegel believes nature does not have beauty independently, and it will only become beautiful when it dimly suggests it to someone's mind. The influence of this idea has persisted up till now such that aesthetics is still called "the philosophy of art" in many works and textbooks today. Looking at it now, this idea is obviously not whole-sided and is even incorrect. With regard to establishing the field of aesthetics, ecological aesthetics is a correction of this one-sided tendency to qualify aesthetics as "the philosophy of art."

So, could we say that ecological aesthetics posits "nature" as its own object of research? We believe that this formulation is missing something, because from the standpoint of the eco-ontological conception of aesthetics, human being and Nature are not diametrically opposed as subject and object, but rather compose a unified whole and singular world with all natural things. Nature also does not have substantive attributes, and there is no "beauty of nature" existing independently of human being. What we call beauty always exists in a unified whole composed of human being and all natural beings; it exists in the long river of time of symbiosis between humans and nature; it exists in the process of gradually manifesting and opening up Being and Truth. Therefore, we cannot simply insist that ecological aesthetics' object of

[1] Hegel [1], (Vol. 1), p. 160.

F. Zeng, *Introduction to Ecological Aesthetics*,
https://doi.org/10.1007/978-981-13-8984-9_11

research is "nature," but should rather regard its object of research as the whole "ecosystem" that includes both all natural things and simultaneously human beings.

"Ernst Haeckel, the German Darwinian, coined the term "oecologie" in 1866. The modern spelling appeared in the 1890s along with the first sophisticated ecological monographs by European botanists. By that time ecology meant the study of how organisms (of any sort) interacted with each other and with their total environment. From its beginnings, ecology concerned communities, systems, wholes".[2] In 1973, Arne Naess applied ecological theory to human society and sphere of ethics, positing "deep ecology." As Naess put it, "[e]cology, as a science, does not ask what kind of a society would be the best for maintaining a particular ecosystem—that is considered a question for value theory, for politics, for ethics." issues, such as natural view and value view; political and ethical issues.... But from the point of view of deep ecology, are we able to satisfy such doubts raised by the fundamental needs of humanity as the right to love, security and proximity to nature, and whilst expressing such doubts, are we also calling into question the fundamental role of society? We seek a type of society, education and religion that as a whole is beneficial to all life on the planet and therefore we are probing further how to realize this necessary transformation, a task that we must complete."[3] "Ecology" as a phenomenon entered society and the social sphere of emotional value judgments from the field of natural science beginning with Arne Naess, which made philosophy of ecology, ecological ethics and ecological aesthetics come into being, and implications like "value," "equality," "justice" and "beauty" were added onto "ecology" beyond the implications of "wholeness" and "systematicity." The object of research into ecological aesthetics is precisely the aesthetic implications of ecosystems. Such aesthetic implications are precisely the manifestation of Being and the opening up of Truth in the four-fold game of "sky, earth, gods and mortals."

Many ecological theorists made proofs of this extremely important idea of ecosystems. What has greatly enlightened us is the famous idea of "land ethic" along the related idea of "ecological community" proposed by the American Leopold in Sand County Almanac. He argues from the perspective of chains of life that soil—plants—animals—humans compose a food pyramid and system of life. He states, "Each successive layer depends on those below it for food and often for other servicès, and each in turn furnishes food and services to those above. Proceeding upward, each successive layer decreases in numerical abundance." He believes, "a land ethic changes the role of Homo sapiens from conqueror of the land-community- to plain member and citizen of it. It implies respect for his fellow=members, and also respect for the community as such." For this he proposed the famous conception of ecological aesthetics, which starts with this ecosystem or community and envelops ecological ethics: "A thing is right when it tends to preserve the integrity, stability, and beauty of the biotic community. It is wrong when it tends otherwise."[4]

[2]Nash [2], pp. 66–67.

[3]Quoted from Yi [3], p. 25.

[4]Leopold [4], pp. 194, 204, 213.

(1) The beauty of ecosystems envelops nature but is not "the perfect beauty of nature" and thus draws a boundary with "ecocentrism"

The important distinction between the beauty of ecosystems and the beauty studied by the aesthetics of traditional rationalism is that it envelops natural factors and affirms the aesthetic value unique to nature. In his famous Philosophy Gone Wild, the ecological ethicist Rolston clearly asserts that nature concretely possesses aesthetic value. He states, "We value the Landscape Arch of the Canyonlands for the same reason that we value the Winged Victory of Samothrace; both have grace. Every admirer of the Tetons or of a columbine admits aesthetic value in nature, and the photographs in Audubon or National Wildlife bring out well this natural aesthetic value."[5] This is to say that in ecological aesthetics the aesthetic value of nature is self-evident. Yet, nature still cannot become an aesthetic object by itself, but rather must rely on human participation. Ecological aesthetics is beauty in relation; it is beauty in an ecosystem. Ecological aesthetics draws a clear boundary between itself and "the perfect beauty of nature" in ecocentrism with this special aesthetic object known as the beauty of ecosystems. As everyone knows, "ecocentrists" brought up the viewpoint of positing nature and moreover all of nature as the object of aesthetic research, and this is fully popular viewpoint among current environmental aesthetics philosophers. Ronald Hepburn, the important starter of environmental aesthetics attacked analytic aesthetics for neglecting the aesthetic beauty of nature while coming up with the concept of natural beauty in that opening article he published in 1966, Contemporary Aesthetics and the Neglect of Natural Beauty. This shows that in his mind, nature as a substance could become a stand-alone object of aesthetic appreciation. In addition, he also raised the problem of the subject's participation in that article. But this participation only referred to the all-round participation of human senses in contradistinction from traditional observational aesthetics, which still fails to break free of the pitfalls of the subject-object dichotomy. The research object that we propose for ecological aesthetics—the beauty of ecosystems dissolves the subject-object dichotomy and does not affirm the independent existence of natural beauty. One of the founders of contemporary Western environmental aesthetics—the Canadian philosopher of aesthetics, Allen Carlson, proposed the famous "perfect aesthetics" thesis that "nature is maximally aesthetically valuable." He argues: "the natural environment, in so far as it is untouched by humans, has mainly positive aesthetic qualities; it is graceful, delicate, intense, unified, orderly, not dull, bland, insipid, incoherent, chaotic. All virgin nature in short is essentially aesthetically good. The appropriate or correct aesthetic appreciation of the natural world is basically positive and negative aesthetic judgments have little or no place."[6] This is the famous "positive aesthetics" theory in contemporary environmental aesthetics represented by Carlson. Its reasoning goes as follows: first, independently standing nature is originally beautiful and only humans are the destroyers of natural beauty; second, any natural thing is valuable or contains value; third, virgin nature like wasteland has a primitive holistic beauty; fourth, even if not all natural things have the value

[5]Rolston [5], pp. 132–133.

[6]Carlson [6], p. 109.

of being beautiful, from the perspective of the whole and connections, nature is perfectly beautiful. For instance, a forest that had been set on fire represents an abridged ecosystem; fifth, people hold different attitudes toward the aesthetic appreciation of nature and art respectively, the former is positive and the latter is critical. In sum, this is a rather self-conscious ecocentric standpoint.

A contemporary Chinese philosopher of aesthetics explained the problem of "nature is maximally aesthetically beautiful" from another perspective, namely made arguments about nature being maximally beautiful from the perspective of uniqueness. He states: "the reason why nature is maximally beautiful is not because every natural thing fits a standard of beauty, but because every natural thing is unique and singular, every natural thing is an unrepeatable and irreplaceable thing in itself. In this sense, natural things are incomparable and cannot be graded, in the sense that they each strive to become themselves, they are equal, but this equality of theirs is expressed precisely in their inequality."[7] The origin of the "nature is maximally beautiful" theory, we believe, should be traced back to the famous biologist Darwin and his On the Origin of Species published in 1859. In the section on "Sexual Selection" in the fourth chapter of On the Origin of Species, "Natural Selection: Survival of the Fittest," he wrote on the problem of the original beauty of nature. When he describes the male striped peacock displays with his best posture, gorgeous plumage and strange antics to attract the female peacock, he writes: "[i]f man can in a short time give elegant carriage and beauty to his bantams, according to his standard of beauty, I can see no good reason to doubt that female birds, by selecting, during thousands of generations, the most melodious or beautiful males, according to their standard of beauty, might produce a marked effect."[8] In China the theory thought to be closest to that of the nature is maximally beautiful theory is the theory put forward by the famous aesthetic philosopher of the older generation, Cai Yi, namely "the example is beautiful." When explaining "the essence of beauty" in his New Aesthetics, Cai Yi remarks, "We believe beauty is objective, not subjective; what makes the beautiful thing beautiful is found in this thing itself and not in the effect of our consciousness." He again states, "We believe beautiful things are exemplary things, namely that among particulars which manifests the general; what is beautiful is the exemplariness of something, namely the general among particulars manifesting the kind."[9]

Since the "nature is maximally beautiful" theory originates from Darwin's On the Origin of Species, we may predict on this basis where the costs and benefits of this theory are. By "benefit" we mean that this theory fully affirms that nature possesses some attributes of value, including aesthetic value, and criticizes the viewpoint that the aesthetic appreciation of nature is entirely "humanized nature." But the costs of this "nature is maximally beautiful" theory clearly protrude into the foreground, most important of which is that it totally expresses the tendencies of "ecocentrism," and absolutizes the value of nature including its aesthetic value, divorcing "the beauty of

[7]Feng [7], pp. 196–197.
[8]Darwin [8], p. 104.
[9]Yi [9], p. 235.

nature" from the "ecosystem" in which nature and humans are closely interconnected, and thereby going down the wrong path of opposing ecology and humanity.

Since we hold to the philosophy and aesthetic viewpoint of "eco-ontology," we must understand and explain the aesthetic value of Nature from within the relationship of being-in-the-world. The facts demonstrate that like what is true and what is good, what is beautiful is not a substance, but is a relational being; it is a process of unfolding being and showing truth in the structure of being-in-the-world, in the ecosystem of nature and humans. Divorced from human being's participation, divorced from the structure of being-in-the-world, divorced from the ecosystem linking humans and nature so tightly together, the aesthetic values of nature will cease to exist. Even the uniqueness of what is natural would not shine forth with the luster of beauty. Even Kant affirms the individuality of the beautiful form, but he still links this unique individuality with its commonality, and insists that its beauty is a bridge growing from nature to the human being, which finally leads to "beauty is a symbol of morality." Therefore, the "nature is maximally beautiful" thesis has no place in Kant's aesthetics.

In summary, beauty gradually grows and manifests in the structure of being-in-the-world, in the "ecosystem" tightly binding human beings to all natural beings, which is not fully explainable in ecocentrism's "nature is maximally beautiful" theory.

(2) The beauty of ecosystems envelops human factors, but differs from beauty in the "empathy" theory, "humanized nature" theory and the "picturesque landscape" theory, and thereby draws a clear boundary with anthropocentrism

The beauty of ecosystems according to ecological aesthetics envelops human factors, which is determined by "the eco-phenomenological method" that ecological aesthetics follows. Eco-phenomenology particularly emphasizes the composing role of the human subject in the process of appreciating beauty, and insists on the explanatory and active nature of "Dasein" in the process of appreciating beauty. As Rolston put it, "There are two sorts of aesthetic qualities: aesthetic capacities, capacities for experience that are only in beholders, and aesthetic properties, which lie objectively in natural things."[10] He also states, "There is aesthetic ignition when humans arrive, the aesthetics emerges relationally with the appearance of the subject-generator."[11] But the beauty of ecosystems also differs from the beauty understood from the perspective of "empathy" theory, "humanized nature" theory and "picturesque landscape" theory, which over-emphasize the role of humans and hence move to anthropocentrism.

"Empathy theory" was first brought up by the German psychologist, Theodor Lipps at the end of the 19th century and beginning of 20th century. He proposed from the perspective of psychology that all aesthetic activity is the result of humans displacing their own emotions and will onto the object. He states that "all aesthetic experiences are all due to us transferring things we have personally experienced, our feelings of force, our efforts, our will, active or passive feelings into something

[10][U.S.] Holmes Rolston: *From Beauty to Duty: Aesthetics of Nature and Environmental Ethics:* Berleant [10], p. 158.

[11]Ibid., p. 156.

external to us, into an event happening in this something or with it".[12] Kant also used empathic transference when explaining the sublime, insisting that the sublime was the result of the subject surreptitiously transferring his own "sense of the sublime" onto the object. The facts show, that this empathy theory completely denies the aesthetic attributes that nature possesses in itself in the aesthetic process of appreciating beauty, which does not fit the facts; it is an expression of both "anthropocentrism" and "idealism."

In the big debates occurring in China during the 1960s and 1980s, the problem of "the beauty of nature" became one of the hot topics of debate. At the time, Li Zehou proposed a very famous viewpoint that the beauty of nature is "humanized nature." As he put it, "natural objects only become beautiful once they have become 'humanized nature," once they 'objectively reveal the richness of human being's essence' in natural objects." When discussing the beauty of the sun, he argued that the beauty of the sun is not in its natural attributes, but in its social attributes. He states, "[o]bviously, the sun as a delightfully bright, beautiful object is found in its own objective social nature, this objective social relation, objective social role and status it has with human life. It is these that cause in people the intense joy and beautiful feeling towards the sun. The sun's objective social attribute is the main condition making up its beauty, and although its natural attributes of emitting heat and light are necessary conditions, they are also secondary conditions."[13] Here, Li Zehou applies Marx's viewpoint about human labor being "humanized nature" in his famous Economic and Philosophic Manuscripts of 1844. But what Marx was clearly talking about was productive labor and not that of appreciating beauty. Marx's viewpoint about aesthetics in the same work combining "the measure of species" and "the intrinsic measure" already clearly tells us that Marx thought aesthetics not only involves human being's "intrinsic measure" but also involves nature's "measure of species." Marx was opposed to "anthropocentrism." "The objectification of man's essential forces" cannot truly reflect Marx's view of aesthetics, and "humanizing nature" also cannot truly reflect Marx's view of the aesthetic beauty of nature.

In the 1960s and 1970s, there emerged a "picturesque landscape" theory of explaining the beauty of nature against the backdrop of Western culture in the process of the Western modern world producing environmental aesthetics. The picturesque landscape theory observes nature with an artistic gaze, and views nature as a picturesque scene. This is the scenic model of environmental aesthetics that Carlson and Sepänmaa brought up. As Yrjö Sepänmaa understands it the starting point is landscape painting or photography, the landscapes we see have frames like paintings. Selection and framing make the landscape. Sepänmaa points out that many historic landscapes made for human appreciation were managed in artistic fashion, including tourism routes, tiny pathways, resting places, route markers, guidebooks, observation towers are all arranged and set up beforehand. Sepänmaa does not agree with this picturesque landscape theory nor the management model associated with

[12] *Edited by* Qi [11] (Vol. 2), p. 841.

[13] Zehou [12], pp. 25, 88.

it, thinking that its fundamental flaw is that "nature is not seen as a whole".[14] In his view, there are five major distinctions between environments and artistic works; the artistic work is the product of human labor, while the environment is pre-given; art is accepted inside of a conventional framework, while the environment is not; artistic works are created for aesthetics pleasure, while the aesthetic quality of the environment is a biproduct; artistic works are fictional, while the environment is non-fictional; artistic products are elliptical, while the environment is just itself. So, she does not approve of the picturesque landscape theory, because it still does not break free of the traditional theoretical constraints of traditional aesthetics as the philosophy of art, and it even more so flatly denies the aesthetic quality of nature itself, which shows a marked anthropocentric tendency. We also naturally disapprove of this picturesque landscape theory for the main reason that this theory still appreciates natural ecology with the "anthropocentric" gaze, and appreciates it as so many scenic paintings presented before man, which actually betrays this implication of aesthetics that the object of research into ecological aesthetics is "ecosystems."

Above, we explained the specific implications of the aesthetic beauty of ecosystems differing from both traditional anthropocentrism and ecocentrism through comparing the beauty of ecosystems with beauty according to the "nature is maximally beautiful" theory, "empathy" theory, "humanized nature" theory and picturesque landscape theory, and thereby distinguished the special significance and value of the object of research into ecological aesthetics. It is knowable from this that the beauty of ecosystems is neither the beauty of pure nature nor the beauty of pure human "empathy, nor "humanized" beauty; it is rather the beauty of the ecosystem and ecological whole in which humans and nature and inseparably connected.

11.2 The Research Method of Ecological Aesthetics—Eco-phenomenology

The basic category of ecological aesthetics is the eco-ontological conception of aesthetics, whose main research method is that of eco-phenomenology. Just as Heidegger said, "Only as phenomenology, is ontology possible."[15] The method of eco-phenomenology is engaging "eidetic intuition" through the "suspension" of material and mental substance, "to the thing itself." By applying the phenomenological method to the field of ecological philosophy and ecological aesthetics, it becomes eco-phenomenology, and the earliest practitioner of eco-phenomenology was Heidegger, who used the phenomenological method to demonstrate the model of the human's being-in-the-world as early as 1927 in his famous work Being and Time. But the formal positing of eco-phenomenology was a later affair. In March of 2003, the German Philosopher Ulrich Melle did the report "Ecophenomenology" in Würzberg at the annual Phenomenology Conference in Germany, where he posed the ques-

[14]Sepanmaa [13], pp. 61, 62.

[15]Heidegger [14], p. 42.

tion of what ecophenomenology is, arguing that ecophenomenology is the attempt to use phenomenology to enrich ecological philosophy that up till now has mainly been reached using the analytic method.[16] As to the concrete implications of ecophenomenology, we will try to make the following several summarizations:

First, abandoning the subject-object dichotomy of instrumental reason and the way of thinking by opposing humans and nature, and "suspending" traditional anthropocentric ideas and the desire to over exploit nature. "In comparison with anthropocentrism which is fundamentally responsible for human plans of self-perfection and world-perfection, ecophenomenology is less likely to allow itself to be established on the dualistic ontological foundation of dichotomizing nature and spirit".[17]

Second, return to things themselves, and first of all returning to the natural foundation of the human spirit in the search for the natural essence of the human spirit and human existence. "For eco-phenomenology, the key to the problem is in further determining the natural foundation of this spirit".[18]

Third, turning around the purely instrumental and arithmetical way that humans handle nature toward an inter-subjective way of interacting with nature in the form of equal dialogue. On the path of eco-phenomenology, "people try to recall and concretely describe another way of experiencing nature and attempt to show that the purely instrumental and calculative way that humans handle nature is a warping of our experiential possibilities and a debasing of our experiential world".[19]

Fourth, eco-phenomenology is only possible under the premise of appropriately affirming the "intrinsic value" of nature. "Only when nature possesses an intrinsic aspect of inexhaustible specifications and riddling self-regulation, that respect for and feeling of awe at non-human nature will only come about and hence nature will only become the object of our care and concern by virtue of itself when the otherness and strangeness of nature possesses an immeasurable characteristic".[20]

Fifth, the philosophical affirmation of the intrinsic value of nature necessarily results in the critique and abandonment of the disenchantment of nature and the mechanistic worldview. As Melle insists, "the affirmation of the intrinsic value of nature is first of all a critique of that disenchantment of nature that occurs through modern natural science and technology".[21]

Sixth, the positing and development of eco-phenomenology could also result in connecting it with deep ecology's thought of "the ecological self." As Muller understands it, "the human other and non-human other are my bigger social self and ecological self, because of which my own self-realization is closely, inseparably and dependently bound to the self-realization of all others, no one is rescued until we are all rescued".[22]

---

[16]Mailer [15].

[17]Ibid.

[18]Ibid.

[19]Ibid.

[20]Ibid.

[21]Mailer [15].

[22]Ibid.

We can only transcend greed toward the aesthetic state of equal dialogue and symbiotic coexistence with all natural beings by following this method of "eco-phenomenology." The ancient Chinese Daoist practices of "purifying the affective mind" (xin zhai 心斋) and "sitting in forgetfulness" (zuowang 坐忘), namely "slacking limbs and body, dismissing smarts and intelligence, distancing from physical form and doing away with knowledge, merging into the greater consistency" (Zhuangzi, 6.9), and the Chan Buddhist method of "suspending" material desire and benevolently treating nature, namely the sixth patriarch of Chan Buddhism, Huineng's lyric "the Bodhi tree is originally not a tree, and the shining mirror is also not a mirror. Originally there is nothing, why stir up dirt, etc., could all be seen as the classical form of eco-phenomenology, and could entirely combine with the construction of contemporary ecology in the effort to make this theoretical method from the West native to the Chinese people.

References

1. [Germany] Hegel. (1979). *Aesthetics*, translated by Zhu Guangqian. Commercial Press.
2. [U.S.] Nash, R. F. (1999). *The rights of nature*, translated by Yang Tongjin. Qingdao Press.
3. Yi, L. (2001). *Deep ecological thought research*. Tsinghua University Press (雷毅:《深层生态学研究》,清华大学出版社2001年版).
4. [U.S.] Leopold, A. (1997). *A Sand County Almanac,* translated by Hou Wenhui.
5. [U.S.] Rolston, H. (2000). *Philosophy gone wild*, translated by Liu Er, Ye Ping. Jilin People's Publishing House.
6. [Canada] Carlson, A. (2005). *Aesthetics and the environment—The appreciation of nature, art and architecture*, translated by Yang Ping. Sichuan People's Publishing House.
7. Feng, P. (2005). *Perfect nature: The philosophical basis of contemporary environmental aesthetics*. Peking University Press.
8. [English] Darwin, C. (1995). *On the origin of species*, translated by Zhou Jianren. Commercial Press.
9. Yi, C. (2002). *Cai Yi's collected works* (1) China Federation of Literary and Art Publishing House.
10. [U.S.] Berleant, A. (2007). *Environment and the arts: A multidimensional perspective on environmental aesthetics*, translated by Liu Yuedi et al. Chongqing Publishing House.
11. Qi, M. (1987). *Selected materials of western aesthetic history*. Shanghai People's Publishing House.
12. Zehou, L. (1980). *The collection of aesthetics*. Shanghai Literature and Art Publishing House.
13. [Finland] Sepanmaa, Y. (2006) *The beauty of environment*, translated by Wu Xiaoxi, Zhang Yi. Hunan Science and Technology Press.
14. [Germany] Heidegger, M. (2006). *Being and time* (revised translation), translated by Chen Jiaying, Wang Qingjie. Sanlian Bookstore.
15. [Germany] Mailer, U. (2004). *Ecological phenomenology*, translated by Ke Xiaogang, Containing in *World Philosophy*, No. 4.

Chapter 12
The Implications of Ecological Aesthetics (Part III): The Basic Categories of Ecological Aesthetics

12.1 The Theory of Natural Disposition in Ecological Aesthetics

An important reason why ecological aesthetics is well-founded is because it largely reflects human being's natural disposition to appreciate the beauty of ecology. Human being's affinity with and aesthetic appreciation of natural ecology are important expressions of human being's natural disposition or human nature, which is an important implication of ecological aesthetics. As the contemporary ecocriticism theorist, Harold From, put it, the problem of ecology is "a philosophical and ontological problem related to the core of the self-definition of contemporary humans".[1]

(1) **Marxism's Exposition about the Ecological Essence of Human Being**

The classical authors of Marxism have discussed the issue of human being's eco-aesthetic nature. In *Economic and Philosophic Manuscripts of 1844*, Marx directly proposed the viewpoint that "man is directly a natural being," and profoundly explained the fact of coexistence between "humanized" nature and "objectified" man through man's practical activity. He states, "To say that man is a corporeal, living, real, sensuous, objective being full of natural vigor is to say that he has real, sensuous objects as the object of his being or of his life, or that he can only express his life in real, sensuous objects. To be objective, natural and sensuous, and at the same time to have object, nature and sense outside oneself, or oneself to be object, nature and sense for a third party, is one and the same thing."[2] In the first volume of *The German Ideology*, he went on to forcefully criticize Feuerbach's historical materialism, and specifically criticized Feuerbach for the practice of intentionally discussing the essence of man in isolation from man's relationship to nature, believing that man unfolds his own essence in relationship to nature. Marx figuratively alludes to the relationship between man and nature with that between fish and water.

[1]Fromm [1], p. 38.

[2]Marx and Engels [2], (Vol. 42) p. 168.

F. Zeng, *Introduction to Ecological Aesthetics*,
https://doi.org/10.1007/978-981-13-8984-9_12

He states, "The "essence" of the fish is its "being," water—to go no further than this one proposition. The "essence" of the freshwater fish is the water of a river. But the latter ceases to be the "essence" of the fish and is no longer a suitable medium of existence as soon as the river is made to serve industry, as soon as it is polluted by dyes and other waste products and navigated by steamboats, or as soon as its water is diverted into canals where simple drainage can deprive the fish of its medium of existence."[3] Marx here suitably alludes to the relationship between man and nature with that between fish and water, thereby explaining the essence of man: just as fish cannot live separately from clean water, how could it be imaginable that man could live separately from a natural ecological environment in good condition? Thus, the essence of man is closely connected to a good, natural, ecological environment. Engels specifically underlined the identity of man and nature in the famous *Dialectics of Nature*. He states, "But the more this happens, the more will men not only feel, but also know, their unity with nature, and thus the more impossible will become the senseless and antinatural idea of a contradiction between mind and matter, man and nature, soul and body…"[4] These materials suggest that the classical theorists of Marxism fully noticed and furthermore proved the original relationship of affinity between man and natural ecology.

(2) **The Concrete Implications of Man's Ecological Essence**

As everyone knows, grasping the essence of human being is the eternal theme of human spiritual life. Hence, the inscription carved into the wall of the temple of Delphi in ancient Greece, "know thyself." Ever since ancient times, there have been two drastically different orientations with respect to grasping the essence of human being. One is the epistemological orientation still popular in many fields today, whose highest aim is to know and grasp the abstract essence of man, in accordance with which there has emerged many formulations from man is a rational animal and man is a sensible animal to man is a political animal and the human essence is humanist "love."[5] The one-sidedness of such formulations is found in grasping the essence of human being in total isolation from the actual facts of life. No such abstract "essence" of human being has ever existed in the real living world. Ernst Cassirer attempted break through the limits of epistemology to think about the essence of human being from the functional perspective. He reduced the essence of human being to an animal that creates and uses signs. He states, "if there is any definition of the nature or "essence" of man, this definition can only be understood as a functional one, not a substantial one. We cannot define man by any inherent principle which constitutes his metaphysical essence-nor can we define him by any inborn faculty of instinct that may be ascertained by empirical observation. Man's outstanding

[3]Ibid., p. 369.

[4]Marx and Engels [2], (Vol. 3), p. 518.

[5]For example, Plato believes that people "divide" the idea; Aristotle believes that "people are political animals"; The British empirical philosophy categorizes the essence of human beings as "sensible" and "feeling"; Feuerbach believes that the essence of human beings is the "love" of humanism.

characteristic, his distinguishing mark, is not his metaphysical or physical nature-but his work. It is this work, it is the system of human activities, which defines and determines the circle of "humanity." Language, myth, religion, art, science, history are the constituents, the various sectors of this circle."[6] Cassirer defining the essence of man from the "functional" perspective of creating and using signs should be lauded as a break-through experiment, but it still does not fundamentally break through the constraints of essentialism, because the capacity to create and use signs is still an abstract description of the essence of man. In actuality, a lively and vitally active subject cannot be completely equated with an abstract subject who creates and uses signs. The former is much richer and much more concrete than the latter.

Opposed to the essentialist orientation of epistemology, the modern Western philosopher, Martin Heidegger, proposed an "ontological and phenomenological" method. He states, "[o]ntology and phenomenology are not two distinct philosophical disciplines among others."[7] To some degree, this is a development of ontological phenomenology, which breaks through old essentialist convention of the subject-object dichotomy in epistemology, and adopts the method of "suspending" every substantial content in order to "return to the thing itself" and hence directly face "being" itself. In this philosophy and worldview, what human being faces is neither a sensible, rational, political, loving nor symbolic substance, but is rather human "existence" itself; it is not the opposition between nature and society, but rather the primordial integration between life and nature. Heidegger's understanding and grasp of the essence of man clearly has an element of this-worldliness and provides rich resources of thought for contemporary eco-ontological philosophy and aesthetics. Heidegger insisted that "[o]ur investigation takes its orientation from Being-in-the-world—that basic state of Dasein by which every mode of its Being gets co-determined."[8] The German philosopher, Wolfgang Welsch also asserted that the definition of human is humans of this world (humans sharing joys and sorrows with the world) and not humans of humankind (humans centered on humankind itself).[9] What he refers to by this world is that all humans are really living humans and not abstract beings. Such really living humans are at every moment inseparable from nature and the ecological environment; they are beings in nature and the ecological environment. There is some integrity to this grasp of the human essence, which is to say that there is no human being in which either sensibility and rationality or society and nature are divided in opposition; rather, all forms of life can only survive in ecosystems that mix them inseparably with all other natural beings. As Rolston pointed out, "our humanity is not absolutely "in" us, but is rather "in" our world dialogue. Our integrity rises from transaction with our opponent-partner and therefore requires a corresponding integrity."[10] There is also a humanistic element to this grasp of humanity, which is to say that true humanity is filled with humanistic feelings, and should not be cold

[6]Cassirer [3], pp. 96–97.

[7]Heidegger [4], p. 45.

[8]Heidegger [4], p. 136.

[9]Welsch [5].

[10]Rolston [6], pp. 92–93.

instrumental reason; what exists deep within humanity is precisely the contemporary Idea of being filled with humanistic feelings. As the first World Environmental Conference in 1972 asserted, there is only one earth, and humanity must show concern for earth, this tiny planet. In 1991, when internationals organizations signed "Protect the Earth—Sustainable Survival Strategy" at UNEP, they aimed at "conservation to keep human actions within the earth's carrying capacity, and development to 'enable people everywhere to enjoy long, healthy and fulfilling lives.'"

From the unique perspective of eco-ontological philosophy, we could summarize contemporary human being's ecological essence into three aspects. First, human being's ecological source. Humans come from nature, nature is the source of human life, and nature is also one of the most important insurers of the continuation of happy human life. This point is extremely important. For a long period of time, what people have insisted upon in idea is the difference between what is human and what is natural, while ignoring the similarity between them, which easily brought about opposition and divergence between the two in practice. As Engels put it, "[i]n particular, after the mighty advances of natural science in the present century, we are more and more getting to know, and hence to control, even the more remote natural consequences at least of our more ordinary productive activities. But the more this happens, the more will men not only feel, but also know, their unity with nature, and thus the more impossible will become the senseless and antinatural idea of a contradiction between mind and matter, man and nature, soul and body."[11]

Second, human being's ecological circularity. Human being's ecological essences envelops an important content, which is that human being is an essential link in the whole ecological chain. Human beings are linked cyclically with ecology, and an individual human being will lose the basic conditions of his own life in isolation from the ecological chain, and will therefor perish. Rachel Carson in *Silent Spring* concretely elucidates the ecological chain as the basic condition of life. She states: "Water must also be thought of in terms of the chains of life it supports—from the small-as-dust green cells of the drifting plant plankton, through the minute water fleas to the fishes that strain plankton from the water and are in turn eaten by other fishes or by birds, mink, raccoons—in an endless cyclic transfer of materials from life to life."[12] The ecological chain is the basic content of human being's ecological essence; in one respect it reflects the commonality and intimate relationship between humans and all natural beings. Humans and other beings are equally links in the ecological chain, and relatively equal, they are inseparable not even for a moment; in another respect, it also involves the aspect of difference between humans and all other natural beings, because humans and other natural beings separately exist in different links in the ecological chain with different positions and functions pertaining to them. For a long period of time, people have defined human being's essence through the difference between what is human and what is natural, while gravely overlooking the commonality and closeness of relation between humans and other natural beings, and industrial civilization, that practice of conquering and plundering

[11]Marx and Engels [2], (Vol. 3), p. 518.

[12]Carson [7], p. 39.

nature is intrinsically founded on this idea of production. As soon as we notice the commonality between what is human and what is natural in the chain of life, and define the essence of human being based on its basic principles, not only will that definition fit more with the essence of human being, but will also make human thought and human activity become more highly scientific.

Third, human being's ecological self-consciousness. As the only animal with reason in the ecological chain, human cannot solely consider their own survival while neglecting all natural beings as other animals do. Humans not only have to maintain their own survival well, but should even more so self-consciously maintain the good links in the ecological chain and sustain the normal existence of other life forms by relying on their own reason. Only in this way can humans ultimately maintain their own survival better. Rolston thinks that the true distinction between humans and non-human beings is that animals and plants only care about their own life, their own posterity and members of the same species, while humans can maintain other life forms and non-human beings with a broader mind. He states that humans are positioned at the top of the food chain and pyramid in biotic systems, and possesses perfection, but precisely for that reason, they show that a path of this perfection is "to care for the earth".[13] The new explanation of the human essence given by eco-ontology will necessarily have an important influence on contemporary human sciences including aesthetics and will provide a new philosophical foundation for these disciplines to readjust the ideas within them and the framework of such disciplines.

(3) **The Expression of Human Being's Ecological Essence in the Field of Humanities**

1. Expanding from Human Equality to Relative Equality between Humans and Nature

"Fairness" and "Equality" are the basic implications of the humanist spirit, and contemporary ecological theory advocates "ecological equality" and expands the humanist principles of "fairness and equality" to the sphere of nature. Some scholars criticize ecological theory on this basis for eliminating basic human rights of survival like eating, drinking, clothing and using, and for having a clearly anti-human hue because of this. This is actually a misunderstanding, because the "ecological equality" that contemporary ecological theory advocates is not absolute equality but relative equality, which is to say equality within the chain of life. What it means is, all beings in the chain of life, including humans, enjoy their own right to survive and develop in their own position in the chain of life, but at the same time should not over step such rights. The founder of deep ecology, Arne Naess spoke of the egalitarianism of the biosphere in principle, which is precisely what we mean here. He states: "To the ecological field-worker, the equal right to live and blossom is an intuitively clear and obvious value axiom. Its restriction to humans is an anthropocentrism with detrimental effects upon the life quality of humans themselves. The quality depends in part upon the deep pleasure and satisfaction we receive from close partnership

[13]Quoted from Mouchang [8], p. 5.

with other forms of life. The attempt to ignore our dependence and to establish a master-slave role has contributed to the alienation of man from himself."[14]

2. Expanding from Human Survival Rights to Environmental Rights

An important component of humanism is the human right to live and advocating that human life have the right to live and develop. Such rights of survival were limited to such aspects as life, work and political rights for a long time, but contemporary ecological theory expands this right to live to the human being's right to an environment, which is the new implication of the humanist spirit contained in contemporary ecological theory. In 1969, the National Environment Policy Act in the USA clearly stipulated that everyone should enjoy a healthy environment and at the same time have the responsibility to maintain and improve the environment. In 1970, the "Declaration of Tokyo" clearly proposed that everyone has the inalienable right to an environment decent for health and well-being and that the heritage that contemporary people pass on to the next generation should be a right to natural resources filled with natural beauty, and as a basic human right, it should established in the body of law. In 1972, the United Nations *Declaration of the United Nations on the Human Environment* asserts, "Man has the fundamental right to freedom, equality and adequate conditions of life, in an environment of a quality that permits a life of dignity and well-being, and he bears a solemn responsibility to protect and improve the environment for present and future generations." In 1973, in a draft written up in Vienna on natural resources and human rights in Europe, the right to a decent environment was affirmed as a new human right. Clearly, contemporary environmental rights includes the right to enjoy a decent environment and protect a decent environment. The former is for the sake of the contemporary age and humanity itself and the latter is for the sake of the next generation, other life forms and non-living matter, but both comprehensively encapsulate human environmental rights.

3. Expanding the value of the human being to the value of nature

Value always indicates the relationship of benefit and interest between objects and human being, and maintaining the value of the human being has always been an indispensable and important element of humanism. But for a long period of time, people have been rather neglectful of the value of nature; it would seem as if rivers, oceans, air and water exist naturally and do not have any value to speak of in themselves. Contemporary ecological theory broke through this notion, expanded the value of the human being to the broader sphere of nature and fully affirmed the major irreplaceable value that nature itself possesses. In 1992, at a lecture at the Chinese Academy of Social Sciences, Institute of Philosophy, Rolston summarized over 13 kinds of values of nature: the value of supporting life, economic value, scientific value, entertainment value, the value of genetic diversity, value for natural history and cultural history, cultural symbolic value, the value of character cultivation, medicinal value, dialectical value, the value of natural stability and openness, the value of respecting life, scientific and religious value, etc.[15] The confirmation of the value of nature has

[14]Quoted from Yi [9], p. 51.

[15]Quoted from Mouchang [8], pp. 66–68.

extremely important significance for further maintaining decent human life; it is a new development of the humanist spirit.

4. Expanding Care for Humanity into Care for Other Species

This is an extension of the humane spirit of human being. Humanism follows the humane spirit and sympathetic feeling of caring for humanity, especially caring for the weak; contemporary ecological theory expands this humane spirit and feeling of sympathy to other species, and forcefully advocates caring about other species in opposition to the inhumane behavior of destroying nature and torturing animals. In "Protect the Earth—Sustainable Survival Strategy" issued by the UN in 1992, a principle concerning environmental ethics is proposed, which states that human development ought not threaten the totality of nature and the survival of other species, and that people should similarly treat all animals, protect them from brutality, torture and unnecessary slaughter.

5. Expanding from the Present Concern for Humankind to the Ultimate Concern for the Future Fate of Humankind

Humanism has maintained an ultimate concern for the future fate of humans, but this did not involve the ecological dimension. Contemporary ecological theory includes the ecological dimension into the ultimate concern, giving it deeper and richer implications. Particularly, the theory of "sustainable development" proposed through the perspective of contemporary ecological civilization is the theory of the ultimate concern that begins with the long-term interests of humankind.

In summary, the philosophy—aesthetics and theory—about the ecological essence of man in eco-ontology has broken with traditional anthropocentrism, but is not reactive toward humanity; rather it is a much deeper and broader concern for the life and development of humans in the new era; it is the new humanist spirit of the new era that envelops the dimension of nature.

(4) **The Expression of the Ecological Essence in the Sphere of Aesthetics**

Historically speaking, China and the West have both had rich in-depth expositions of human being's essence to appreciate the beauty of ecology. Particularly in ancient China, prior peoples lived in agricultural society for a long period of time, and multiplied dwelling in the vast loess plateau. So, humanity's relationship to nature was the most important relationship that China's prior peoples encountered, from which took shape the ancient Chinese orientation of thought to "investigate the verge between nature and humanity, and make sense of the transformation between past and present. It is precisely this cultural context that gave birth to ancient China's aesthetic thought, which mainly consists in reflecting human being's essence to appreciate the beauty of ecology and life. *The Zhou Book of Changes* in Ancient China provided this philosophical and aesthetic thought that "producing life is what we call change." "Producing life is what we call change, becoming signs is what we call Qian, emulating patterns is what we call Kun, exhausting the numerology to know the future is what we call prognostication, making sense of changes is what we call the event, the incalculability of *yin* and *yang* is what we call mysterious"

(The *Zhou Book of Changes,* "Xici I"). This means that the fundamental point of changes is "producing life," that is the life and survival of humans and other beings under the vast background of "the relationship between nature and man." The *Zhou Book of Changes* furthermore states "the great virtue of heaven and earth is called life, the great treasure of the sage is called position" ("Xici II"), and views ecology, survival and life as the highest favor that Nature bestows upon humankind; according to which, treasuring the life of all things is seen as the highest standard of human behavior; the *Zhou Book of Changes* insists that the best state of life for humans and all things is a vital state of thriving growth and decent life characterized by "beginning, growing, expanding and rectifying." Reaching this state demands that heaven and earth, Qian and Kun, *yin* and yang can all correspond with the being and working of ecological laws. Zhou Book of Changes, the tai 泰 hexagram, "Commentary on the Decision," states: "Heaven and Earth intercross and all creatures interlink, high and low pass into one another and wills converge." The tai 泰 hexagram has Qian on top and Kun on bottom, Qian symbolizes Heaven and signifies male, Kun symbolizes Earth and signifies female, the energies of Heaven and Earth, *yin* and *yang* intercross and communicate, fostering the vital growth of all natural beings in the sky and on earth along with the fluid development of vitality. The *tai* hexagram is precisely the symbol of this harmonious relationship between man and nature. In the diametrically opposed *pi* 否 hexagram, Qian is on bottom and Kun is on top, the energies of Heaven and Earth, *yin* and *yang* do not affectively mix, "Heaven and Earth do not intercross and all beings fail to interlink, high and low do not cross into one another and the world under Heaven fails to consolidate," which can only lead to disaster in human society plight with nature.

The *Zhou Book of Changes* asserts that the natural workings of Heaven and Earth in accordance with ecological laws is a kind of "beauty." The judgment of line 3 of the *kun* 坤 trigram in the *Zhou Book of Changes* reads "Implies grace that may persist, sometimes doing things following the King's intention, unsuccess will end," and the judgment of line 5 reads "[w]earing a yellow skirt [the color of the earth] initiates auspiciousness." "Grace" and "yellow" are both [considered] beautiful, so the commentary on the *kun* trigram states: "*yin* 阴 has beauty, contain it by doing things following the King's intention, but do not dare succeed. The *Dao* of Earth is the way of the wife, the way of the servant. The *Dao* of Earth is unsuccessful and is compensated for with an end;" it also states: "The authoritative person (*junzi* 君子) in yellow links the patterns, each in the proper position of the body, beauty is found therein, flowing out to the four limbs, and executed in his affairs, it is the consummation of beauty." These are the two places where the *Commentary on the Zhou Book of Changes* mentions the character for beauty, *mei* 美, but to understand the *Zhou Book of Changes* on "beauty," we need to combine the judgments of the lines in the *kun* trigram, "*kun* 坤, beginning expansion, benefit from the loyalty of the female horse. The authoritative person has outgoing conditions, if he pushes forth he will get lost, if he follows he will gain mastery, beneficial. If in the Southwest, he will get a friend, if in the Northeast he will lose a friend. If he rests safely [where he is], it will be auspicious and right." Integrating these statements, "beauty" in *Zhou Book of Changes* firstly refers to the combination of *yin* and *yang*; the "grace" and "yellow skirt" corresponding to this "beauty" appear in the judgments of lines 3 and 5 of the *kun* hexagram, and these two lines have the broken *yin* line occupying the *yang* position, which has the meaning of *yin* and *yang* integrating. But "beauty" in the *Zhou Book of Changes* refers to is the supple and soft beauty of introverted implications, so *The Commentary on the Book of Changes* states, "*yin* has beauty, contain it," in yellow links the patterns" refers to the beauty of soft on the outside but firm on the inside." Second, the important implication of "beauty" in *Zhou Book of Changes* is "in the proper position occupying due place," meaning that the energies of *yin* and *yang* are each occupying their rightful position, and thereby each rests in its own place, each exercises its own function, and thereby fosters the production of all things in Heaven and Earth along with the thriving growth of vitality in the world. This is what is meant by the *tai* 泰 hexagram in stating "Heaven and Earth intercross and all creatures interlink, high and low pass into one another and their wills converge. *Yang* on the inside and *yin* on the outside, firm on the inside and yielding on the outside, authoritative person on the inside and petty man on the outside, the authoritative person's path is stretches long and the petty man's path disappears quick." What the *kun* hexagram symbolizes is the beauty of supple softness, therefore, line 3 of *Commentary on Xiang* states ""*yin* 阴 has beauty, contain it by doing things following the King's intention, but do not dare succeed. The *Dao* of Earth is the way of the wife, the way of the servant. The *Dao* of Earth is unsuccessful and is compensated for with an end." The beauty of supple softness is the auxiliary role of "being in the proper position occupying the due place," "resting in its own place," resting in "the *Dao* of Earth," "the way of the wife," "the way of the minister," but in combination with the energy of firm *yang*, it makes "Heaven and Earth transform, grasses and trees grow luxuriantly" (*Commentary on Kun*), namely success but without basking in achievement. This is what is meant by "beauty is found therein, and flowing out to the four limbs, executed in affairs, it is the consummation of beauty."

Summarily speaking, "beauty" in the *Zhou Book of Changes* is the early ancient Chinese formulation of the aesthetic essence of ecology and life; it is China's representative notion of "the beauty of striking harmony" (*zhonghemei* 中和美). By "striking harmony" what is meant is Heaven and Earth, Qian and Kun each occupying their proper positions, "Heaven and Earth intercross and all creatures interlink," so the weather comes timely and all creatures thrive, full of vitality. All creatures can only thrive and prosper when the heavens and the earth, *yin* and *yang* enter the right positions. It becomes clear from this that the beauty of striking harmony in ancient China is fundamentally speaking the ecological and biological beauty of all creatures prospering and thriving. The consistency of this ecological conception of the beauty of life has had wide and profound impact. Zhuangzi championed "nurturing life," including "maintaining the body," "fully actualizing oneself," "nurturing one's kind" and "completing one's life"; it is also intimately related to the aesthetics of life. Cao Pi's writings championed "the matter-energy of art," namely "art is guided by vital energy, there is substance to the clarity or turbidity of flowing energy, which you cannot reach by forcing it. For instance, in music, although the measure of the tune is balanced, the rhythm always checks it to the point where the flowing energy put into it is uneven, and there is material to the smoothness or clumsiness of flowing energy, and although it is present in the father or older brother, it cannot be transmitted to the son or younger brother." (*dianlun, lunwen*). Here, flowing energy (*qi* 气) refers to the rhythm of breath, and the clarity and strength of the flowing energy is directly related to the smoothness or clumsiness of musical art, which is an ancient Chinese theory of the aesthetics of life. Liu Xie theorized on the relationship between literature and nature, and proposed the theory of "the color of things," namely "the poet senses things, and associates analogies endlessly. At the verge of fluidly connecting countless images, he submerges to recite at the interval between seeing and hearing: the flowing energy of the writing and the likeness of the picture [drawn] yields flexibly to the thing itself; attributing the image and attaching the sound vacillates with the affective mind." (*wenxin diaolong,* "The Color of Things"), and the vividly moving sounds used in the *Book of Poems* like *shuo-shuo* 灼灼, *yi-yi* 依依, *gao-gao* 杲杲 and *jie-jie* 喈喈 illustrate the condition of the poet imbuing the object with vitality in the creative process of artistically sensing things and associating analogies, fluidly connecting countless images, writing rhythm and drawing likenesses. Xie He of Southern Qi clearly put forward the theory of "lively rhythm" (*qiyun shengdong* 气韵生动) in painting, positing the expression of the vital forces of things as one of the important criteria for successful painting. Wang Changling of the Tang dynasty put forward the deeply and widely influential notion of "the poetic state of mind" (*yijing* 意境) in *The Laws of Poetry* (*shige* 诗格), namely "the search is sought in images, the affective mind enters the state, and the spirit gathers on the thing, which is gotten due to the mind;" Wang Changling here fully expresses the condition of creating the artistic product through fusing and converging one's affective state with images, one's intention with the state of affairs, and one's spirit with the material thing itself and thereby filling the latter in each with vitality. The contemporary aesthetic philosopher Liu Baihua forcefully advocated "vital aesthetics" (*shengming meixue* 生命美学) on the basis of summarizing ancient tradition, asserting that "[a]rt is originally the outward development of the human's—the artist's—spiritual life, concentrating in the natural material object, spiritualizing it, idealizing it." He then states: the characteristic state of mind expressed by Chinese painting, one could say, is the basic philosophy rooted in the Chinese people, namely the cosmology of *Book of Changes*; the two kinds of *qi*, *yin* and *yang* transform to produce all phenomena, which come alive due to possessing the *qi* of Heaven and Earth, every entity could be called "a collection of *qi*" (Zhuangzi: Heaven is accumulating *qi*). These ceaselessly producing and reproducing energies of *yin* and *yang* interweave into a life with rhythm."[16] From this it is clear enough to see that ancient China possesses a rather long-lasting ecological and vital aesthetic tradition.

The developed nature of abstract logical reasoning in the West was determined by its particular background of natural historical culture, and so aesthetics was pushed towards purely rational thought divorced from natural ecology from the very beginning. Thus, there emerged such definitions of aesthetics divorced from human being's essence to appreciate the beauty of ecology as "beauty is the Idea," beauty is symmetry, beauty is the perfection of sensuous knowledge, beauty is the sensuous manifesta-

[16]Baihua [10], pp. 7, 118.

tion of the Idea, beauty is the form of purposeless purposiveness, etc. But even under the constraint of the rule of reason, Western thought also still had some theoretical expositions about the vital and ecological properties of aesthetic beauty. In 1725, the Italian philosopher of aesthetics, Vico, explored the primitive form of "poetic thought." As his master key to the new science in *The New Science*. In our view, this "poetic thought" is substantively speaking a primitive thought of ecological aesthetics. When Vico explains the primitive "poetic metaphysics" and "poetic thought," he states: "Hence poetic wisdom, the first wisdom of the gentile world, must have begun with a metaphysic not rational and abstract like that of learned men now, but felt and imagined as that of these first men must have been, who, without power of ratiocination, were all robust sense and vigorous imagination, as established in the Axioms [185]. This metaphysics was their poetry, a faculty born with them (for they were furnished by nature with these senses and imaginations)"[17] Here, Vico specifically hammers on the vital properties of primitive human nature, "all robust with senses and vigorous imagination," "a faculty born with them." In 1795, Schiller saw aesthetics as the original nature of man in his famous *Letters on the Aesthetic Education of Man*, but, constrained by rationalism, he never succeeded in explaining aesthetics from the ecological perspective. Beginning in 1936, Heidegger attempted to build his ecological conception of aesthetics in "the fourfold game of earth, sky, gods and mortals, viewing the relationship between natural ecology and mortals as a relationship of chance between Dasein and the world, Dasein's authentic state of being-in-the-world, whose vitality unfolds as Dasein's, i.e. man's, being in space and time, which is an aesthetics of ecology and life.

Here, we specifically want to introduce two Western theorists, who made outstanding contributions among those researching human being's essence to appreciate the beauty of ecology. The first is the famous philosopher and educator, Dewey, who, in the book *Art as Experience* (1934), brought up the need to overcome the universal tendency to divide subject from object and spirit from flesh at the time. He asks, "Why is the attempt to connect the higher and ideal things of experience with the basic vital roots so often regarded as betrayal of their nature and denial of their value?...A complete answer to the question would involve the writing of a history of morals that would set forth the conditions that have brought about the contempt for the body, fear of the senses, and the opposition of flesh to spirit." (p. 20) Dewey asserts that the aesthetic subject is the new concept of "the live creature," and that aesthetics is an experience of satisfaction generated by the interaction between this "live creature" that is man and the world he lives in. He concretely describes it in the following way: "The existence of art is the concrete proof of what has just been stated abstractly. It is proof that man uses the materials and energies of nature with intent to expand his own life, and that he does so in accord with the structure of his organism—brain, sense-organs, and muscular system. Art is the living and concrete proof that man is capable of restoring consciously, and thus on the plane of meaning, the union of sense, need, impulse and action characteristics of the live creature." (p. 25) "Nature is the mother and the habitat of man, even if sometimes a stepmother

[17] Vico [11], pp. 181–182.

and an unfriendly home. The fact that civilization endures and culture continues—and sometimes advances—is evidence that human hopes and purposes find a basis and support in nature."[18]

The second theorist is the American philosopher of neo-pragmatist aesthetics, Richard Shusterman, who in *Pragmatist Aesthetics*, brought up the concept of "the aesthetics of the body," which is an aesthetics of life for the new era unifying sensation and philosophy. He states, Somaesthetics, roughly defined, concerns the body as a locus of sensory-aesthetic appreciation (*aisthesis*) and creative self-fashioning. As an ameliorative discipline of both theory and practice, it aims to enrich not only our abstract, discursive knowledge of the body but also our lived somatic experience and performance."[19] Here, he does not use the term "body-aesthetics," but the ancient Greek term "somaesthetics," which implies the union of mind and body. So, what is the intention behind him putting forward the concept of somaesthetics? He states that one of the reasons is for the sake of revitalizing Baumgarten's idea of regarding aesthetics as a cognitive science of improving life that includes both theory and practical training; the second reason is for ending Baumgarten's catastrophic bringing of bodily denial into aesthetics; the third reason is that somaesthetics can make important contributions to many vital philosophical concerns, and thereby restore philosophy its initial role as a living art. Quite obviously, Shusterman's somaesthetics implies the ecological aesthetics of life.

12.2 "Poetically Dwelling"

(1) The Formulation and Implications of "Poetically Dwelling"

"Poetically dwelling" was first put forward by Heidegger in *Remembrance*, a question and answer piece on the origins of poetry and the poet. This is question asked universally for a long period of time: who is man and where does man dwell peacefully? Why art and why the poet?—the truth of poetry and the poet makes man poetically dwell on this earth between the existence of gods and the real life of mortals, the poet conveys the meaning of being to mortals who have lost being in boundless darkness, and makes godly luminescence shine upon calm and poor reality, thereby building a better spiritual home. This is one of the most important ecological conceptions of aesthetics put forward by Heidegger; it is another more poeticized formulations of his ontological aesthetics without extremely important value and significance.

For a long period of time, people have only spoken of pleasure, eye-pleasing delight and at most molding in aesthetics, but exceptionally few people have explained aesthetics from the perspective of aesthetically living, and specifically "poetically dwelling." Dwelling itself necessarily touches on human being's relationship of affinity and friendship with nature, and hence is an important category

[18] Dewey [12], pp. 20, 29, 31–32.

[19] Schusterman [13], p. 354.

of ecological aesthetics. In Remembrance, Heidegger first begins by introducing Hölderlin's poem, "Full of merit, yet poetically, man dwells upon this earth," and continues, stating that "every work and activity, every building and caring are a kind of 'culture,' but cultura is always will forever be a consequence of dwelling, which is poetic".[20] In actuality, "poetically dwelling" is a necessary implication of Heidegger's ontological aesthetics. When formulating his own structure of being-in-the-world, he explicates the implications of being-in-the-world, and makes the distinction between epistemological "being-in" and ontological "being-in," pointing out that ontological "being-in" implies the meaning of dwelling and inhabiting. He states, "Nor does the term "Being-in" mean a spatial 'in-one-another-ness' of things present-at-hand, any more than the word 'in' primordially signifies a spatial relationship of this kind. 'In' is derived from "innan"—"to reside," habitare," "to dwell" [sich auf halten], "An' signifies "I am accustomed," "I am familiar with," "I look after something." It has the signification of "colo" in the senses of "habito" and "diligo." The entity to which Being-in in this signification belongs is one which we have characterized as that entity which in each case I myself am [bin]. The expression 'bin' is connected with 'bei,' and so 'ich bin' ['I am'] means in its turn "I reside" or "dwell alongside" the world, as that which is familiar to me in such and such a way."[21] Clearly, "being-in-the-world" is human being residing in, inhabiting and dwelling in the world. But can one succeed in "poetically dwelling"? An essential aspect of it is the need to care for nature and rescue the earth. In Building, Dwelling, Thinking, Heidegger remarks: "Mortals dwell in that they save the earth-taking the word in the old sense still known to Lessing. Saving does not only snatch something from a danger. To save really means to set something free into its own presencing. To save the earth is more than to exploit it or even wear it out. Saving the earth does not master the earth and does not subjugate it, which is merely one step from spoliation."[22] "Poetically dwelling" as "saving the earth," breaks free of subjugating and controlling the earth, and makes it return to its ownness, and thereby allows humans to live better upon the earth, in the world. This is an essential element of contemporary ecological aesthetics. What especially requires explanation here is that Heidgger's notion of "poetically dwelling" had a very clear signification at the time, which pointed to the ever more grave tendency toward human being's "technological dwelling" under the control of instrumental reason in industrial society. In the 20th century in which Heidegger lived, capitalism already entered the age of imperialism. Due to industrial capitalists striving for the maximization of profit and superstitiously acquiring surplus value through technology, mass deforestation, destruction of resources and invasion of weaker countries have become the malady of the entire age. Heidegger deeply sensed this, calling it technology's violence upon the human being, a "technological dwelling" that betrays humanity. He attempts to lead humans toward dwelling poetically through aesthetic means: "Europe's technological-industrial domination has already covered the entire earth. On the other hand, the earth, as a planet, has already

---

[20]Heidegger [14], p. 105.

[21]Heidegger [4], pp. 63–64.

[22]Heidegger [15], p. 1193.

been included in the interstellarcosmic space which is placed at man's disposal by the planned projects of man. The earth and heaven of the poem have vanished. Who would day say whither? The in-finite relation of earth and heaven, man and God, seems to have been destroyed." Heidegger questions this situation: "Must Europe, as this cape and brain, first become a land of an evening from which another morning of world-destiny prepares its rise?"[23] Visibly, he already view "poetically dwelling" as the rise of another morning of the world-destiny. In that long and dark night, the utopian character of this notion is hard to miss. But Heidegger is not unique here. Around nearly the same time as Heidegger, the British novelist Lawrence wrote Lady Chatterley's Lover, which through intense comparisons lashes out at the pollution engendering coalmines and calculative mine owners in capitalist society, and praises ecologically thriving forests and the ranger striving for natural life, all of which expresses the yearning for human being's "poetic dwelling" in harmony with nature.

(2) **Ancient Chinese Aesthetic Wisdom Concerning Poetic Dwelling**

The ecological aesthetics that this proposition of "dwelling poetically" implies shares profound connections with Eastern culture, especially the cultural tradition of China. As everyone knows, ancient Western aesthetics is a classical "harmony theory" of aesthetics, which is based on something's self-symmetry and proportions. Actually, what this "harmonious beauty" emphasizes is the sculpted beauty of static objects, namely "noble purity and silent greatness." But in the East, and especially in ancient China there grew a "beauty of striking harmony" quite different from that in the ancient West. This beauty of striking harmony is an aesthetics of human life, of ethics and of life, which emphasizes the auspiciousness of life and soundness of living, namely human being's "dwelling poetically." The earliest book to mention "the beauty of striking harmony" is Book of Documents "Canon of Shun," "The Di said, "'Kui, I appoint you to be Director of Music, and to teach our sons, so that the straightforward shall yet be mild; the gentle, dignified: the strong, not tyrannical: and the impetuous, not arrogant. Poetry is the expression of earnest thought; singing is the prolonged utterance of that expression; the notes accompany that utterance, and they are harmonized themselves by the standard tubes. (In this way) the eight different kinds of musical instruments can be adjusted so that one shall not take from or interfere with another; and spirits and men are brought into harmony.'" This passage of text clearly asserts the function of such synthetic art forms as poetry, music and dance in "brining spirits and men into harmony" and insists on cultivating human character to be straightforward yet mild, gentle yet dignified, strong yet not tyrannical through musical education, all of which implies the notion of "poetically dwelling." After this, Record of Rites "Record of Music" furthermore explained the ontological conception of aesthetics centered on "striking harmony," "In music of the grandest style there is the same harmony that prevails between heaven and earth; in ceremonies of the grandest form there is the same graduation that exists between heaven and earth. Through the harmony, things do not fail (to fulfil their ends); through the graduation we have the sacrifices to heaven and those to earth. In the

[23] Heidegger [14], pp. 215, 216.

visible sphere there are ceremonies and music; in the invisible, the spiritual agencies. These things being so, in all within the four seas, there must be mutual respect and love." It also states: "Therefore, when the music has full course, the different relations are clearly defined by it; the perceptions of the ears and eyes become sharp and distinct; the action of the blood and physical energies is harmonious and calm; (bad) influences are removed, and manners changed; and all under heaven there is entire repose." These expositions summarily reveal the ancient Chinese cultural tradition's goal of reaching the social peace and living health of "integrating nature and humans into one" and "mutually respecting and loving one another" through "ritual and musical cultivation." Actually, the substance of "the beauty of striking harmony" in ancient China was reaching mutual harmony between nature and man, yin and yang and male and female, in order to secure the goal of fortunate peace in society and human life. This beauty of striking harmony expected human being to dwell poetically not unlike Heidegger's concept of poetic dwelling in the "fourfold" world of eco-ontological aesthetics. All of these convergences will become important resources for the building of contemporary ecological aesthetics.

12.3 The Fourfold

(1) The Posing of the Fourfold

"The fourfold game of earth, sky gods and mortals" is an extremely important idea of ecological aesthetics proposed by Heidegger in his late phase; it is the only way for the Being of "Dasein" to unfold and attain aesthetic life in the fourfold world-structure of sky, earth, gods and mortals. As is widely known, there was a process to Heidegger's positing of the fourfold. His unfolding of "Dasein" is realized in the strife between the world and the earth. Here, the world possesses openness while the earth is self-enclosing, and the revealing of the world is prior to the revealing of the earth, or in Heidegger's words: "The world is the self-opening openness of the broad paths of the simple and essential decisions in the destiny of a historical people. The earth is the unforced coming forth of the continually self-closing, and in that way, self-sheltering. World and earth are essentially different and yet never separated from one another." Moreover, "The opposition of world and earth is strife."[24] Beginning in 1936, Heidegger gradually realized the relationship that ought to take hold between humans and all nature beings through critiquing the perils of the "anthropocentrism" expressed by the abuse of modern instrumental reason. It was in this context that he put forward the famous fourfold game of sky, earth, gods and mortals. In 1938, he forcefully criticized the proposition that "man is the measure of all things" in the lecture The Age of the World Picture: "Man does not set the measure to which all beings in their being here have to accommodate themselves, out of a detached

[24] Heidegger [16], p. 30.

I-ness."[25] In the famous lecture The Thing in 1950, Heidegger expounded the fourfold game of sky, earth, gods and mortals. Here, he mainly explains the relationship between humans and all natural beings beginning with the abuse of instrumental reason in the process of modernization and its concealing of the thinghood of the thing. Heidegger insists that instrumental reason follows the subject-object dichotomy in the theoretical viewpoint of philosophy, and views all entities as external and objective; especially representative of this is Kant in viewing the thing as passive "thing in-itself" and the scientific experiment in likewise viewing the object as nothing but the object of scientific observation. The thinghood of the thing is concealed because of this as humans only see separate beings, but cannot see the Being of the thing. He cites the example of a jug, insisting that the thinghood of the jug is not in the original material used to make the jug nor in the specific usefulness of the jug, but rather in the jug qua vessel, namely in the true meaning of Being contained in the jug's pouring-offering that unifies sky, earth, gods and mortals. He states, "[b]ut the gift of the outpouring is what makes the jug a jug." "The spring stays on in the water of the gift. In the spring the rock dwells, and in the rock dwells the dark slumber of the earth, which receives the rain and dew of the sky. In the water of the spring dwells the marriage of sky and earth. It stays in the wine given by the fruit of the vine, the fruit in which the earth's nourishment and the sky's sun are betrothed to one another. In the gift of water, in the gift of wine, sky and earth dwell."[26] It is clear from this that Heidegger's fourfold was put forward through the critique of the excessive abuse of instrumental reason and the excessive advocacy of anthropocentrism.

(2) **The Aesthetic Implications of "The Fourfold Game"**

Heidegger's fourfold game is an important proposition of ecological aesthetics, which contains extremely rich implications:

1. As an important component of eco-ontological aesthetics, the fourfold game builds the structure of "the world" in which Being opens up and Truth emerges

In eco-ontological aesthetics, beauty is not a substance, but a relation; beauty is not static, but is a process, a transition from concealing to clearing; beauty is not a single element, but is rather gradually unfolded in the world-structure of humanity's relationship to nature. The fourfold game builds a necessary precondition for the unfolding of beauty and the manifesting of Truth, namely the world-structure of the harmonious oneness of the fourfold whole of sky, earth deities and mortals. As Heidegger put it, "This appropriating mirror-play of the simple onefold of earth and sky, divinities and mortals, we call the world: The world presences by worlding. That means: the world's worlding cannot be explained by anything else nor can it be fathomed through anything else."[27] This is to say, the fourfold of sky, earth, deities and mortals constitutes the simple play of the whole, and constitutes the world that makes worlding become what it is. It is precisely in this world-structure fusing together

[25] Heidegger [15], p. 915.

[26] Heidegger [15], pp. 1172, 1172–1173.

[27] Heidegger [15], p. 1180.

mortals and all natural beings, which cannot be proved through traditional logic or calculated using instrumental reason, where Truth appears and Being opens up. The "world" that the fourfold play constructs becomes the precondition of appreciating beauty.

2. The play of the fourfold is the overcoming of the strife between world and earth in earlier Heidegger, and furthermore moves toward equality and harmony between humans and all natural beings

After 1936, Heidegger gradually overcame the earlier theory of strife between earth and world while putting forward the fourfold play of sky, earth, deities and mortals. He states, "[e]arth and sky, divinities and mortals—being at one with one another of their own accord—belong together by way of the simpleness of the united fourfold. Each of the four mirrors in its own way the presence of the others. Each therewith reflects itself in its own way into its own, within the simpleness of the four. This mirroring does not portray a likeness. The mirroring, lightning each of the four, appropriates their own presencing into simple belonging to one another. Mirroring in this appropriating-lightening way, each of the four plays to each of the others. The appropriative mirroring sets each of the four free into its own, but it binds these free ones into the simplicity of their essential being toward one another."[28] This is extremely rich content in this. First of all, this is a conception of ecological equality through the harmonious coexistence between humans and all natural beings that thoroughly destroys anthropocentrism. In Heidegger's view, man is but one equal member of the fourfold that like the other three takes on the task of lightening the others and lightening itself, neither special nor unique. Moreover, the unity of the fourfold constitutes the inseparable wholeness of the fourfold, which ought to be considered an entirely new conception of ecological equality; second, the fourfold is a continuation and development of "the theory of play" in traditional Western aesthetics. The theory of play sprouted forth in the Western enlightenment and classical German aesthetics, it refers to "the free harmony" between the imagination, understanding, reason and judgment. Heidegger's theory of play takes on this meaning of freedom in the classical theory of play, but it becomes the free harmony between sky, earth, deities and mortals; he even uses the betrothal and the metaphor of utmost intimacy between lovers, but overcomes the absolute subjectivism in the traditional theory of freedom, giving it instead an entirely new dimension of meaning in the free equality between humans and all natural beings; finally, he further clarifies the active constructing role that man plays in the opening up of the beauty of ecological being. Heidegger insists that humans are completely equal to all natural beings on one hand, but on the other hand insists that human being is the only being among all natural beings with rational consciousness, and hence, human being has a special constructive role in the manifesting of being, the opening up of truth and the lightening of beauty. He states, "When and in what way do things appear as things? They do not appear by means of human making. But neither do they appear without the vigilance of mortals. The first step toward such vigilance is the step back from the

[28] Heidegger [15], pp. 1179–1180.

thinking that merely represents—that is, explains—the thinking that responds and recalls."[29] In Heidegger's view, although the unfolding of the essence of the thing's existence is not the result of human being's theoretical explanation and scientific research, it is human being's vigilance and thinking that responds and recalls, which is the constructive role that human being plays in the opening up of truth and the constructive role that the subject in phenomenology plays through "intentionality." But the aesthetic relationship between human being and natural ecosystems only becomes possible through a relationship of involvement between natural beings and humans when they become handy. So, why would human being have this constructive capacity and role? This is related to the "Being-there" unique to human being among all other beings. Heidegger calls "Dasein" or "Being-there" the "mortal" who can presence the Being of beings, he states, "The mortals are human beings. They are called mortals because they can die. To die means to be capable of death as death. Only man dies. The animal perishes. It has neither death ahead of itself nor behind it. Death is the shrine of Nothing, that is, of that which in every respect is never something that merely exists, but which nevertheless presences, even as the mystery of Being itself. As the shrine of Nothing, death harbors within itself the presencing of Being. As the shrine of Nothing, death is the shelter of Being. We now call mortals mortals—not because their earthly life comes to an end, but because they are capable of death as death. Mortals are who they are, as mortals, present in the shelter of Being. They are the presencing relation to Being as Being."[30]

3. Heidegger's Late Turn and Comparing the "Fourfold Play" to Ancient Chinese Daoism

There are many facts that can demonstrate how must ancient Chinese Daoism inspired Heidegger's late turn and his positing of the fourfold play. The dialogue between Heidegger and Daoism has become "the foreign explanation of Laozi's theory of Dao"; they are songs gushing forth from a common origin. As is widely known, Heidegger's early thought took the world-structure of the presencing of truth to be the strife of world and earth. Although this theory progresses greatly with respect to breaking through the subject-object dichotomy model of thought, it is still clearly anthropocentric. After the 1930s, Heidegger began to turn from anthropocentrism toward eco-holism in the positing of his famous fourfold play of sky, earth, deities and mortals. We have sufficient materials to prove that Heidegger's ecological turn was the result of dialogue with the ecological wisdom of Chinese Daoism. In the 1930's, he began adeptly utilizing the thought of Laozi and Zhuangzi. He used two German translations on the works of Laozi and Zhuangzi and translated chapter 8 of the Daodejing with the Taiwanese scholar Xiao Shiyi in 1946. During this time period, he used Lao-Zhuang thought quite a bit to argue his own viewpoints. His "fourfold of sky, earth, gods and mortals" is fully consistent with the theory that "there are four greats in the realm and human being is one among them" in Laozi's thought; he also used Laozi's "know the white and hold the black" to explain his thought of moving

[29]Heidegger [15], p. 1182.

[30]Heidegger [15], p. 1179.

from concealing to clearing; he used Laozi's "thirteen spokes converge on one hub, but it is its nothingness that enables the cart to function" to explain the distinction between Being and beings; he used "the Dao that can be spoken is not the eternal Dao to explain his viewpoint that "saying is different from speaking"; he used Zhuangzi's "greater use of the useless" to explain that human dwelling is not utilitarian; he used the dialogue about the "joy of the fish" between Zhuangzi and Huizi on the bridge to explain the distinction between ontology and epistemology, and so on. The ecological turn of Heidegger's thought was greatly informed by Daoism, which is sufficient to prove that the wisdom of ecological aesthetics in ancient China still has important theoretical value in the contemporary age.

12.4 Sense of Home

In modern society, people have been struck universally with a lost sense of homelessness due to the destruction of the natural environment and the intensifying spiritual anxiety that comes from it. Consciousness of home is posited through contemporary ecological aesthetics in this crisis as one of its important implications. Consciousness of home not only envelops human being's relationship to natural ecology, it also implies an even deeper truth of Being in the poetic dwelling of authentic human being.

(1) **Consciousness of home in Heidegger's ontological philosophy—aesthetics**

Heidegger was the earliest to mention consciousness of home in philosophical aesthetics, and in a certain sense, this consciousness of home is an organic component of his ontological philosophy. In 1927, when Heidegger expounded on the ontological model of being-in-the-world in Being and Time, he explicated the implications of Dasein in the world, insisting that it implies the meanings of "to inhabit," "to dwell," and "to reside" as "home." He states, "Nor does the term "Being-in" mean a spatial 'in-one-another-ness' of things present-at-hand, any more than the word 'in' primordially signifies a spatial relationship of this kind. 'In' is derived from "innan"—"to reside," habitare," "to dwell" [sich auf halten], "An' signifies "I am accustomed," "I am familiar with," "I look after something." It has the signification of "colo" in the senses of "habito" and "diligo." The entity to which Being-in in this signification belongs is one which we have characterized as that entity which in each case I myself am [bin]. The expression 'bin' is connected with 'bei,' and so 'ich bin' ['I am'] means in its turn "I reside" or "dwell alongside" the world, as that which is familiar to me in such and such a way."[31] Clearly, the relationship of being-in-the-world in Heidegger's ontological philosophy implies this strong consciousness of home in "one being at home" where human being is intimately and inseparably bound into one with all beings in the world including natural ecosystems.

---

[31] Heidegger [4], pp. 63–64.

But in modern society dominated by instrumental reason, the relationship between human being and the world that includes all natural beings—the authentic relationship of "being-at-home" is warped, and human being finds herself in an uncanny state of anxiety at losing the home she had. Heidegger states, "[i]n anxiety one feels 'uncanny'. Here the peculiar indefiniteness of that which Dasein finds itself alongside in anxiety, comes proximally to expression: the 'nothing and nowhere.' But here "uncanniness" also means "not-being-at-home" [das Nicht-zuhause-sein]."[32] "Uncanniness is the basic kind of Being-in-the-world, even though in an everyday way it has been covered up."[33] "In uncanniness Dasein stands together with itself primordially. Uncanniness brings this entity face to face with its undisguised nullity, which belongs to the possibility of its own most potentiality-for-Being."[34] This shows that uncanniness is not only an affect unique to humans in modern society; "anxiety" as the basic state of Dasein's unfolding is also "primordial." Moreover, "uncanniness" and "unhomeliness" as essential contents of "anxiety" are similarly primordial such that one could say they are innate in man qua man. Of course, under the domination and assault of all of the modern social factors, this sense of uncanniness becomes ever more intense. For this reason, "sense of home" is necessarily an important implication of contemporary eco-ontological philosophy—aesthetics.

On the 6th of June of 1943, Heidegger explicitly brought up the sense of home in aesthetics in his lecture entitled "Homecoming" commemorating 100 years of Hölderlin's passing. The lecture is an explanation of Hölderlin's poem "Homecoming"; it is a dialogue between thought and poetry. He tries to enter the historical uniqueness of the poem through a thought dialogue, and thereby find the aesthetic implications of the poem. The poem Homecoming emphatically expresses the aesthetic implications of sense of home. Heidegger underscores that: "'The house' is intended here to mean the space which for men houses that wherein alone they can be 'at home' and so fulfill their destiny. This space is given by the immaculate earth. The earth houses the people in its historical space. The earth serenifies 'the house.' And the earth, which thus serenifies is the first angel 'of the house.'"[35] "Home" here for Heidegger is actually a philosophical and aesthetic relationship with a primordial character in ontology; it is the manifestation of Dasein's involvement with the world and human being's involvement with Nature. In this "home" Truth appears and Being bursts forth. Heidegger says two very interesting things about his. In one passage, he states: "[b]oth earth and light, the 'angels of the house', and 'the angels of the year' are called 'preservers' because as the greeting ones they bring to light the gaiety in whose clarity the 'nature' of things and people is safely preserved."[36] Here, the "earth" and the "angels of the house" are houses of the world and nature and the "light" and "angels of the year" refers to "human being" and "Dasein." In this being-with or involving and preserving between Dasein and the world or between human

[32]Ibid., p. 218.

[33]Ibid., p. 318.

[34]Ibid., p. 328.

[35]Heidegger [14], p. 15.

[36]Heidegger [14], p. 15.

being and nature, the "light as being" shines forth and clarifies, which is what the consciousness of home implies. In another passage, Heidegger states: "[t]he poet's vocation is homecoming, by which the homeland is first made ready as the land of proximity to the most joyous, and in the process of guarding it to unfold it—that is the care of homecoming'".[37] The goal of the poet's aesthetic yearnings is "homecoming," namely entering the proximity of "the angels of the house." This primordial path of homecoming is the path of the unfolding of the mystery as being; this movement from mysterious secrecy to opening forth, from concealing to clearing realized through historical preserving and unfolding is path to presenting the aesthetic "consciousness of home."

After the middle of the 20th century, the industrial revolution only deepened and environmental destruction only worsened as instrumental reason strengthened human being's feeling of homelessness, in which case, how do we counter the ever vitalizing spirit of science and technology and the ever strengthening feeling of homelessness? In 1955 Heidegger wrote answers this question in Gelassenheit, in which he first describes the massive pressure brought upon people behind the excessive expansion of instrumental reason. Under the pressure of the ever-strengthening world-view of instrumental reason, "Nature becomes a gigantic gasoline station, an energy source for modern technology and industry. This relation of man to the world as such, in principle a technical one, developed in the seventeenth century first and only in Europe." "The power concealed in modern technology determines the relation of man to that which exists. It rules the whole earth."[38] Its concrete manifestation is, "[m]any Germans have lost their homeland, have had to leave their villages and towns, have been driven from their native soil. Countless others whose homeland was saved, have yet wandered off. They have been caught up in the turmoil of the big cities, and have resettled in the wastelands of industrial districts. They are strangers now to their former homeland. And those who have stayed on in their homeland? Often they are still more homeless than those who have been driven from their homeland."[39] Modern technology stirs, harasses and torments human being, fatally threatening human being's basis of survival, doubly throwing human being into the depths of homelessness. So, how do we cope with this grave situation. Heidegger proposes the method of Gelassenheit (releasement). He believes the blind suppression of science and technology is perfectly idiotic, but being enslaved by science and technology is even more tragic. He states, "[w]e can use technical devices, and yet with the proper use also keep ourselves so free of them, that we may let go of them at any time."[40] At the same time, he also insists that we should uphold ecological holism and solidly find a foothold on the earth. He borrows the words of John Stewart Bell, stating that "we are plants, and whether we admit it or not we must grow up rooted in the earth for the sake of being able to blossom in the sky".[41] In an interview in his late years with

[37]Ibid., p. 31.

[38]Heidegger [15], p. 1236.

[39]Ibid., pp. 1234–1235.

[40]Ibid., p. 1239.

[41]Heidegger [15], p. 1241.

a reporter at Der Spiegel in 1966, when Heidegger speaks about humanity's way out of the ever-worsening crisis, he again says that humans should firmly protect their own "homeland" which alone can bring about something great enough to turn fate around. He states, "I know that, according to our human experience and history, everything essential and of great magnitude has arisen only out of the fact that man had a home and was rooted in a tradition."[42] This further highlights the important status Heidegger's ontological philosophy gives to the awareness of home.

(2) **Sense of Home in Contemporary Western Ecological and Environmental Theory**

In 1972, to arrange the UN Conference on the Human Environment and the Declaration of the United Nations Conference on the Human Environment, over 70 famous scientists and intellectuals from 58 countries organized an advisory committee tasked with providing the conference with detailed written materials. In the same year, the first UN Conference on the Human Environment in Stockholm was led by the UN General Secretary Maurice Strong, and the economist, and the economist Barabara Ward co-wrote Only One Earth: The Care and Maintenance of a Small Planet with the biologist Rene Dubos, in which they clearly put forward the important viewpoint that the earth is humanity's only home. The report states: "As we enter the global phase of human evolution it becomes obvious that each man has two countries, his own and the planet Earth."[43] In the age of globalization, everyone has a national home as the root of his or her culture, and at the same time an earthly home as the root of his or her survival. At the end of the paper, the authors even more clearly assert: "[a]lone in space, alone in its life-supporting systems, powered by inconceivable energies, mediating them to us through the most delicate adjustments, wayward, unlikely, unpredictable, but nourishing, enlivening and enriching in the largest degree—is this [Earth] not a precious home for all of us earthlings? Is it not worth our love? Does it not deserve all the inventiveness and courage and generosity of which we are capable to preserve it from degradation and destruction, and by doing so, to secure our own survival?"[44]

In 1978, the American scholar, William Rueckert first brought up the concepts of "ecocriticism" and "ecopoetics" in Literature and Ecology, where he clearly states the viewpoint that the "ecosphere" is humanity's home. When listing the grave problems of environmental pollution that humans have been wreaking on the earth, he states that "these problems are currently destroying the ecosphere, which is our home".[45] The famous British historian Arnold Toynbee wrote at the end of chapter 82 of Mankind and Mother Earth in 1973, "Will mankind murder Mother Earth or will he redeem her? He could murder her by misusing his increasing technological potency. Alternatively he could redeem her by overcoming the suicidal, aggressive greed that, in all living creatures, including Man himself, has been the price of the Great

[42]Ibid., p. 1305.

[43]Ward and Dubos [17], "Foreword", p. 17.

[44]Ward and Dubos [17], "Foreword", p. 260.

[45]Ruckert [18], p. 115.

Mother's gift of life. This is the enigmatic question which now confronts Man."[46] Today's ecosphere is the one and only habitable space we have—or once had.

Since entering the 21st century, humans have begun focusing on the issue of the ecological environment more intensively. In 2002, the famous American environmentalist Arnold Berleant edited Environment and The Arts: Perspectives on Environmental Aesthetics, which collates relevant works from many important contemporary environmental theorists, among which Holmes Rolston III clearly expounded on the problem of home from the perspective of aesthetics in From Beauty to Duty: Aesthetics of Nature and Environmental Ethics: When nature is nearer at hand and must be managed on our inhabited landscapes, we might first say that natural beauty is an amenity—only an amenity—and injunctions to its care would seem less urgent. But this gestalt changes with the perception that the ground is under our feet and the sky over our heads, that on Earth we are at home. Disinterest is not self-interest, but the self is not disembodied. Rather the self is incarnate and emplaced. This is ecological aesthetics, and ecology is vital relationships, a self at home in its world, I identify with the landscape on which I reside, my home territory. This 'interest' does lead me to care about its integrity, stability and beauty." He furthermore says: "The whole Earth, not just the marsh, is a kind of wonderland, and we humans – we modern humans more than ever before– put so much of this grandeur in jeopardy. No one, well placed in the world, can be either logically or psychologically uninterested in that."[47] Here, Rolston explicates the sense of home in ecological aesthetics from the more modern perspective that "earth is humankind's home." He asserts that humankind only has one earth, that earth is the home of humankind's prospering survival, and only the earth enables humans to have a "self." Therefore, protecting our own "home," putting it in order, making it stable and beautiful are the needs of human survival, which is ecological aesthetics.

(3) Cultural Resources about "Sense of Home" in the West and Ancient China

It is the primordial nature of the sense of home that not only gives it an extremely important modern significance and value, but also makes it become the mother problem of human literary arts for thousands of generations. The west as maritime countries are simultaneously also the countries where capitalism developed the earliest, and their culture and literary resources mostly emphasize traveling and expanding abroad like Robinson Crusoe; but "sense of home" as humankind's need for authentic existence was also the mother problem or archetype problem of culture and literature. The earliest historical poem in the West The Homeric Hymns and Odysseus writes of the Greek hero, Odysseus, who, after the end of the battle of Troy, conquered many forces and obstacles from giants, fairies, wind gods and sea monsters to water sprites, and ultimately returned home, metaphorically expressing the cultural "mother problem" that humans must return to a spiritual home despite having to undergo all hazards and dangers. The description of "The Garden of Eden" in The Bible is another explanation of Western culture on the sense of home since ancient

46Toynbee [19], p. 529.

47Berleant [20], pp. 167–168.

Greece. According to the record of Genesis, God made a garden in Eastern Eden, the garden had fertile soils and nourishing rivers, all kinds of trees, flowers and edible nuts of mystifying beauty and plenty. God made Adam out of the soil and made Eve out of his rib, placing both of them securely in the Garden of Eden. At this point, man, God and nature were harmoniously one, and man lived in an incomparably beautiful home. But God warned Adam and Even, "you may freely eat the fruit of every tree in the garden, but you must not eat from the tree of the knowledge of good and evil; for in the day that you eat of it, you will surely die." But Eve was seduced by the sly snake, "When the woman saw that the tree was good for food and pleasing to the eyes, and that it was desirable for obtaining wisdom, she took the fruit and ate it. She also gave some to her husband, and he ate it…" God knew that after all of this, he would drive Adam and Eve out of the Garden of Eden, from which point on they were homeless across the world. Moreover, since Adam and Eve committed the sin of greed, God made them plow the earth and toil for their whole lives. While Odysseus's long homecoming in ancient Greece was brought about by the fatal cause of the battle of Troy, man being driven out of the Garden of Eden in The Bible was brought about by the cause of the "original sin" of greed. We should say that the cause of original sin plays the role of warning humankind in The Bible. After this in Western literature, the loss and reconstruction of the Garden of Eden becomes one of the eternal themes through which we see the important status of "sense of home" in Western literature.

As an ancient agricultural state, China has had a strong sense of home running through each successive dynasty of cultural and literary works, all of which provide extremely valuable resources for constructing the "sense of home" in contemporary ecological aesthetics and ecological literature. Book of Poems records China's prior peoples selecting land to settle on and the history of choosing a resting place beneficial to the people's thriving prosperity. For instance, the third chapter of the famous "Major Court Hymns" records the process of King Tai of Zhou leading his people out of Bin, crossing the river Qizu, passing mount Liang, and finally settling in the fertile plane of the Zhou dynasty. "The plain of the Zhou looked beautiful and rich, with its violets and sowthistles [sweet] and dumplings. There he began with consulting [his followers]; There he singed the tortoise-shell, [and divined]. The responses were – there to stay, and then; And they proceeded there to build their houses." ("Mian" trans. James Legge) The plane of Zhou was so fertile that even the bitter vegetables that grew out of it were as sweet as sugar. That they underwent careful and detailed planning, debate and prognostication, illustrates that this was a suitable place to reside and build a homeland. "Odes of Wei, He Guang" specifically describes the Song people living as guests in the state of Wei, confronting the homesickness incited by the river water. "Who says that the He is wide? It will not admit a little boat. Who says that Song is distant? It would not take a whole morning to reach it." The main protagonist walks to and fro the riverside with his native state on the other side, but could not cross the river over to it, and anxious in his hearth with worries pressing on his mind for so long like billowing waves, the lines in the ode gush forth out of his mouth. "The Minor Court Hymns, Cai Wei" writes of the traveling master returning home: "At first, we we set out, the willows were fresh and green; Now, when we shall be returning, the snow will be falling in the clouds," which became a long renown verse from early times.

Book of Changes is an important ancient Chinese classic, whose core concerns the relationship between humans and nature; it is the ecological wisdom of ancient China's classical theory of life that explains "production and reproduction is what we mean by change," and it contains a strong "sense of home" with philosophical implications. The Tuan Commentary on the qian 乾 hexagram states "[v]ast is the 'great and originating (power) indicated by Qian! All things owe to it their beginning: - it encompasses all of the heavens"; on the kun 坤 hexagram, it states: "[c]omplete is the 'great and originating (capacity)' indicated by Kun! All things owe to it their birth; it receives obediently the influences of Heaven"; the four virtues of "originating, expanding, harmonizing and rectifying" in the judgment of the Qian hexagram, and "pacifying, rectifying and benefacting" in the judgment of the Kun hexagram—all of this reveals the consciousness of home wherein the natural ecology of sky and earth act as the roots of human survival. In the "Jia Ren" hexagram in the Tuan Commentary on the Book of Changes, it states that only by first rectifying the way of the home and enacting it to rule the realm can one "stabilize the realm," which speaks of the harmonious relationship between ruling the home and pacifying the realm. The Lu 旅 hexagram in the Book of Changes is composed of Li 离 on top and Gen 艮 on bottom, Gen stands for mountain and stopping, Li stands for fire and brightness. Mountain stopping below suggests that those travelling abroad should stay still, but should also rise up and rely on the light. Leaving home and travelling to live outside of home comes with many inconveniences, so the judgment of the hexagram states "Lu (travelling) indicates minor attainments and progress." This clearly shows the important position that sense of home occupies in Chinese culture and literature. The Fu 复 hexagram in the Zhou Book of Changes expresses the meaning of going back to the root and returning, the image of the hexagram is Zhen (thunder) below and Kun above, and one unbroken yang line produced below five broken yin lines symbolizes yang energy returning from disappearing and means that things return back when they reach their extreme pole, which not only expresses the law of the cyclical transformation of things, but also reveals the consciousness of human being wanting to return home. The judgment of the second SIX of the Fu hexagram is "the admirable return (of its subject). There will be good fortune," that is the auspicious image of yang energy coming back and the traveler returning home. The argumentation of the Fu hexagram is actually an elevated summary of the traditional philosophical ideas of ancient China in "what changes transforms," "things necessarily return when they develop to the extreme pole," and "good weather comes along at the pinnacle of the storm," which explicate the law that every phenomenon and all things necessarily return to their origin. This is the philosophical basis of the sense of home in traditional Chinese culture and arts. The traditional Chinese "sense of home" not only implies the shallow sense of "returning home," but even more so implies the deeper sense of yin and yang returning to their original position and their authentic existence, which is deep in philosophical implications. In Li Bai's poem "Quiet Night Thought," "Lift my head and see the moon, Lower my head and I miss my home," became a famous world-renown line expressing the traveler's recollection of his or her native land—the home becames an aesthetic proposition moving everyone's heart.

In summary, sense of home in the shallow sense has the meaning of the home safeguarding human survival and protecting the environment. At the current moment as environmental pollution gets worse and worse, the development of a sense of home becomes ever more urgent. According to the statistics, in the current age of mass consumption where it is fashionable to throw away everything after one use, the entire world each year throws out no less than 2 trillion, bottles, cans, plastic cartons, paper cups and plastic cups, and that isn't even counting plastic bags. Our home is increasingly becoming a wasteland filled with abandoned trash, and the environment of human survival is worsening by the day. Early in 1975, the American Happiness magazine published an advertisement in a development zone of the Philippines territory that states "To attract companies like yours, we have already deforested mountains, cleared forests, filled-in marshlands, redirected rivers and moved towns all so that you and your business can operate a little easier." This is but a miniature of the severe environmental destruction that has resulted from exploiting all developing countries, including China. Treasuring and protecting our already completely degraded home of survival is today's humanity's shared responsibility; but looking at it from a deeper level, sense of home even more so means the return to and liberation of one's authentic being, namely one returning one's mind and spirit to authentic being and clearing through a path of suspending and overcoming.

12.5 Sense of Place

While sense of home is a great human being's primordial sense of being, sense of region is intimately related to the concrete environment of human life and human experience.

(1) Heidegger on Sense of Region

"Sense of region" was also first brought up by Heidegger. He states, "This 'whither', which makes it possible for equipment to belong somewhere, and which we circumspectively keep in view ahead of us in our concernful dealings, we call the 'region'." "The regional orientation of the multiplicity of places belonging to the ready-to-hand goes to make up the aroundness—the 'round-about-us' [das-Um-uns-herum]—of those entites which we encounter as closest environmentally." "The discovery of regions beforehand is co-determined [mitbestimmt] by the totality of involvements for which the ready-to-hand, as something encountered, is freed."[48] For Heidegger, "region" refers to the location and condition of articles intimately related to human being's existence. This is actually a "totality of involvement" of "things ready-to-hand." This is to say, in human being's everyday life and labor, surrounding articles form relationships of involvement with human being, and thereby become "ready-

[48]Heidegger [4], pp. 120, 121.

to-hand." But, such problems as being "handy" and "unhandy" occur to what is "ready-to-hand" as well as being "a good involvement" and "bad involvement." For instance, the surrounding environment in which one lives is polluted, and the surrounding nature is destroyed, there is all kinds of gases and noises harming oneself, which are precisely cases of them being "unhandy." Such surrounding things are involvements that are no good for human beings; they are "regions" that are bad for human existence. In Heidegger's treatment of "sense of region," he touches upon the extremely important problem of "spatiality." Wuej Heidegger expounds upon sense of region and things ready-to-hand, they are unfolded in the explanation of "spatiality," so the title of the section about sense of region is "The Spatiality of the Ready-to-hand Within-the-world."[49] In actuality, what Heidegger calls "space" is the "world," which is to say "region," in which Dasein in each case finds itself. As he himself says, "In concerning itself circumspectively with the world, Dasein can move things around or out of the way or 'make room' for them [um-, weg-, und 'einräumen'] only because making-room—understood as an existentiale—belongs to its Being-in-the-world."[50] And every one of his explanations of "spatiality" begins with Dasein's Being-in-the-world—which is actually "a region." When he explains the "ready-to-hand" within a region, he uses the formulation of "close by" or "closeness," but "[e]very entity that is 'to hand' has a different closeness, which is not to be ascertained by measuring distances. This closeness regulates itself in terms of circumspectively 'calculative' manipulating and using."[51] This is to say "closeness" is not numerical length of distance, but is whether or not something has a relationship of involvement with human being in the existential sense, so it is always in the scope of one's "circumspection." Thus, "region" comes with the property of "a totality of places for a context of equipment"[52] In 1969, wrote a sketch in his later years entitled—Art and Space, where he further explains his own view of space, or rather, his own "sense of region" from the existential perspective. Because capitalist modernization was already developing deeply at the time, technology's compressing and destroying of "space" or "region" was getting increasingly severe. Heidegger exposed and criticized this fully in the article, stating: "Space—is it that which, since that time (Newton), challenges modern man increasingly and ever more obstinately to its utter control?"[53] Technology's "challenging" and "controlling" of space causes modern man's pollution at the material level and pressure at the mental level to reach the gravest phase. So, what should ideal "clearing-away of space" or "sense of region" be? Heidegger gives an answer in the article, which is the free of dwelling and the opening up of Truth—"Clearing-away brings forth the free, the openness for man's settling and dwelling."[54]

---

[49] Ibid., p. 119.

[50] Ibid., p. 129.

[51] Ibid., p. 119.

[52] Ibid., p. 120.

[53] Heidegger [15], p. 482.

[54] Ibid., p. 484.

(2) **Arnold Berleant on "Sense of Region"**

Arnold Berleant is the main representative of contemporary Western environmental aesthetics. He explores the problem of environmental aesthetics from the perspective of the phenomenology of aesthetic experience, and his viewpoints are largely in agreement with ecological aesthetics with respect to basic theory. He explicates the issue of "sense of region" in accordance with his environmental aesthetics, and first gives his own explanation of "region," which he calls "place." He states, "[w]hat is basic here is the fact that places are the product of a dynamic juxtaposition of many factors: inhabitants, structures replete with meanings, perceptual involvement, and a common spatiality… There is rather the interpenetration, indeed the continuity, of person and place."[55] "Place" is composed by such multiple factors as human perception and space, and is dynamically processual. When expounding on "sense of place," Berleant gives a concrete definition to "place": "Here is familiar ground, personal ground, whose streets and buildings are unified by customary association, recognized with relief, experienced with ease, and remembered with affection. Whatever aids a neighborhood in achieving identity and in generating the warmth of personality helps turn it into a place with which we can associate ourselves and be at ease."[56] Here, he specifically emphasizes the emotionality of places in people's associations, namely places that make people feel free and at ease. Because people have strong associations with "spaces," Berleant insists, "aesthetics of place" is a sensuous phenomenology of environment, an aesthetics of reciprocity in environmental experience. He states that, "we are concerned here not with a psychology of place but an aesthetics of place. In his now-classic work, The Phenomenology of Perception, Merleau-Ponty argued for a unified collaboration of all the senses, including touch, since we do not perceive through separate, differentiated channels of sensation.8 A phenomenology of environmental perception must likewise be more inclusive and take perceptual synaesthesia as basic, beginning, as Merleau-Ponty holds, with the body."[57] From this point, he further argues that "reciprocity is a constant feature of environmental experience."[58] He criticizes the traditional idea of disinterestedness in static aesthetics, and advocates the aesthetics of engagement involving the joint participation of all the senses. Berleant points out: "More forcefully than in any other situation, environmental perception engages the entire, functionally interactive human sensorium in a process through which we become part of our environment in an interpenetration of body and place." "We not only see our living world; we move with it, we act upon and in response to it. We grasp places not just through color, texture, and shape, but with the breath, by smell, with our skin, through our muscular action and skeletal position, in the sounds of wind, water, and traffic. Those major dimensions of environment—space, mass, volume, and depth—are encountered not primarily by the eye but with the body in our movements and actions."[59] This is the new Idea of

---

[55]Berleant [21], p. 135.

[56]Ibid., p. 66.

[57]Ibid., p. 136.

[58]Ibid., p. 139.

[59]Berleant [20], p. 10.

aesthetics in ecological aesthetics, which differs from the traditional conceptions of aesthetics relying on higher level organs like vision and hearing. Berleant insists that "sense of place" in contemporary ecological aesthetics is not just visual and auditory awareness, but is furthermore olfactory, tactile and motile awareness, including taste. He distinguishes the human senses into receptors that keep distance like vision and hearing on one hand and receptors of direct contact like smell, taste, touch and motility on the other, but both types of receptors play a role in aesthetics. This is not just a new development, it also fits with the actual state of contemporary ecological aesthetics. From the perspective of ontological aesthetics, the influence of the natural environment upon human being is far from just visual and auditory, it also includes fragrance, touch and motility. Not only noise and noxious gases bring harm to human being, dust storms and SARS are even more dangerous to people's well-being. Of course, from another perspective at the higher mental levels, the rapid growth of urbanization, the erection of high rises, the acceleration of the rhythm of life, the isolation between people, people's remoteness from nature, cramped living conditions and cookie-cutter housing models all make people actually lose the genuine places of better life. This dimension of ecological aesthetics will become an important dimension in the building of contemporary culture and urban planning, and will simultaneously give prominence to the idea of "putting people first."

(3) **"Sense of Place" in Ancient Chinese Philosophy**

Sense of place in ancient Chinese philosophy could be understood through "the spatial consciousness" in ancient Chinese philosophy. Looking at the Zhou Book of Changes, the spatial consciousness in ancient Chinese philosophy is three-dimensional as the theory of "heaven, earth and man." Zhou Book of Changes "Xici Part II" states: Changes as a book is greatly all-encompassing: there is the way of Heaven, there is the way of man, and there is the way of earth, it combines [these] three natures and doubles them, hence six." Commentary on Characters states: "The great man integrates his virtue with Heaven and Earth, fuses his brightness with the sun and moon, integrates his order with the four seasons, and integrates his fortune and misfortune with ghosts and spirits." All of these statements intend to reveal the meaning of human being living in "heaven and earth" and harmonizing with heaven and earth. This is precisely the thought in the Tuan Zhuan on the Qian hexagram, "holding together the supreme harmony is harmonizing and rectifying." Moreover, spatial consciousness in ancient Chinese philosophy is dynamic as well, as in the statement "Heaven and Earth share fortunes." In ancient Chinese philosophy, a sound place suited to human survival is the product of harmony between nature and humans, yin and yang, Qian and Kun, which is dynamic and filled with living vitality. Ancient Chinese Fengshui is actually an art of choosing habitats based on the cosmological conception of "heaven and earth." Even though it is shrouded in thick superstitious overtones, it does still largely encompass an ancient rational "idea of habitat selection" and "sense of place." For instance, in the selection of a residence, there is the expression, "in the southwest of the residence have a pond, in the northwest an imposing hill complements it even more. Mountainous earth with a ridge means more wealth and nobility, your descendants will gain gifts from heaven marked in satin clothing.[60] Even though getting "wealth and nobility" and "gifts from heaven" are groundless superstitions, having the residence in the north facing the south with mountains in the back and water in the front amounts to a healthy natural environment that is beneficial to human beings, and is worth advocating; also, as the hand copy of the rare edition of Residence Essentials (yangzhai cuoyao 阳宅撮要) made by Guag Xu of the Qing dynasty states: "[a]ny residence must be grounded in a foundation without irregularities, the framework must be well-balanced, the human eye likes to see regularity as auspicious. If it is too high, too wide, too lowly and petty, or if it is pulling to the east or dragging to the west, if the east end is overfull and the west end is foreshortened, it will harm your wealth." "If the shape of the house is tight and upright, the atmosphere is imposing, the sand barrier neatly lines up, and it is unimpeachably dignified, it is a noble residence; if the walls are sealed all-around, the four sides are well-lit, the courtyard is bright and clean, and dividing lines are tightly contracted, it is a noble residence." Here, although the formulation of "noble residence" is tinted by feudal consciousness, the height and grandness, brightness and cleanliness of the house, the solidity, durability and stability of the architecture are however features certainly beneficial to the life of human dwelling. The Qing dynasty Ten Manuals on Residences (yangzhai shishu 阳宅十书) also says: "human being's place of residence should have important features like mass of land, mountains and rivers. The imposing manner of the arteries feeding it should be the grandest." "The outflowing veins of the residence should be no different, the place of residence must use broad and even propensity. The courtyard must be capacious enough to fit 10,000 horses, the hall gates and side rooms must first establish the position. The east wing and classroom in the west along with the kitchen, the front courtyard, balcony and garden, whether residing along a mountain or a flat plain, having a moat all around it embraces nobility. Having paths on the left and right is no different, but if you have to backtrack you must avoid it."[61] All of this emphasizes the natural environment in which the residence is placed, and the belief in the "importance of the mass of land, mountains and rivers," and moreover "connecting it to human misfortune and fortune is most vital," demands the place of residence grounded on broad and flat earth, the courtyard open, and the residence following along the mountain with a water moat all-around and paths to the left and right for convenience of transportation. The famous Qing dynasty playwright Li Yu also used a considerable amount of space to talk about the environmental problem of human habitation in Occasional Enjoyment (xianqing ouji 闲情偶寄), stating: "[h]uman being's inability to live without a house is like the body's inability to survive without clothing. Good clothing keeps cool in the summer and warm in the winter, and a house is no different." On the issue of the direction the house faces, Li Yu proposed "the house should face the south as the proper direction, but if that is impossible to get, the north facing house should be open in the back to soak up exposure to the south; the east facing house should be open on the right, and the west facing house should be open on the left for the same

[60] *Yangzhai Ten books, Kan-yu Integration* (2), Gu Yu Editor, Chongqing Publishing House 1994 edition, p. 199.

[61] Ibid., p. 191.

reason. If there is no extra leeway in the east, west or north, remedy it by opening a window to the sky." When touching on "pathways," he states "for pathways, the most convenient is the quick shortcut, and none is more mystifying than the circuitous"; when discussing the depth of projections and recesses," he states "whether the house is refined or crude, always of value is the ability to protect from wind and rain."; when touching on "bricks," he states "if you do not cover the soil with brick, you will try to suffer its dampness, and it will also be prone to producing dust. Using floorboards to make a ground also faults in one's steps making sound, it is noisy and never silent. If you use concrete bricks, the building will be extremely firm, and making it perfectly like stone is most greatly economical and well-tailored."[62] All of these statements strongly emphasize operationality and are worth referencing.

12.6 Aesthetics of Engagement

This is an extremely important aesthetic notion in ecological aesthetics, which we have already touched on much when discussing "sense of place."

(1) **The Value and Significance of "Aesthetics of Engagement"**

Aesthetics of engagement was clearly brought up by Arnold Berleant, who states, "first, the inadequacy for architecture of the theory of disinterestedness and the need for what I call an aesthetics of engagement."[63] Then, he states, "Thus both aesthetics and environment must be thought of in a new, expanded sense. In both art and environment, we can no longer stand apart but join in as active participants."[64] The positing of "aesthetics of engagement" breaks with the traditional "aesthetics of silent observation" continuously praised since Kant, and seeks to establish a completely different subject as well as idea of aesthetic appreciation based on the active participation of all the senses in that subject. This is a break-through and positive construction in the field of aesthetics with important value and significance. As Berleant puts it, if we were to take aesthetic experience of the environment as the standard, we would throw out the conception of disinterested aesthetic appreciation and support a model of aesthetic engagement. He states, "[a]esthetic engagement is not only illuminating, then, for architecture and environment; it can also be applied with remarkable consequences to the other arts, traditional as well as contemporary."[65] Carlson further makes an appraisal of the value of the aesthetics of engagement from the perspective of constructing the field of aesthetics, arguing that the key to shaping environmental aesthetics into a field is not only concern with the aesthetic appreciation of natural environments, but is even more so the concern the aesthetic appreciation of our

---

[62]Yu [22], pp. 303, 308–309, 310, 423, 424.

[63]Berleant [21], p. 134.

[64]Berleant [20], p. 9.

[65]Berleant [21], p. 142.

entire surrounding world.[66] These expositions reveal the important implications that environmental aesthetics has for the field of aesthetics in the universal sense. This universal significance Berleant calls the reconstruction of the way of doing artistic research.[67]

(2) **The Implications of the Aesthetics of Engagement**

What is the aesthetics of engagement? First of all, this is a model of aesthetics that is differs from the traditional aesthetics of disinterested contemplation; it is an aesthetics of all the senses of the human body actively engaging in the process of aesthetic appreciation. As Berleant puts it, "The aesthetic mark of all such times is not disinterested contemplation but total engagement, a sensory immersion in the natural world that reaches the still-uncommon experience of unity. Joined with acute perceptual consciousness and enhanced by the felt understanding of assimilated knowledge, such occasions can become clear peaks in a cloudy world, high points in a life dulled by habit and defensive disregard."[68] This is actually an ontological conception of environment and aesthetics, because contemporary ontological philosophy already overcomes the model of being and thinking in the world through the dualistic opposition between subject and object, man and nature, mind and body, and what is layed out instead is a model of being-in-the-world: Dasein is "in the world" and fuses into one with the world. Berleant is a famous phenomenological philosopher of aesthetics, who continuously upholds the phenomenological method of contemporary ontology as his philosophical standpoint in the belief that there is nothing outside of nature, everything is involved in it. He states, "Environment, in the large sense, is not a domain separate and distinct from ourselves as human inhabitants. We are rather continuous with environment, an integral part of its processes. The usual tradition in aesthetics has difficulty with this, for it claims that appreciation requires a receptive, contemplative attitude. Such an attitude befits an observer, but nature admits of no such observer, for nothing can remain apart and uninvolved."[69] The nature of which he speaks here is not nature in the simple sense, but nature in which there is human participation. Berleant points out that "we can rediscover Spinoza's insight that there is no nature apart from our human presence and nothing human separate from nature."[70] He insists on opposing the dualistic tendency of opposing human being and environment, "in the sense of regarding humans as placed in them rather than as continuous with them."[71]

From this we can see that what Berleant calls "nature" or "environment" is actually "the totality of life," and what is he calls "aesthetics of engagement" is actually the "aesthetics of ecological holism," or rather we should say, it is "the aesthetics of ecological ontology." This conception of aesthetics is made up of the aesthetic sub-

---

[66]Carlson [23], p. 7.

[67]Berleant [21], p. 155.

[68]Berleant [21], p. 154.

[69]Ibid., p. 12.

[70]Ibid., p. 36.

[71]Ibid., p. 11.

ject and aesthetic object in the process of appreciating beauty, which in one respect insists on the subject's role of subjectively constituting in the process of appreciating beauty, but also does not deny the aesthetic properties latent in nature itself. Rolston reduces the aesthetic appreciation of nature to two related conditions, which is the combination of the aesthetic capacities of human being and the aesthetic properties of nature, thinking that the aesthetic appreciation of nature only becomes possible in the unity of the two and the active participation of human being. He states, "There are two sorts of aesthetic qualities: aesthetic capacities, capacities for experience that are only in beholders, and aesthetic properties, which lie objectively in natural things. The experience of beauty does arise in the beholder, but what is this experience of? It is of form, structure, integrity, order, competence, muscular strength, endurance, dynamic movement, symmetry, diversity, unity spontaneity, interdependence, lives defended, coded in genomes, regenerative power, speciation, and so on. These events are there before humans arrive, the products of a creative evolutionary and ecosystemic nature; and when we humans value them aesthetically, our experience is being superposed on natural properties."[72] He draws the example of people's enjoyment of the grace of the impala's leap, insisting that the impala acquired a muscular force of bodily movement through the long period of biological evolution, because of which it is capable of gracefully leaping—the aesthetic experience of this enjoyment only becomes possible in the human act of appreciation, when this leap encounters human being's subjective aesthetic capacity.

(3) The Costs and Benefits of "Aesthetics of Engagement"

The positing of "aesthetics of engagement" undoubtedly broke-threw the traditional aesthetics of disinterested contemplation, and brought the long-neglected appreciation of nature and environment into the field of aesthetics, which is of fully important significance; it not only broke with the framework of the uniqueness of art in terms of the aesthetic object; it also broke with the dualistic model of the subject-object opposition in terms of the manner of aesthetic appreciation. What should be specifically highlighted here is "aesthetics of engagement" elevates aesthetic experience to a considerable height in the belief that purely "contemplative" observation of "picturesque" scenery is far from adequate when facing natural landscapes filled with life and vitality, which is to say the "engagement" of all the senses are necessary. As Rolston put it, "We initially may think of forests as scenery to be looked upon. But a forest is entered, not viewed. It is doubtful that one can experience a forest from a roadside pullover, any more than on television. The forest attacks all our senses: sight, hearing, smell, feeling, even taste. Visual experience is critical, but no forest is adequately experienced without the odor of the pines or of the wild roses."[73]

The British philosopher, Malcolm Budd, pointed out the three principal defects of the aesthetics of engagement in The Aesthetic Appreciation of Nature: first, that we are essentially in, not over against, the landscape, does not prevent our aesthetic experience from being contemplative, which often it properly is. Second, disinter-

[72]Berleant [20], p. 158.

[73]Berleant [20], p. 166.

estedness of response in this sense is not only compatible with the various aspects of engagement that Berleant articulates that are aesthetics, but is a condition that, it seems, any satisfactory understanding of the notion of an aesthetic response must satisfy. Third, Berleant's failure to replace them disqualifies his aesthetics of engagement with nature from being acceptable either as an account of nature appreciation or as a conception of aesthetic experience of nature.[74] Budd's criticism does make some sense, but overall, he still examines "the aesthetics of engagement" from the standpoint of the traditional aesthetics of contemplation, and even more importantly, he ignored the fact that the aesthetics of engagement is a contemporary ontological aesthetics—whose definition of beauty is no longer directly related to the traditional form of beauty, but is rather a sort of poetic dwelling and clearing of truth.

Of course, there at least a few theoretical difficulties to the aesthetics of engagement at present. While the role of the aesthetics of engagement in the aesthetic appreciation of the natural environment is clear, using the total engagement of all the senses (eyes, ears, nose, tongue, body) in the aesthetic appreciation of art is pushing it a little too far. Berleant obviously notices this point, and tries to mount some kind of defense with respect to the complexity of modern art. He points out in the recently written work Recent Developments in Environmental Aesthetics: "Throughout the century that has just ended, the arts have not only flourished in their customary forms as painting, sculpture, architecture, music, theater, literature, and dance. They have pressed with growing insistence beyond their traditional boundaries. In dada and the many innovative movements that followed, painting has incorporated taboo materials and subject matters, employed texts in conjunction with images, broken out of the canvas and extended beyond the frame. Sculptures have enlarged their size and form so that we can walk on, through, and into them, and they have expanded into environments, both enclosed and out-of-doors. Architecture has gone beyond monumental edifices, challenging conventional shapes and structures and merging with the site. Music has employed new modes and ranges of tone production, both in the synthesizer and in the use of noise and other nonmusical sounds, and has explored different performance venues, such as environmental music. Poetry has relinquished rhyme and meter, while fiction has transformed the novel and other traditional forms. Dance has not only developed into different styles of modern dance but has broken away from conventional standards of gesture, lightness, and grace. Theater, along with other arts, has developed forms that require active audience participation."[75] (Environment and the Arts; Perspectives on Art and Environment, 2002) The growth of modern art toward active art has indeed prepared the conditions for the engagement of the whole body's eyes, ears, nose, tongue and body in the aesthetics of engagement. But, when faced with traditional forms of art, the absolute validity of the aesthetics of engagement is dubitable, in which case, it would not hurt for us to expand "engagement" to the subject's active participation, firstly the participation of the subject's aesthetic perceptual capacity in the constitution of the aesthetic object, and when faced with a natural environment, it involves the participation of all the

[74]Bard [24].

[75]Berleant [25].

senses. Aesthetics of engagement as the contemporary aesthetics of ecological ontology highlighting the readiness-to-hand and unreadiness-to-hand of the object with respect to its involvement with the subject certainly breaks through the aesthetics of contemplation and its subject-object dichotomy. In this expansion, the aesthetics of engagement as an aesthetics of ecological ontology that uses the phenomenological method has great universality and theoretical self-evidence.

12.7 Ecological Arts

(1) **Why Propose Ecological Literary Arts**

The positing of ecological literary arts is the real need to take ecological aesthetics from theory and put it into practice. As is widely known, the biggest feature of contemporary ecological aesthetics is that it is not entirely a purely academic form of theory; one of its basic or principal properties is that it has the greatest practicality, which is of course firstly exhibited in the literary arts, because the literary arts is one of the most important practical forms of aesthetics. In addition, the positing of ecological literary arts is also the manifestation of the ecological aesthetics researcher's social duty. Why do we have to research ecological aesthetics? Aside from the theoretical and academic impulse, there is also a strong sense of social responsibility, a strong social responsibility to save the environment, save the earth and save humanity. The founder of ecocriticism and eco-poetics—William Rueckert asks, how can people engage in the great duty of creative and cooperative activity in the ecosphere as readers, teachers and critics of literature? He think that we need to start by answering these questions, and continue to do the work we have always been doing, seeking help from poets and then seeking help from ecologists. We need to establish an ecological poetics.[76] The positing of ecological literary arts is also related to the current state of our literary arts. The notion that humans will triumph over Nature and the thought of natural beauty belonging to "humanized nature" still persists strongly. The continuous worsening of the environment in the top-down nature of real life makes us aesthetics workers need not be silent and absent, but rather let our voices be heard, and take steps into changing the reality at hand, especially the reality of the literary arts.

(2) **The Implications of Ecological Literature**

1. "The Green Principle" of Literary Activity

The positing of the "green principle" in literary arts drew lessons from the American ecocritic Lawrence Buell point of view on "Green Reading." Buell once remarked that there is no green literature without green thinking and green reading. Here, Buell continues to use these three important concepts: green thinking, green reading and green literature. "Green" is the color of the earth, representing life, ecology and

[76]Ruckert [18], p. 107.

growth. "The green principle" is the "symbiotic" principle of "the biosphere," that of the ecological whole, which differs from both traditional "anthropocentrism" and extreme "ecocentrism." What it advocates is the theoretical principle of an ecological whole in which "all creatures nurture one another without destroying one another." Just as Leopold's Sand County Almanac states: "A thing is right when it tends to preserve the integrity, stability, and beauty of the biotic community. It is wrong when it tends otherwise."[77]

2. Ecocriticism

In 1978, William Rueckert, a researcher of American literature, published an article entitled "Literature and Ecology: An Experiment in Ecocriticism" in The Iowa Review (Winter, 1978), where he first used the concept of "ecocriticism," from which point onward, ecocriticism became an extremely important form of literary criticism after social criticism, aesthetic criticism, psychoanalytic criticism and archetype criticism; it also became an important component and practical form of contemporary ecological aesthetics, and very quickly became a vigorously developing "dominant theory."

"Ecocriticism" is first of all a cultural criticism; it is a cultural criticism that unfolds from the perspective unique to ecology; it is a noble attempt to combine ecological responsibility with literature and aesthetics by workers in literature and aesthetics in the face of increasingly severe environmental pollution. When Rueckert states his own reasons for putting forward the idea of "ecocriticism," he states, "[t]rue to the experimental motive intrinsic to me as a human being, I have taken on the question of how reading, teaching, and writing about literature might function creatively in the biosphere, to the ends of biospheric purgation, redemption from human intrusions, and health." Faced with such severe environmental pollution, he yells loudly at workers in literature and aesthetics that people have to start doing something![78] The philosopher of environmental aesthetics, Berleant, further emphasizes "the fundamental coalescence of aesthetics and ethics," and Rolston advocates "the beauty of the ecosphere,"[79] asserting that Grand Canyon protections have been undertaken out of considerations of its beauty and splendor. From this perspective, he argues that the ultimate historical basis of environmental protection is aesthetics.[80] From the perspective of contemporary ecological aesthetics, the unity of ethics and aesthetics is also the most fundamental principle. Because the environment is not always positive for humanity, for instance, because noise is disruptive to human perception and damaging to human health, noise is therefore neither good nor beautiful.

From this we can see that the unity of environmental ethics and aesthetics is the most basic principle of ecocriticism. The theorists of ecocriticism believe: art possesses an energy that can change humanity, an energy that expresses itself in changing people's minds and hence changing their attitudes, thereby enabling them

---

[77]Leopold [26], p. 213.

[78]Grottefelti etc. [27], p. 40.

[79]Berleant [20], p. 24.

[80]Berleant [20], p. 151.

to turn from destroying nature to protecting nature. As Francis Ponge, states in The Voice of Things, "Hope therefore lies in a poetry through which the world so invades the spirit of man that he becomes almost speechless, and later reinvents language."[81] Perhaps, the ecocritics overestimate the role of literary art, but changing people's cultural attitude through aesthetic education, and thereby making people gradually succeed in treating nature with an aesthetic attitude is still possible. We should continue our efforts in this direction in the expectation that we can succeed.

3. The Re-evaluation of the Value of Art

We are currently faced with a massive shift in social and cultural values from "anthropocentrism" to "ecoholism," and since ecological literature must make a huge adjustment in the position of philosophical culture and the principle of artistic creation, it necessarily faces the problem of re-evaluating the value of literary arts. The facts suggest that any massive shift in economic, social and cultural values necessarily results in the re-evaluation of values in the sphere of literature and arts. Early on at the end of the 19th century and the beginning of the 20th following the end of classical German philosophy, the traditional principle of subjectivity gradually exposed its negative side, from which came Nietzsche's important propositions in The Birth of Tragedy, "the re-evaluation of all values" and "God is dead." Today, the massive socio-economic shift from industrial civilization to ecological civilization will necessarily precipitate a reassessment of literature and art. In Revaluing Nature, the American scholar, Glen A. Love acutely observes, "so must it happen that our critical and aesthetic faculties will come to reassess those texts—literary and critical—which ignore any values save for an earth-denying and ultimately destructive anthropocentrism."[82] This reassessment is an important component of the "green reading" that Buell advocates, namely revaluing past works of literature using the "green principle" of "the ecological whole." On the issue of this revaluing in the field of literary arts, the academic world is currently debating. The American, Jay Parini gives his own viewpoint on "revaluing" through ecocriticism in The Greening Humanities. He states, ecocriticism "marks a return to activism and social responsibility; it also signals a dismissal of theory's more solipsistic tendencies. From a literary aspect, it marks a re-engagement with realism, with the actual universe of rocks, trees and rivers that lies behind the wilderness of signs."[83] Here, Parini speaks of a few principles of literary revaluation, some of which are desirable, for instance, the return to social responsibility and the abandonment of solipsistic tendencies. Social responsibility here naturally refers to the social responsibility of ecological protection, which is the essential duty of every member of humanity on earth; and the "solipsistic tendencies" that naturally ought to be abandoned here refer to the "anthropocentric tendencies" that have been growing ever since the industrial revolution, under whose guidance humankind has greatly intensified its destruction of the earth and pollution of the environment. But a return to the behaviorist psychology

[81] Ruckert [18], p. 105.

[82] Love [28], p. 235.

[83] See Ning's [29], p. 289.

of a fundamental distinction between humans and animals would be inappropriate, same with a "re-engagement with realism." Although realism follows the principles of faithfully reflecting reality, the flourishing period of critical realism was precisely the time for the deepening of the industrial revolution in the 19th century. Even if it is critical realism, it still follows the principle of "anthropocentrism"; the "anthropocentric" thought that runs through it along with its praise of industrialization and the newly arising bourgeois class still have not broken free of the old conventions of anthropocentrism; and even though the modern schools opposed to it, whose distinctive feature is exaggerated transfiguration, are characterized by different variations of nature, they do imply the deeper significance of criticizing the overgrowth of instrumental reason. The famous mural Guernica by Picasso uses extreme exaggerations and transfigured pictures of humans, bulls and other animals as a denunciation of and protest against the savage behavior of fascism, in order to shock people.

(3) **Eastern Resources of Ecological Literature**

Ancient China possesses exceptionally fruitful resources of ecological literature, which are very much worth our efforts to uncover them. From the perspective of philosophical thought, ancient China followed the theory of "the integral oneness of nature and humans," and placed Heaven and Earth, the Universe and Nature in the position of humans and all things all being on the same flat plane; with respect to art, Chinese arts turned nature into the object of expression early on. The graph inscription huodan 火旦 (flaming dawn) on the earthenware statue unearthed among the Dawenkou cultural artefacts of Shandong, according to archaeological analysis, is assumed to be ancient people's depiction of the natural scene of the sun rising over a mountain temple that they observed. Book of Poetry and Encountering Sorrow (li sao 离骚) have many depictions of natural mountain waters. Mountain water painting vigorously erupted in the Wei-Jin era to become a painting form unique to China. The Southern dynasties were even more so an era of vigorous growth in mountain water poetry and landscape poetry. Ancient Chinese arts did not grow like ancient Western arts in statue carving focally centered on imitating the human form as lifelike as possible, but rather were founded on the poetic and painted depictions of natural mountain waters guided by the idea of "the integral oneness of nature and humans," an idea which puts nature on equal and friendly footing with humankind.

1. Analogy and Incitement (bixing 比兴)

"Analogy and incitement" is the principal artistic technique in the *Book of Poetry*. In our view today, analogy and incitement is really the ancient principle of "green aesthetics" that puts nature on equal footing with humanity. The Chinese character bi 比 is described in Explaining Graphs and Analyzing Characters (shuowen jiezi 说文解字) as two human beings depending each other. The explanation goes: "bi 比 means close. Two humans (ren 人) is two people following [the same direction] cong 从, two people following the reverse direction (fancong 反从) is bi 比."Duan Yu of the Qing dynasty comments on this saying: "its basic meaning is mutual dearness and closeness." Visibly, the basic meaning of "bi 比" is two people living intimately with one another. In the *Book of Poetry* is, bi 比 is used as an expressive technique to "analogize with things." Zheng Zhong of Eastern Han says of xing 兴: "xing 兴 is setting affairs off against things." Duan Yu comments in the Commentary on Explaining Graphs and Analyzing Characters: "xing 兴 means incite." The guangyun 广韵 (Broad Rhymes) states: 'flourishing, means to raise, to perfect.'" "Xing 兴" in Book of Poetry is the using of natural things to incite the words the poet writes. Liu Xie of the Southern dynasties wrote The Literary Mind and the Carving of Dragons (wenxin diaolong 文心雕龙), in which the section entitled "Analogy and Incitement" (bixing 比兴) states "the poet analogizes and incites, triggering the all-around view of things. Although the things are as distant as the state of Hu and the state of Yue, they come as close together as the liver and gall bladder. Description takes the affective mind, making a judgment necessitates daring. Putting together the disparate in poetic form is like the vanishing of a river." Clearly, "analogizing and inciting" is the expression of the human being's emotion with natural things. Although the mind and things are greatly different from one another, as long as they are used precisely, they match up and come together with one another like the liver and gall bladder work with one another. The application of analogy and incitement needs both description and the affective mind, namely both writing the likeness of natural things and manifesting one's inner spiritual rhythm. The "likeness" and "mentality" of "things" is closely and intrinsically bonded with the emotions and consciousness expressed by the poet.

2. Analogizing Virtue (bide 比德)

"Analogizing virtue" is metaphorically illustrating the virtuous characteristic of people's noble purity with natural things. The first to employ this technique was Confucius, who in Analects "Yong Ye" states: "the wise enjoy waters, the humane enjoy mountains. The wise are active, the humane are still. The wise find delight, the humane find longevity." Xunzi inherits this theory and develops it, saying in "The Proper Model and Proper Conduct" (faxing 法行) that "[a]s for jade, the gentleman compares it with being-virtuous. The way it is warm, smooth, and lustrous resembles being-humane. The way it has structure and contains patterns resembles being-wise. The way it is firm and unyielding resembles being-righteous. The way it is sharp but does not cut resembles acting-appropriately. The way it can be broken but does not bend resembles acting-courageously. The way that even its flaws are visible resembles being-genuine. In the way that when struck, its sound is pure, rises high, and can be heard far away, but when it stops, it finishes completely, it resembles speaking-properly." (Xunzi: The Complete Text, Eric Hulton, p. 331 [modified]). Here, Xunzi uses the natural qualities of this natural thing, jade, its warmth, smoothness and lustrousness, its firmness and unyieldingness, its patterns and structure, its being-sharp but not-cutting, its breaking but not bending, its being-flawed but visible, its pure rising sound that when stops finishes completely to metaphorically illustrate the outstanding virtue of someone being-humane, being-wise, being-righteous, acting-appropriately, acting-courageously, being-genuine and speaking-properly. Afterward, the thought of "analogizing virtue" grew further in the world of Chinese literary arts. For instance, "bamboo" was used to illustrate the nobility and incorrupt nature of conduct, "Chinese plum tree and the orchid" was used to illustrate the character needed to survive difficult conditions, and "lotus" was used to illustrate the virtue of staying clean through dirtying circumstances, "pine, bamboo and plum" are "the three winter companions," which metaphorically illustrate someone's will to survive difficulties without yielding to them.

3. zaohua 造化 (Creator)

"Creator" (zaohua 造化) originally refers to the greatness of Heaven and Earth or Nature creating all beings. Du Fu's Gazing at the Mountain has the verse, "Here the Creator gathers wonders divine; Its northern and southern slopes divide dusk and dawn," after which zaohua (造化) has referred to greater Nature, which nourishes all beings. As stated previously, one of the most important characteristics of Chinese literary arts is that of turning Heaven and Earth or Nature into the object of expression, because of which zaohua also become an important concept in the philosophy of Chinese art. The famous Tang dynasty painter Zhang Zao enunciated the immortal statement about painting: "Outside the teacher is the natural Creator, inside the origin is the affective mind," which reveals the basic principle of ancient Chinese mountain water painting creation. Wang Wei then points out in Knack of Mountain Waters (shanshui jue 山水诀): "In the way of painting, the water ink is esteemed the highest. Commence the tendency of what is naturally so, complete the effort of the natural Creator. In a picture of eight cun by one chi, draw a landscape of one thousand li. East, West, North and South seemingly just before the eyes: Spring, Summer, Autumn and Winter, created right under the brush." In Wang Wei's view, water ink paintings can better depict the natural "Creator" as the source of all things in Heaven and Earth, because of which, ancient Chinese painting has always turned to the idea of "the natural Creator as teacher." Luo Dajing of the Song dynasty recorded two perfectly vivid examples of "taking the natural Creator as teacher" in Theory of Painting (huashuo 画说): one is Emperor Ming of Tang ordering the painter Han Gan to paint horses; he first made him observe the paintings depicting horses he had archived, but Han Gan said "the horses in the yard are all teachers." The painter, Li Boshi of the Northern Song painted horses, spending all day observing imperial horses without any time to talk with guests to reach the effect of "having his heart become full horse in his breast, trusting desire to drop the brush, naturally beyond subtle." Zeng Yunchao was good at painting grass and insects, reaching the acme of perfection, and surpassing refined in old age. Someone asked Zeng, do you have something to teach on painting grass and insects? Zeng stated: "[e]ver since I was young I grabbed grass insects, put them in a cage and observed them all day and night without getting sick of it. But then I feared their spirits were not whole, and proceeded to put them amongst the grass on the ground to observe them, so they would begin to acquire their Nature. At the brink of setting down my brush, I did not know whether I was a grass insect or the grass insect was me, which is so close to the trigger between the natural Creator and creature there is no difference between them." These two examples both vividly illustrate the marvel of ancient Chinese arts "taking the natural Creator as teacher."

4. Intensity of Spirit (qiyun shengdong 气韵生动)

Qi Xiehe of the Southern dynasty brought up the famous theory of "six methods in painting" in Catalogue of Ancient Paintings (guhua pinlu 古画品录), and gave primary status to "intensity of spirit." Zhang Yanyuan of the Tang dynasty contrasted ancient painting and the painting of his day in On Painting (lun hua 论画), offering a further explanation "on the six methods of painting," and asserting that ancient painting should be the model. He states: "Some paintings of ancient times could change the likeness of their shapes but elevate the intensity in their bones, and seek their painting outside of the likeness of shapes, which is hard to pair with the customary way of people [today]." In his view, an outstanding work of painting cannot simply seek the likeness of shapes, but should seek the "intensity in the bones" outside of the likeness of their shapes. Here, "bones" means "the repository of marrow," referring to "the original energy of Heaven and Earth," but "intensity in the bones" refers to the convergence of human being's most basic and most principal vital spirit. Zhang Yanyuan uses "intensity in the bones" to explain "the intensity of spirit," highlighting the nature of "intensity" reflected in human being's vital spirit. Deng Chun of the Song dynasty explains "intensity of spirit" as "transmitting spirit" in Continuation of Painting (huaji 画继). He states, "the function of painting is grand. What fills Heaven and Earth is beings, everyone with outstanding use of thought curves [the plane] to fully express his poise. There is but one method that makes possible the curving of [the plane] to fully express. Which is the one [method]? It is called transmitting spirit and that is all." Because of this, "[t]he first method of painting is intensity of spirit." Deng Chun points out: "the people of this age are aware there is spirit in human being, but are not aware there is spirit in other creatures." Obviously, what he calls "chuan shen 传神" is transmitting the vital spirit of all creatures of Heaven and Earth.

5. "Constant Joy and Constant Satisfaction"

Guo Xi of the Song dynasty explained the cause behind the flourishing of mountain water painting in China in the text The Lofty Message of Forests and Springs (linquan gaozhi 林泉高致), reducing that cause to the natural disposition of human being to find constant joy and constant satisfaction in mountain waters. He states: "[t]he reason why junzi love mountain waters, where does it lie? It is for these reasons: that in a rustic retreat he may nourish his nature; that amid the carefree play of streams and rocks, he may take delight; that he may constantly meet in the country fishermen, woodcutters, and hermits, and see the soaring of the cranes, and hear the crying of monkeys. The din of the dusty world and the locked-in-ness of human habitations are what human nature habitually abhors; while, on the contrary, haze, mist, and the haunting spirits of the mountains are what human nature seeks, and yet can rarely find. It is positive with great peace and flourishing days, one can both serve one's ruler and one's kin for one's whole life carefree and unsullied, and yet exit the social world with integrity and righteousness intact without having to engage in the hermit's behavior of divorcing from the world, climbing high and withdrawing far away." The creation and appreciation of mountain water paintings originated largely from people's notion to detach from secular life in the desire to tread high and far away from society and social conventions, and simultaneously, because natural mountain waters were the places where superior human beings would constantly dwell, enjoy, adapt to and observe. Here, "constantly dwelling," "constantly enjoying," "constantly fitting into" and "constantly observing" reveal human being's essence to get closer to mountain waters. Of course, people getting closer to natural mountain waters, craving the art of mountain water paintings, and striving for transcendent concerns beyond material profit was also related to the dark turmoil in society at the time, and the yearnings for hermetic Daoist thought brought about by the tumultuous changes in official career life at the time. This was perhaps also an important reason behind the rise of mountain water painting during the Wei-Jin period of China. Guo Xi bringing up the constant dwelling among, constant joy of, constant satisfaction with, constant observation of, and constant desire for natural mountain waters reflects the philosophical idea of "nature and human being merging into one" deeply fusing into people's minds in China as a major agricultural state. Ancient Chinese people in general, and Chinese artists in particular, naturally possessed the artistic thought and feeling of being amongst mountain waters and befriending Nature.

12.8 The Two Forms of Ecological Aesthetics: The Beauty of the Health of Supple Yin and the Beauty of the Strength of Firm Yang

(1) The Beauty of the Health of Supple yin and the Beauty of the Strength of Firm Yang in the ancient Chinese Zhou Book of Changes

We think that there are two different forms of ecological aesthetics, one of which is the beauty of the health of supple yin and the beauty of the strength of firm yang. The earliest to bring up these two forms and conduct deep investigations of them was the ancient Chinese classic the *Zhou Book of Changes*. In fact, the 64 hexagrams of the *Zhou Book of Changes* revealed the variety of human being's existential conditions in the "relationship between Nature and humans," and these states of existence are states of human life as well as aesthetics states of humanity's ecological being, indicated by the two basic affects of yin and yang and their relationship to one another. People are more familiar with the beauty of supple yin in human being's ecological aesthetics as explicated in the *Zhou Book of Changes*. This beauty of supple yin revealed in the *Zhou Book of Changes* is the beauty of Earth, which was comprehensively expounded in the judgments of the lines of the Kun hexagram and the Tuan Zhuan and Xiang Guan commentaries on characters and signs. Its main characteristics are "occupying the substance of the proper position," and "complete suppleness and complete stillness"; its basic quality is "the thickness of Kun to bear creatures"; its basic forms are "the linking the patterns in the Yellow Center" and "the way of Earth and the way of the wife"; its basic duty is to nourish the life of all creatures. This is the beauty of the health of striking harmony characterized by Heaven and Earth intercrossing, all creatures interlinking and favorable weather. Just as the Tuan Zhuan commentary on the tai 泰hexagram in the *Zhou Book of Changes* puts it, "In 'Tai' (we see) the petty pass away and the great come along, good fortune with progress and success. It is thus Heaven and Earth intercrossing, and all creatures interlinking; the higher and lower (in society) cross paths, and their wills converge. Yang on the inside and yin on the inside, assertive within and compliant without, superior master within and minor human being without."

Bumper harvest with much millet and much rice,
And with a granary filled high,
One trillion and one thousand that.
Offering wine and offering sweet wine,
Many conferences on ancestors.
To fit hundreds of rituals appropriately,
Raining fortune all around with much fortune.

But, Heaven and Earth intercrossing, yin and yang harmonizing, and weather coming favorably are only cases that do not occur frequently in agricultural society; it much rather represents people's ideal of life. Most harvests are natural and human disasters one on top of another, and especially when productivity was extremely underdeveloped, what nature brought to people was most frequently not bumper harvests, and perhaps more frequent was major disastrous and minor difficulties, in which case the reason why humans could continue living and reproducing without withdrawing in the face of difficulty generally had to do with the noble spirit of tirelessly striving to be stronger expressed in a complex emotion of resistance to Nature, respect for Nature and prayer to Nature. It was this spirit that enabled humans to brave the storms and ceaselessly plough forward all the way to today. *Zhou Book of Changes* is the crystallization of all of the wisdom accumulated by ancient Chinese people to cope with the difficulties in the matching of Nature and human society. Just as *Zhou Book of Changes* "Xici Part II" states, "Was it not in the middle period of antiquity that the Book of Changes began to flourish? Was not he who made it familiar with anxiety and calamity?" The *Zhou Book of Changes* is the product of humans coping with anxiety and calamity. The ancient Chinese people's most effective means of coping with anxiety and calamity was the unyielding spirit to tirelessly strive to become stronger. The judgment of the third NINE line of the Qian hexagram in the *Zhou Book of Changes* states "In the third NINE, undivided, (we see its subject as) the superior man active and vigilant all day, and in the evening still careful and apprehensive. The position is dangerous, but there will still be no mistake." In the face of such predicaments as calamity, natural disaster and imminent danger, in those brutal natural conditions, people never slacked, but rather maintained a state of tireless fight all day, vigilant at every moment, taking preventive measures. This is the spirit expressed in the Xiang Zhuan commentary on the Qian hexagram of the *Zhou Book of Changes*: "Heaven, in its motion, (gives the idea of) strength. The superior man, in accordance with this, nerves himself to ceaseless activity." The *Zhou Book of Changes* insists that the way of the heavens is the power of unyielding yang. The Qian hexagram is made up of 6 unbroken yang lines symbolizing the unbreakable assertiveness of the way of the heavens. People emulate the firm power of the way of the heavens, and therefor tireless strive to become stronger. It is the virtue of this spirit to tirelessly become stronger that has enabled humans to cope with the variety of disasters, overcome all difficulties and dangers, and develop on forward the *Zhou Book of Changes* concentrates and expresses the ancient dialectical spirit of the East, namely the spirt of "good weather comes at the peak of the nasty storm"; it is also the main spiritual power of the Chinese people as enveloped in the *Zhou Book of Changes*.

The Pi 否 hexagram in the *Zhou Book of Changes* is originally an extremely inauspicious sign, meaning "Heaven and Earth do not interlock and all creatures interlink, the higher and lower do not cross paths and the realm fails to unite." But the line judgments of the Pi hexagram similarly inform people that as long as they tirelessly try to become stronger amongst difficulties and dangers, they can eliminate all difficulties and move to the light. The line judgment of the first nine of the Pi hexagram states, "The first NINE, undivided, suggests the idea of grass pulled up, and bringing with it other stalks with whose roots it is connected. Advance (on the part of its subject) will be fortunate," informing the superior human being that difficulty can be overcome and an auspicious state of fortune reached if he or she tirelessly tries to become stronger. The judgment of the line of the sixth NINE of the final line of the Pi hexagram states, "The sixth NINE, undivided, shows the overthrow (and removal of) the condition of distress and obstruction. Before this there was that condition. Afterward there will be joy." The Xiang Zhuan commentary on the said line states, "The distress and obstruction having reached its end, it is overthrown and removed. – How could it be prolonged?" This is the meaning of the famous saying that "good weather comes at the peak of the distressing storm." Then look at the Zhen 震 hexagram of *Zhou Book of Changes*, the hexagram sign is Zhen on top and Zhen on bottom, whose sign is lightning, symbolizing the unbroken string of terrifying lightning, startling Heaven and moving earth, great force and massive destruction, startling the human heart. But the hexagram judgment of the said hexagram informs people that as long as they "tirelessly try to become stronger" and seriously deal with it, even in this dangerous circumstance, they can still render danger into safety and calm the crisis. The saying goes, "thunder comes growling, laughter speaks cawing, auspicious," which is to say, as long as one maintains calm amongst intense tumult and fear, and moves with ease, dealing with it in laughter, one can still obtain an auspicious result. The *Zhou Book of Changes* places the Qian hexagram symbolizing "the assertive power of heavenly movements" at the very beginning of the 64 hexagram cycle, not only implying the praise of "the great virtue" of "Heaven" in creating all creatures, but also expressing respect for the spirit of the tireless self-empowerment coming from the yang force of Qian. The Wen Yan Zhuan commentary on the Qian hexagram points out: "Qian, (thus) originating, is able with its admirable benefits to benefit all under the sky. We are not told how its benefits are conferred; but how great is (its operation)! How great is (what is emblemed by) Qian!—Strong, vigorous, undeflected, correct, and (in all these qualities) pure, unmixed, exquisite!" This is to say, the Qian emblem not only "nourishes the beginning of all things with its yang energy, but also feeds humans with its spirit of "assertive power of heavenly movement" and "undeflected correctness of strong vigor," making humans tirelessly try to become stronger and surmount difficulties. This is truly a uniquely virtuous power and beautiful action.

(2) **Nietzsche's Expositions on the Dionysian Spirit and Apollonian Spirit**

The famous German philosopher, Nietzsche, wrote the renown *Birth of Tragedy* in 1872, and brought up the Dionysian spirit and Apollonian spirit. Even though what Nietzsche was discussing was still not eco-ontological aesthetics, one point is very clear, which is that the Dionysian spirit and Apollonian spirit do not entirely belong purely to the philosophy of art; it was rather a deconstruction of Hegel's traditional philosophy of art, discussing life aesthetics in the broader sense and the tragedy of human life including the tragedy of human being in Nature and society. In addition, what is perfectly clear is that even though Nietzsche mainly used Western resources when discussing the tragic spirit, he also reflected on Eastern resources. Because of this, the two forms of the tragic spirit that Nietzsche expounded upon were not entirely the sublimity of language that the ancient Greek Longinus wrote on in On the Sublime nor the sublimity of reason in Kant's *Critique of Judgment*. Nietzsche states: "We shall have gained much for the science of æsthetics, when once we have perceived not only by logical inference, but by the immediate certainty of intuition, that the continuous development of art is bound up with the duplexity of the Apollonian and the Dionysian: in like manner as procreation is dependent on the duality of the sexes, involving perpetual conflicts with only periodically intervening reconciliations." And further, "In order to bring these two tendencies within closer range, let us conceive them first of all as the separate art-worlds of dreamland and drunkenness; between which physiological phenomena a contrast may be observed analogous to that existing between the Apollonian and the Dionysian."[84] We can see from Nietzsche's exposition of the origin of tragedy that what he discovered was ancient Greek art in the primitive state and the esoteric rituals of ancient Greek tribes, through which the implications of tragedy along with the implications of man's ecological being in the Dionysian spirit become much clearer. Nietzsche argues that the Dionysian spirit is a set of primordial forces that after being dismembered again transform into air, earth, fire and water to become reborn. He states, "this hero is the suffering Dionysus of the mysteries, a god experiencing in himself the sufferings of individuation, of whom wonderful myths tell that as a boy he was dismembered by the Titans and has been worshipped in this state as Zagreus: whereby is intimated that this dismemberment, the properly Dionysian suffering, is like a transformation into air, water, earth, and fire…" Lastly, "we have the mystery doctrine of tragedy: the fundamental knowledge of the oneness of all existing things, the consideration of individuation as the primal cause of evil, and art as the joyous hope that the spell of individuation may be broken, as the augury of a restored oneness."[85] Nietzsche distinguishes the two types of tragedy along the lines of the two sexes, masculinity and femininity, which

[84] Wilhelm Nietzsche [30], pp. 2, 3.

[85] Ibid., pp. 41, 42

obviously relates his account to the influence of the yin-yang doctrine in Eastern culture. The primordial social background of tragedy's birth in ancient Greece that he discovered also reveals the eco-ontological implications communicating human being with Nature involved in the Dionysian spirit of tragedy.

Nietzsche insists that Apollo rules over appearance, moderation, simplicity and dream. Apollo is extremely important to humankind. He states, "Apollo, as the god of all shaping energies, is also the soothsaying god. He, who (as the etymology of the name indicates) is the 'shining one,' the deity of light, also rules over the fair appearance of the inner world of fantasies. The higher truth, the perfection of these states in contrast to the only partially intelligible everyday world, ay, the deep consciousness of nature, healing and helping in sleep and dream, is at the same time the symbolical analogue of the faculty of soothsaying and, in general, of the arts, through which life is made possible and worth living."[86] The shine, harmony, appearance and fairness of the Apollonian spirit gives human life a goal to strive for, but the greater worth of human life is found in the Dionysian spirit. Nietzsche insists that the Dionysian spirit is a release, madness, intoxication and emotional outburst, where the true worth and meaning of human life lies. He states, "we shall gain an insight into the being of the Dionysian, which is brought within closest ken perhaps by the analogy of drunkenness. It is either under the influence of the narcotic draught, of which the hymns of all primitive men and peoples tell us, or by the powerful approach of spring penetrating all nature with joy, that those Dionysian emotions awake, in the augmentation of which the subjective vanishes to complete self-forgetfulness." He then states, "Under the charm of the Dionysian not only is the covenant between man and man again established, but also estranged, hostile or subjugated nature again celebrates her reconciliation with her lost son, man. Of her own accord earth proffers her gifts, and peacefully the beasts of prey approach from the desert and the rocks. The chariot of Dionysus is bedecked with flowers and garlands: panthers and tigers pass beneath his yoke."[87] Nietzsche here explicates the unique reconciling role that the Dionysian spirit plays in the relationship between man and man as well as that between man and nature, once again illustrating the implications of ecological aesthetics in the Dionysian spirit.

Nietzsche also particularly highlighted the effect of "metaphysical comfort" that tragedy uniquely possesses. He states, The metaphysical comfort,—with which, as I have here intimated, every true tragedy dismisses us—that, in spite of the perpetual change of phenomena, life at bottom is indestructibly powerful and pleasurable."[88] This metaphysical comfort is not the comfort of reason nor the comfort of sensibility, but rather a higher more vital aesthetic comfort, which calls upon man and the vitality of the earth with this spirit of Dionysis.

---

[86] Wilhelm Nietzsche [30], p. 4.

[87] Ibid., pp. 5, 6.

[88] Ibid., p. 28.

(3) Heidegger's Explanation of the Two Forms of Human Dwelling

We have already conducted an in-depth exploration of the implications of eco-ontological aesthetics in Heidegger's philosophy and aesthetic thought. The problem now is: which forms of human dwelling did Heidegger bring up? Without question, Heidegger forcefully advocated realizing "poetically dwelling" of human being in the fourfold world-structure. But is it really possible that truly poetically dwelling is the normal state of human being? In real life in normal conditions, human being actually finds it very difficult to poetically dwell. Heidegger without any doubt sees this clearly. In his explanations of the work of art, he reveals his viewpoint on man's way of dwelling.

First look at the analysis everyone is very familiar with, Heidegger's analysis of Van Gogh's oil painting, A Pair of Shoes. We should first be clear that Heidegger's analysis of A Pair of Shoes is not really an analysis of oil painting. The oil painting is only a stage proper for the purpose of explaining his own philosophical viewpoint on aesthetics. Heidegger uses Van Gogh's oil painting here only to explain that the essence of art is the Truth of beings freely setting itself into the work,[89] but he thereby explains his viewpoint on human being's dwelling. He states, "From the dark opening of the worn insides of the shoes the toilsome tread of the worker stares forth. In the stiffly rugged heaviness of the shoes there is the accumulated tenacity of her slow trudge through the far-spreading and ever-uniform furrows of the field swept by a raw wind. On the leather lie the dampness and richness of the soil. Under the soles slides the loneliness of the field-path as evening falls. In the shoes vibrates the silent call of the earth, its quiet gift of the ripening grain and its unexplained self-refusal in the fallow desolation of the wintry field. This equipment is pervaded by uncomplaining anxiety as to the certainty of bread, the wordless joy of having once more withstood want, the trembling before the impending childbed and shivering at the surrounding menace of death."[90] Here, what we see is not any sort of "dwelling poetically," but rather "toil, trudge, anxiety, trembling, and shivering," the toil and tread of the Truth and Being of beings setting into this painting.

Let us take another look at Heidegger's interpretation of the ancient Greek temple. He states, "It is the temple-work that first fits together and at the same time gather around itself the unity of those paths and relations in which birth and death, disaster and blessing, victory and disgrace, endurance and decline acquire the shape of destiny for human being. The all-governing expanse of this open relational context is the world of this historical people. Only from and in this expanse does the nation first return to itself for the fulfillment of its vocation."[91] The states of human being's survival that Heidegger touches upon in his exposition of the ancient Greek temple is not "poetically dwelling" whatsoever, but is rather birth and death, disaster and blessing, victory and disgrace, endurance and decline. In sum, it is the world of this "historical people." This world that we see is filled with the gun-powder smoke of history; it is the co-existence of honor and disgrace and relentlessness, from which

---

[89] Heidegger [15], p. 256.

[90] Ibid., p. 254.

[91] Ibid., p. 262.

we can see that what Heidegger presents to us is actually this aesthetic state of simultaneous coexistence of "poetic dwelling" in ideality and "the forward trudge of toil" in everydayness.

There are currently theorists who have already touched upon modes of appreciating beauty in ecological aesthetics. For instance, the contemporary American ecocritic, Scott Slovic, brings it up in Going Away to Think, and the American post-modern ecocritic Li Rozelle wrote about it in the book *Ecosublime: Environmental Awe and Terror from New World to Oddworld.* Rozelle states: "There is no affective difference between the natural sublime and the rhetorical ecosublime; both have the power to bring the viewer, reader, or player to heightened awareness of real natural environments. Both can promote advocacy. My work thus argues that mountain peaks, ozone holes, books, DVDs, advertisements, and even video games have the potential to spark environmental awe and terror."[92] Rozelle already carries the "ecosublime" into our everyday life, advocating the use of every modern means to explicate environmental terror and awe at it. We can see from here that the main implications of the "ecosublime" in his understanding include "environmental awe and terror." Of course, simply awe and terror cannot make up the aesthetic characteristics of the ecosublime; it also must include the "the free presencing of Truth" advocated by eco-ontological aesthetics, which is to say it must express the brilliance of contemporary humanity.

The modes of appreciating beauty in eco-ontological aesthetics marks an entirely new topic. We believe that as ecological literary arts develop further and as ecological aesthetics, ecological literature and environmental aesthetics grow and mature, it will inevitably draw more and more attention and boast continuous developments.

12.9 Ecological Aesthetics Education

(1) What is Ecological Aesthetics Education

Ecological aesthetics education is the educating of the broader masses with the ideas of ecological aesthetics, especially the younger generation of the broader masses, in the effort to make them learn how to treat Nature, care for life and protect the earth with the ecological aesthetics attitude. It is an important component of ecological aesthetics and an important channel through which this theoretical formation of ecological aesthetics plays a role. Ecological aesthetics education should become one of the most important forms of cultural education for contemporary citizens and especially contemporary youth; it is the education of important habits that people must learn from childhood on.

Ecological aesthetics education has been an important component of the environmental education that has been strengthening internationally since the 1970s. One could even say that it is one of the main theoretical positions of environmen-

---

[92] Slovic [31], p. 197. See Ping [32], p. 295.

tal education. Aesthetically treating Nature becomes one of the important reasons for humans to care for the environment. In 1970, the IUCN pointed out: "Environmental education is the process of recognizing values and clarifying concepts in order to develop skills and attitudes necessary to understand and appreciate the inter-relatedness among man, his culture and his biophysical surroundings. Environmental education also entails practice in decision making and self-formulation of a code of behavior about issues concerning environmental quality." In 1972, the United Nations held the Conference on the Human Environment in Stockholm, where they formally established the name of "environmental education"; In 1975, the United Nations formally set up the International Environmental Education Project. In the same year, the United Nations, Educational, Scientific and Cultural Organization published the famous Belgrade Charter, which states in line with the nature and goals of environmental education that: environmental education is "1. To foster clear awareness of, and concern about, economic, social, political, and ecological interdependence in urban and rural areas; 2. To provide every person with opportunities to acquire the knowledge, values, attitudes, commitment, and skills needed to protect and improve the environment; 3. To create new patterns of behavior in individuals, groups and society as a whole toward the environment."[93] From this we can see that the aim of environmental education is to establish the right attitude of human being toward the environment along with the right behavior standards and make every person obtain the knowledge, values, commitment and skills to protect and improve the environment in the hope of establishing new patterns of human development harmoniously coordinated with the environment. The aesthetic attitude toward the natural ecological environment has also become one of the most important and most basic attitude of "symbiotic affinity" between contemporary humans and the natural environment.

Ecological aesthetics education is one of the most important avenues through which every citizen enjoys environmental rights and the right to environmental education. In 1972, the UN Conference on the Human Environment stipulated that every person has the right to live a life with dignity in a good environment. In 1975 The Belgrade Charter stipulated that everyone has the right to receive environmental education. Considering the connotations of "rights", fir there is the right to know, which is to say know that one has this right; second is the right to understand, which is to say understand what the intension of this right is. From the perspective of understanding, the role of ecological aesthetics education is a big one. Putting into practice "environmental rights" makes everyone "live aesthetically" and "dwell poetically," which alone is a "life with dignity"; "the right to environmental education" is to make everyone understand the main contents of ecological aesthetics education that environmental education must involve. The right to environmental education is incomplete or defective without ecological aesthetics education.

[93] See Ping [32], p. 295.

(2) The Contents of Ecological Aesthetics Education

The most basic foothold of ecological aesthetics education is that of contemporary eco-ontology, namely making the basic content of education be the ideas of ecological aesthetics about the eco-ontological conception of aesthetics, the method of eco-phenomenology, the research object of ecological aesthetics, the beauty of ecosystems, the essence of human being to appreciate the beauty of ecology as well as poetically dwelling, the play of the fourfold, sense of home, sense of place, the aesthetics of engagement and ecological literature on the different foundations of social-economy, philosophical culture and aesthetic arts guided by the Marxist ontology of practical materialism; in terms of the goal of ecological aesthetics education, it should involve enabling the broader masses, and especially the younger generation, to establish the attitude needed to appreciate the beauty of natural ecology and the ecological and aesthetic consciousness needed to poeticize dwelling.

To establish the ecological and aesthetic attitude needed to appreciate Nature, what is first necessary is to establish the right conception of Nature and to uphold a way of being-in-the-world through the eco-ontological conception of aesthetics, namely the conception that Nature is not external to humans, nor opposed to humans, but is rather one with and inseparable from humans. On these grounds, what is more important is the need to learn how to appreciate the beauty of Nature. As Santayana put it, aesthetic education is training us to see the greatest possible beauty, to see it in the material world—the material world surrounding us, is to walk the path of marriage between imagination and reality, which is our ultimate hope.[94] Taking aim at the total erasure of the beauty of Nature against the background of the industrial revolution for so long, what Santayana here proposes is making us see the greatest possible beauty, the beauty of nature, which is beauty "in the material world." Ever since Hegel called aesthetics "the philosophy of art" and henceforth erased the beauty duly possessed by Nature itself, many of our aesthetics works still continue to simply reduce natural beauty to "humanized nature," which is extremely one-sided. Even though we never deny the role duly possessed by human being in the aesthetic appreciation of beauty, we should also recognize and accept the aesthetic potential possessed by Nature itself. Aesthetic beauty is precisely the tapping, exhibiting or on such grounds creating of this aesthetic potential of nature itself. Of course, the discovering or creating of the aesthetic potential of the natural or material world admits of the distinction between shallow and deep, right and wrong. The Canadian environmental aesthetics philosopher, Allen Carson, made a specialized exposition of such cases. He states that the aesthetics of Nature admits of a distinction between the thin sense and the thick sense, "Hospers describes it as a distinction between the 'thin sense' and the 'thick sense' of 'aesthetic,' labels that serve for our purposes. The thin sense is relevant when we aesthetically enjoy an object primarily in virtue of the physical appearance of the object, including not only its surface physical properties, but also its formal properties having to do with line, shape, and color. The thick sense, on the other hand, involves not merely the physical appearance of the object, but also certain qualities and values that the object expresses or conveys to the

[94] Quoted from Carlson [23], p. 126.

viewer. Prall calls this the 'expressive beauty' of the object, while Hospers speaks of objects expressing 'life values.'"[95] This is to say, Carlson takes some "physical appearances" of natural materials are the "thin" potentially aesthetic characteristics, while "expressive beauty" and "life values" are their "thick" potentially aesthetic characteristics. He judges natural material objects with this viewpoint, insisting that "it seems to me that while farms along the road may express determination, discarded car bodies do not, and while the skyline of a city may express vision, a strip mine does not."[96] Aesthetic attributes "in the thick sense" of "expressive beauty" and "life values" as Carson argues largely fall under the scope of eco-ontological aesthetics.

Second is the eco-aesthetic way of life. The founder of the Club of Rome, Peccei, insist that the current environmental problem today is ultimately a cultural problem, a problem of what way of life people choose. For China and Chinese people currently engaging in the building of modernization, this problem is likewise fully important. Some people think that China's current problem is that of "development first, environment later," and we could even say the path of "pollute first, clean up later." This is an extremely irresponsible attitude. The facts show that this path will not only pay a high price, but will also be basically impracticable for such a high population and resource scarce country as China. We have to place development and environment in equally important positions, and take the win-win path of building "environmentally-friendly society." This demands our people to choose and cultivate the ecologically aesthetic way of life, which importantly consists of building modern "ecological civilization," and fundamentally speaking, it is also an aesthetic way of life in affinity with Nature, or rather, the lifestyle of "poetically dwelling." Concretely speaking, this way of life includes content like "enough is enough," environmental protection, recycling and treasuring life. By "enough is enough" we mean a comfortable and economical way of life, opposing the extravagant and ostentiatious wasting of resources and energy. Instead of the bigger the better when buying houses, and the more the merrier in buying cars, furniture and clothing, what is praised is frugality, and enough is enough; by environmental protection and recycling, we mean self-consciously not polluting the environment and choosing biofuels as much as possible along with a way of life that enables the cyclical utilization of matter and energy; by "treasuring life," we mean self-consciously protecting every living thing, plant or animal in Nature, protecting wild species, rare species, and caring for animals with the spirit of treasuring them. As the American ecologist David Griffin put it, we must read lighter on this earth, only use the things we need to use and maintain an ecological balance for our neighbors and next generation, which will become "common sense."[97]

Last is cultivating an aesthetic attitude toward Nature, which is the most central content of ecological aesthetics education. Ultimately, ecological aesthetics education is an education of an aesthetic view and even more so of a worldview. The cultivation and establishment of this aesthetic attitude toward Nature has to undergo

---

[95]Carlson [33], pp. 206–207.

[96]Carlson [33], pp. 214–215.

[97]Griffin [34], p. 227.

the following three conversions: first, in philosophical viewpoint, we have to convert from anthropocentrism to eco-holism. Only by undergoing this conversion can the aesthetic attitude toward Nature truly be established. For a long period of time, anthropocentrism has occupied an unwaveringly important position in philosophy, namely the omni-suppressive idea of humans conquering nature and controlling nature, under whose guidance humans have always seen themselves as the masters of Nature incapable of establishing an equal and harmonious aesthetic relationship with Nature; second, in terms of aesthetic viewpoint, we have to convert from the idea that the beauty of nature is "humanized nature" to the idea that the beauty of nature is the symbiosis of nature and humans. The traditional conception of aesthetics regarded natural beauty as humanized nature, which still only sees human being while ignoring Nature and covering up the presentation of natural beauty. But the viewpoint of contemporary eco-ontological aesthetics insists that the beauty of nature is the beauty of "Nature and humans reaching symbiosis by virtue of their co-existence." Because Nature and humans co-exist inseparably as equals, Nature and humans can collectively aspire to prosperous growth and full vitality, and present the beauty of living and Being; third, in terms of the nature of aesthetic conceptions, human being's aesthetic attitude toward Nature must convert from the purely aesthetic conception to a conception of human life and the world, an attitude toward life demanded by the new era. The new era of China has the goal of establishing "harmonious society," and harmony between humans and nature correspondingly has become the important foundation for socio-economic development and the beautified survival of the people. Thus, human being's aesthetic attitude toward nature has become the precondition of harmonious coexistence between nature and humans; it is the view of human life and the world view that this new era must champion and establish.

References

1. [U.S.] Fromm, H. (1996). *From transcendence to degeneration: A road map*, Containing in *Ecological Criticism Reader*. University of Georgia Press.
2. Marx and Engels. (1979). *Collected works*. People's Publishing House.
3. [Germany] Cassirer, E. (2004). *An essay on man*, translated by Ganyang. Shanghai Translation Publishing House.
4. [Germany] Heidegger, M. (2006). *Being and time* (revised translation), translated by Chen Jiaying, Wang Qingjie. Sanlian Bookstore.
5. [Germany] Welsch, W. (2004). *How to transcend anthropocentrism?* Containing in *the National Art Research*, No. 5.
6. [U.S.] Rolston, H. (2000). *Philophy gone wild*, translated by Liu Er, Ye Ping. Jilin People's Publishing House.
7. [U.S.] Carson, R. (1997). *Silent spring*, translated by Lu Ruilan, Li Changsheng. Jilin People's Publishing House.
8. Mouchang, Y. (1999). *Ecological ethics,* Capital Normal University Press (余谋昌:《生态伦理学》,首都师范大学出版社1999年版).
9. Yi, L. (2001). *Deep ecological thought research*. Tsinghua University Press (雷毅:《深层生态学研究》,清华大学出版社2001年版).

10. Baihua, Z. (1987). *Yi Jing*. Peking University Press.
11. [Italian] Vico. (1986). *New science*, translated by Zhu Guangqian. People's Literature Publishing House.
12. [U. S.] Dewey, J. (2010). *Art as experience*, translated by Gao Jianping. The Commercial Press.
13. [U.S.] Schusterman, R. (2002). *Pragmatism aesthetics—The beauty of life, the thinking of art*, translated by Peng Feng. The Commercial Press.
14. [Germany] Heidegger, M. (2014). *Interpretation of Holderlin's poetry*, translated by Sun Zhouxing. The Commercial Press.
15. [Germany] Heidegger. (1996). *Selected works ofHeidegger*, Selected by Sun Zhouxing, Shanghai Sanlian Bookstore.
16. [Germany] Heidegger, M. (2008). *Off the beaten track*, translated by Sun Zhouxing. Shanghai Translation Publishing House.
17. [U. S.] Ward, B., & Dubos, R. (1997). *Only one earth,* Translated and edited by Editorial Board of *Foreign Public Welfare Series*. Jilin People's Publishing House.
18. [U.S.] Ruckert, W. (1996). *Literature and ecology: An experiment of ecological criticism*, Containing in *The Ecocriticism Reader*, University of Georgia Press.
19. [English] Toynbee, A. (2001). *Mankind and mother earth*, translated by Xu Bo et al. Shanghai People's Publishing House.
20. [U.S.] Berleant, A. (2007). *Environment and the arts: A multidimensional perspective on environmental aesthetics*, translated by Liu Yuedi et al. Chongqing Publishing House.
21. [U.S.] Berleant, A. (2006). *The aesthetics of environment*, translated by Zhang Min, Zhou Yu. Hunan Science and Technology Press.
22. [Qing] Yu, L. (1998). *Xian Qing Ou Ji*, Du Shuzhen's commentary, Xueyuan Publishing House, 1998.
23. [Canada] Carlson, A. (2006). *Nature and landscape: A collection of environmental aesthetics*, translated by Chen Libo. Hunan Science and Technology Press.
24. [English] Bard, M. (2008). *The basic pedigree of aesthetics of natur*e, translated by Liu Yuedi, see *World Philosophy,* No. 3, 2008.
25. [U.S.]Berleant, A. (2008). *The development of environmental aesthetics and its recent issues*, translated by Liu Yuedi, see *World Philosophy*, No. 3.
26. [U.S.] Leopold, A. (1997). *A Sand County Almanac,* translated by Hou Wenhui. Jilin People's Publishing House.
27. [U.S.] Grottefelti, C. etc. (1996). *Ecological criticism reader*. University of Georgia Press.
28. [U.S.] Love, G. (1996). *Reevaluating nature*, Containing in *The Ecocriticism Reader*. University of Georgia Press.
29. Ning's, W. (2001). Compilation of *New Literature History*. Tsinghua University Press.
30. [Germany] Wilhelm Nietzsche, F. (1986). *The birth of tragedy*, translated by Zhou Guoping, Sanlian Bookstore.
31. [U. S.] Slovic, S. (2010). *Going away to think*, translated by Wei Qingqi, Peking University Press.
32. Ping, Y. (2007). *The pedigree of environmental aesthetics*. Nanjing Press.
33. [Canada] Carlson, A. (2005). *Aesthetics and the environment—The appreciation of nature, art and architecture*, translated by Yang Ping. Sichuan People's Publishing House.
34. [U.S.] Ray Griffin, D. (1998). *Peace and postmodern paradigm, postmodern spirituality and society*, translated byWang Chengbing. Central Compiler Publishing House.

Part VI
An Ecological Aesthetics Interpretation of Literary Works

Chapter 13
An Ecological Aesthetics Interpretation of Chinese Writings

13.1 An Ecological Aesthetics Interpretation of *Book of Poetry*

At present, research on eco-ontological aesthetics has already deepened to the stage of exploring its specific aesthetic implications. One of the main avenues of this exploration is conducting an aesthetic interpretation of some classical works from the perspective of eco-ontological aesthetics in order to find some things of regularity in them. An interpretation of the ecological aesthetics of ancient China's famous collection of songs *Book of Poetry* is one of such attempts.

(1) .

The reason why we choose *Book of Poetry* to begin with is because it emerged in the 500 plus years from 11th century BCE to 5th century BCE, which was the time at which ancient China's eco-ontological philosophy of "integral oneness of Nature and humans" gradually took shape. Some important literature and written contributions of thought that emerged during this time were tightly connected to the religious beliefs, mythology, tradition of life, mode of production and human being's relationship to the natural environment in that era. *Zhou Book of Changes* was also largely produced and completed during this time, or even earlier. China's early peoples found a place to dwell and thrive on Chinese soil with agriculture as the principal mode of production in the cultural climate of this ancient thought of "gods and men finding harmony." They cultivated the earth in extremely backward conditions of production, harvested the crops, reproduced the next generation, and resisted foreign invaders. At the same time, in the state of a tight combination of ritual sacrifices with music and dance, they prayed to the heavens, commemorated ancestors, praised the harvest and expressed emotions. The *Book of Poetry* was produces under such conditions; it was the proto-ecological work of China's early people, and the true expression of their authentic form of life; it was the place of origin of the uniquely Chinese ancient arts. Its primordial nature is what makes it extremely valuable; it was basically untouched by the later Confucian thought of the *rui*, and still maintained the ancient Chinese

F. Zeng, *Introduction to Ecological Aesthetics*,
https://doi.org/10.1007/978-981-13-8984-9_13

artistic plea for the "integral oneness of Nature and humans" and the unique style of "the beauty of striking harmony." The facts prove that *Book of Poetry* was produced in the pre-Confucian age; it was the ancient Chinese people's ode to life and song of survival.

Of course, later explanations of *Book of Poetry* inevitably came with distortions based on other thoughts, particularly after the emergence of the Confucian classics, at which point many implications of ancient eco-ontological aesthetics were either intentionally or unintentionally covered up. Confucius' discussion of *Book of Poetry* rather comprehensively and not without profundity revealed its nature along with its status and significance in the Chinese cultural tradition. For instance, "stimulate oneself in poetry, establish oneself in ritual and accomplish oneself in music" (*Analects*, 8.8) which combines the tradition of cultivation in ritual and music to demonstrate poetry's role in aesthetic education; another example is, "poetry may be used to stimulate, to contemplate, to socially consolidate, and deal with resentment." "[From poems] we largely become acquainted with names of animals and plants…" (*Analects,* 17.9), which, concentrating on the aesthetic dimension of moving people emotionally, demonstrates the social role of poetry; again, for example, "[The Guan Ju] is expressive of enjoyment without being licentious, and of grief without being hurtfully excessive…" (*Analects*, 3.20), which demonstrates the beauty of poetry in "striking the harmonious medium." But Confucius' treatment of *Book of Poetry* is largely based on the distortions of Confucian thought and the feudal institution of ritual propriety. For instance, "In *Book of Poetry*, there are three hundred pieces, but the design of all of them may be embraced in one sentence, namely 'thinking without depravity'" (*Analects*, 2.2), which could be accused of distorting *Book of Poetry* with the moral consciousness of Confucianism; For instance, there are statements advocating "the right songs" and "elegant music" and denying the value of "the songs of Zheng," like "Banish the songs of Zheng, and keep far away from specious talkers. The songs of Zheng are licentious; specious talkers are dangerous…" (*Analects*, 15.11), "I hate the manner in which purple takes away the luster of vermillion. I hate the way in which the songs of Zheng confound the music of Ya. I hate those who with their sharp mouths overthrow kingdoms and families" (*Analects*, 17.18); then, for example, there are statements that overemphasize the political and ethical role of poetry like "the Master said: having learned all three hundred odes by heart, if someone is entrusted with a governmental charge, but does not reach the goal, or if, when sent to any quarter on a mission, he cannot independently secure diplomacy, no matter how many he has learned, what would still be the use of them?" (*Analects*, 13.5), and "near at hand in serving one's father, and more remote in serving one's prince." *The Greater Order of the Mao Commentary on the Odes* (*maoshi daxu* 毛诗大序) from the Han dynasty and later Confucian research on *Book of Poetry* generally unfolded in line with Confucius' trains of thought like "thinking without depravity," "banishing the songs of Zheng," "expressive of enjoyment without being licentious, and of grief without being hurtfully excessive." This is always the case, for instance, in the *Mao Commentary*, "The *Guan Ju* is the virtue of the empress and concubines," "prior kings used this poem to set norms for marriage, to complete filial respect, to deepen human relationships, to refine cultivation and to change the customs." Traditional *Book of Poetry* research is of course valuable, but does not however exhaust the implications of the *Book of Poetry*.

So, what are the core implications of *Book of Poetry*? Why do we also say that *Book of Poetry* encapsulates content of eco-ontological aesthetic thought? We believe that it is unproblematic to reduce the core implication of *Book of Poetry* to "poetry is the expression of earnest thought." *Book of Documents* "Canon of Yao" records, "The Di said, 'Kui, I appoint you to be Director of Music, and to teach our sons, so that the straightforward shall yet be mild; the gentle, dignified: the strong, not tyrannical: and the impetuous, not arrogant. Poetry is the expression of earnest thought; singing is the prolonged utterance of that expression; the notes accompany that utterance, and they are harmonized themselves by the standard tubes. (In this way) the eight different kinds of musical instruments can be adjusted so that one shall not take from or interfere with another; and spirits and men are brought into harmony.' Kui said, 'I smite the (sounding-) stone, I gently strike it, and the various animals lead on one another to dance.'" This passage rather comprehensively records the actual situation of ancient Chinese people's artistic creation. Firstly, art at the time was the unity of music, dance and poetry; secondly, the pursuit of art was "harmonizing with the standard tubes," and "adjusting the eight different kinds of musical instruments so that one shall not take from or interfere with another" such that "spirits and men" ultimately reach harmony; thirdly, the core implication of art was "poetry is the expression of earnest thought." The key to the problem is, what is "earnest thought"? The *Mao Commentary* states: "the poem is what holds earnest thought, in the mind it is earnest thought, expressed in words it is poetry. Affects stir within, and take shape in words. If wording it falls short, one therefore sighs it; if sighing it falls short, one therefore chants in song; if chanting it falls short, who knows the waving of hands and shaking of feet will do," "the affect springs into song, and the song becoming writing is called notes." Clearly, what we call "earnest thought" is mainly what is stored in the feelings of the inner mind. In *Comprehensive Account of Chinese Poetry Scholarship* (*zhongguo shixue tonglun* 中国诗学通论) Yuan Xingpei pointed out after detailed analysis that, "in these explanations and understandings, the intension of "earnest thought" (*zhi* 志) is "emotion" and "intention," which is to say, the emotion and will inside the affective mind of the poet."[1] A certain thinking consciousness is the reflection of a certain social being, and the "affective will" of the people in those times was the reflection of social life in those times. China has been an agricultural society, and our ancient people were a primarily agricultural people with extreme reliance on the land, Nature and climate. Because of this, worship of and affinity with Heaven, Earth and Nature constituted an important content of our ancient people's "affective will." It was precisely against this backdrop of agricultural society that our ancient people created theory own unique "Book of Changes Culture" in "the integral oneness of Nature and humans." The *Zhou Book of Changes* "Commentary Explaining Hexagrams" (*shuo gua zhuan* 说卦传) states: "the sages of the past in creating *Book of Changes* had to do so with the patterns following the natural tendencies of life. Therefore in establishing the way of Heaven, they spoke of *yin* and

[1] Xingpei et al. [1], p. 19.

yang, in establishing the way of Earth, they spoke of soft and hard, in establishing the way of human being, they spoke of being-humane and being-righteous." "the great man merges his virtue with Heaven and Earth, fuses his light with the sun and the moon, merges his order with the four seasons, and marries his fortunes and misfortunes with demons and gods. Innately in no contradiction to the natural order, learning in reverence of the heavenly seasons." This is to say, the "*Book of Changes* Culture" in ancient China had profound implications of a culture of "integral oneness of Nature and humans"; it included the principles of *yin* and *yang*, that of soft and hard, and that of humane and righteous, and forcefully advocated an ancient "eco-humanist spirit" that championed integrating with the virtues of the heavens and the earth and fusing with the light of the sun and the moon and merging with the order of the four seasons. This was also the necessary implication of the "affective will" of the people in those times. Primitive Chinese art was an art form unifying ritual, music, dance and poetry that originated in the activity of sacrifices, whose basic aim was the pursuit of "the great music converging in harmony." As the *Record of Rites* "Record of Music" puts it, "the great music converges in harmony with Heaven and Earth, and the great ritual converges in rhythm with Heaven and Earth. Because of harmony, there is no loss in everything that happens, because of rhythm sacrifice to Heaven and offer to Earth. In the light, there is ritual and music; in the dark, there are ghosts and spirits. If so, within the four seas, there is merging in respect and converging in love. The ritual is particular affairs combining in respect; music is different patterns combining in love. The affects of ritual and music converge, and thus the kings of light follow one another." Clearly, the "affective will" of poets in those times was the "affective will" of a "greater music combining in harmony with Heaven and Earth."

Summarizing what has been previously stated, the affinity of humanity with Nature, humans merging their virtue with Heaven and earth, and "the greater music of converging in harmony with Heaven and Earth," namely the "affective will" of the poets in those times, are expositions of the beauty of "striking the harmonious medium" of "Nature and humans fusing into one." These all constitute eco-ontological aesthetic thought encapsulating the classical eco-humanist spirit of "integrating human virtue with that of Heaven and Earth." The core of *Book of Poetry* is precisely the pursuit of this beauty of striking the harmonious medium, which is the ancient spirit of ontological aesthetics with ecological implications unique to China constituting the aesthetics and artistic spirit of ancient China; worlds apart from the ancient aesthetics and artistic spirit of the Western world; it is an extremely macroscopic aesthetic spirit of "Nature and humans integrating into one." As *Record of Rites* "Striking the Medium" (*Doctrine of the Mean*) puts it, "the middle is the expanding root of the realm; harmony is the realm reaching the way. Striking the harmonious medium, Heaven and Earth come into phase and all creatures grow." This is to say, "striking the harmonious medium" is the basic law of evolutionary development for all creatures between Heaven and Earth relating to the workings of Heaven and Earth and the breeding of all creatures, namely "the expanding root" and "reaching the way." The concentrated manifestation of this beauty of striking the harmonious medium in art is the *Book of Poetry*, which is the artistic expression of the "affective will" of the ancient Chinese people; it is the concretion of the Chinese people's aesthetic spirit. The "harmony" championed by ancient Westerners was a ratio and proportionality of the material world whose substance was the "Idea" or "number." Aristotle thought that the most important demand on beauty and artistic works was wholeness, concretely unfolding as the "completion," "order," "proportionality" and "clarity" of the plot and the characters' actions,[2] the aesthetic idea behind which was "imitation," and the representative arts of this were sculpture, tragedy and epic poetry. Particularly ancient Greek sculpture illuminated the world through its "proportionality, symmetry and harmony," exhibiting a noble purity and solemn greatness.[3] What the *Book of Poetry* expressed was a very different beauty of "striking the harmonious medium" that was both dynamic and macroscopic in scale. Presently we will cite the *Odes of Wei, He Guang* for example:

Who says that the He is wide?
With [a bundle of] reeds I can cross it.
Who says that Song is distant?
On tiptoe I can see it.

Who says that the He is wide?
It will not admit a little boat.
Who says that Song is distant?
It would not take a whole morning to reach it.

This is a famous poem about the thought of homecoming. The poet was a person of Song living as a guest in the state of Wei. Facing the span before him, he becomes homesick thinking of the gushing Huang He river separating him from home, and lets out the call "Who says that the He is wide? With [a bundle of] reeds I can cross it," "Who says that Song is distant? It would not take a whole morning to reach it." In the artistic world of the poet, the gushing Huang He is no longer an obstacle to returning home in the eagerness to cross it with a tiny bundle of reeds and immediately rush home to the state of Song and the company of dear family and friends. The expression of this passion to eagerly return home is so pronounced! "Homecoming" has been the "mother theme" of literature present in both China and the West since ancient times; it has thick implications of eco-ontological aesthetics, but the ode *He Guang* handles it in the unique manner of treating Nature as friendly equal. Through artistic imagination, the poet imagines a bundle of reeds as natural objects as being a small boat capable of helping the traveler cross the flooding Huang He. In the traveler's eager mood to return home, this artistic treatment seems till potentially inadequate, but by deploying the imagination he suddenly narrows the wide Huang He, and seems to stand on the tips of his toes to see home as if he could quickly return home and unite with dear friends and family. At this point in time, the reeds and even the flooding Huang He became the traveler's friends, who help him realize his wish to return to his home. This is "affective will" to befriend Nature through art is uniquely different

[2]Aristotle [2], p. 26.

[3]Ephraim Lessing [3], p. 215.

from the affective state of the Homeric Hymns when describing the return home overseas by boat following victory in battle. What the Homeric Hymn *The Odyssey* speaks of is the story of the Greek hero, Odysseus, returning home at the conclusion of the battle of Troy, which metaphorically expresses the battle between man and Nature and describes the process of Odysseus conquering such natural forces as the sea god Poseidon and the mythical titans, fairies, wind spirits and water sprites, and finally smoothly returning home. This is a picture of man battling with Nature, a praise of human being conquering Nature, which is worlds apart from the aesthetic implications of "He Guang" in *Book of Poetry*.

(2) .

If we open up the *Book of Poetry* and interpret it from the perspective of eco-ontological aesthetics, we will discover extremely rich content in it. What should be reiterated again is, the eco-ontological aesthetics of which we speak is an ontological aesthetic thought that encapsulates the ecological dimension, but which far exceeds the simple aesthetic relationship between man and Nature, and ultimately rests in the better survival and poetic dwelling of human being.

1. **"Odes" with Eco-humanist Implications**

The *Mao Commentary* states, "Thus there are six sense of poetry: one is the ode, two is the rhapsody, three is the analogy, four is the incitement, five is the hymn, and six is the sacrificial song." Kong Yingda of Tang pointed out in *The Correct Meaning of the Mao Commentary*, "the ode, the hymn and the sacrificial song are the different styles of poem; the rhapsody, the analogy and the incitement are the different verses of poetry. They differ in size, but combine to form six senses of poetry, the rhapsody, analogy and incitement are what the poem uses; the ode, the hymn and the sacrificial song are the complete formations of the poem." Later researchers of *Book of Poetry* continued to use this formulation. We believe that among the "six senses" in *Book of Poetry*, the most important are the ode, analogy and incitement. Let us first discuss "the ode." The *feng* 风 (ode) is a poetic style unique to the *Book of Poetry* with singular characteristics containing implications of ecological humanity; it is not only a treasure in the bank of Chinese literature, but also shines a unique light in world literature. The ode style poem is a principal component of *Book of Poetry*, which includes 305 poems, among which 160 are "national odes," mainly local ethnic songs of feudal territories, and 105 are *Major Hymns* and *Minor Hymns*. Gao Heng insisted that "[t]he hymns made use of the Xia character script, so the editors of the *Book of Poetry* used Xia characters to mark the poetry of the imperial capital of Western Zhou."[4] In this way, we could say the *Minor Hymns* and *Major Hymns* are also "odes." Because of this, aside from the 40 "sacrificial songs" (*song* 颂) used in temple sacrifices, the "ode style poems" amount to 265 of the total, and are the main part of the *Book of Poetry*.

So, what is an "ode"? The *Mao Commentary* argues, "the *feng* 风 (ode) means style, means cultivation; the ode is to move you, the cultivation is to transform you."

[4] *Heng* [4], p. 4.

It also states: "those above use the ode to cultivate those below, those below use the ode to satirically critique those above. Mastering writing to craft an admonishment, the one who speaks it is guiltless, and the one who listens to it finds it enough to take precaution, thus we call it *feng*." This principally explains the "ode style poem's" characteristic of political cultivation from the perspective of traditional "poetic instruction." Because they could observe the political passions of the people through the "ode style poem's" function of satirically critiquing the higher class, the rulers established the institution of "collecting folk songs." It is said, that the Zhou dynasty preserves this institution of collecting odes passed on since high antiquity. *Record of Rites* "Royal Regulations" records: "The son of Heaven, every 5 years, made a tour of inspection through the fiefs. In the second month of the year, he visited those on the East......ordered the Grand music-master to bring him the poems (current in the different states), that he might see the manners of the people..." There already existed this institution of collecting poems by the music-master and Grand music-master. *Book of Han* "Annals of Shi Huo" records: "In the first month of spring, those living together in concentration will disperse, and passers-by ring wooden bells, and follow the route to collect poems, and present them to the Grand music-master, who compares the temperament of their notes, to play them for the son of Heaven." He Lin commented on the *Gong Yang Commentary of the Spring and Autumn Period* "Year 15 of Xuan Gong": "Men and women had some resentment, heard one another and sang. The hungry sang about their food, the over-worked sang about their choirs. Those 60 men and 50 women without children, the officials clothed and fed them, and sent them among the people to collect poems. From the countryside the poems moved to the county center, from the county center the poems moved to the state capital, and the state capital played them to the son of Heaven. Thus, the kings did not need to step foot outside of the house, and still heard exhaustively of the bitter struggles in the realm." Gao Heng explained the implications of *feng* 风 (odes) from the perspective of the likeness of music and odes to Nature and their reflection of customs and manners. He states, "*feng* was originally the common name for musical tunes," "why would musical tunes be called *feng* then? Mainly because *feng* music admitted differences in height, volume, clarity and curvature, and the pitch of musical tunes also had differences in height, volume, clarity and curvature. Musical tunes had similarities to *feng* (odes), so the ancients called music *feng* (odes). At the same time, the content and form of musical tunes were generally the reflection of custom, so musical tunes called *feng* and were also connected to the manner of customs. IN this respect, what we call national odes were actually the musical tunes of each state." Ancient China also began with the cultural idea of "integrating with the virtues of Heaven and Earth," thinking that "music" could integrate Heaven and Earth. *Record of Rites* "Record of Music" states: playing music "rouses the light of ultimate virtue and activates the blending of the four atmospheric conditions (*si qi* 四气), in order to show the patterning principle of everything that happens. Thus, clear and light like Heaven, broad and massive like earth, beginning and ending like the four seasons, periodic and returning like the winds and rain. The five colors become patterned and do not blur, the eight winds follow rhythm and do not pervert it, the hundred degrees are numbered and possess constancy." This here explicates musical tunes as if they were natural winds coming from the eight directions with their own rhythms. *Explaining Graphs and Analyzing Characters* explains the implications of *feng* from the perspective of Chinese character composition, "[the character] *feng* 風 comes from bug (*chong* 虫), sounds like *fan*," "the wind moves and bugs are born, thus after eight days bugs transform." This could demonstrate that naming musical tunes "*feng*" positively takes from them reflecting the most primitive meaning of living activity, and already contains implications of ancient eco-humanism. The most typical expressions of the ancient Chinese thought of "integral oneness of Nature and humans" was "the reproduction of life is what change means," the fluid connection of *yin* and *yang* affecting one another and being affected by one another reproducing all creatures between Heaven and Earth. *Yin and Yang* are the foundations of life, and the flow of air (*feng* 风) is the result of *yin* and *yang* affecting one another, and generating a charge of energy until it is equalized, which is the driving force behind the life of all creatures. The winds activate and bugs are born, there is only life with the movement of air. Thus, the most primitive artistic *feng* is the same as natural wind, which is the representation of the authentic state of human life. The "ode style poem" of *feng* is this most primitive artistic movement similar to natural wind, which is a proto-ecological rhythm of life reflecting the most authentic state of human existence. The content of "ode style poems" is principally the expression of the most basic needs and conditions of human life. "Alimentary and sexual drives is human nature," satisfying alimentary and sexual drives, working and reproducing life are the most basic conditions of living beings. The artistic expression of these most authentic needs and conditions of human being is precisely a return to human being's ecological essence, which is where the value of the "ode style poetry" in *Book of Poetry* lies.

Of course, the *Book of Poetry*'s expressions of human being's most authentic ecological essence are exceptionally plentiful and diverse such that we can only cite some of the essentials to explain it. *Minor Hymns* "Tiao Zhi Hua" is a famous verse of "the hungry singing about their food." Let us take a look at what the poem concretely describes:

The flowers of the bignonia,
Are of a deep yellow.
My heart is sad;
I feel its wound.

The flowers of the bignonia [are gone],
[There are only] its leaves all-green.
If I had known it would be thus with me,
I had better not have been born."

The ewes have large heads;
The Three stars are [seen] in the fish-trap.
If some men can get enough to eat,
Few can get their fill.

This is a starving people's deep-cutting description of the disastrous harvests brought about by years of military campaigns in the Zhou dynasty; it is especially a deep and figurative expression of unprecedented famine. The poem first incites the tone with the blossoming of yellow bignonia flowers in the summer season to show in reverse the pain and worry of the human heart in hunger; it then says, if such suffering in famine could have been known earlier, the poet would have chosen not to be born; finally through the gaunt body and big head of the lamb and the empty fish-trap through which only the star light shines, it illustrates how there is nothing edible left, and if a tiny bit of food is wrested from nothing, few can eat their full. This poem powerfully expresses the state of human survival in the major famines of the Zhou dynasty with vivid imagery. Especially the verses "If I had known it would but thus with me, I had better not have been born," "if some men can get enough to eat, few can get their fill," are calls for life emanating from the bottom of people's hearts in extreme predicaments, which is the most basic please for the dignity of life. If human being is incomparably worse than the bignonia, starving all day, of what value is there still in human life? The famous *Odes of Wei* "Fan Tan":

Kan-kan go his blows on the sandal trees,
And he places what he hews on the river's bank,
Whose waters flow clear and rippling.
You sow not nor reap; -
How do you get the produce of those three hundred farms?
You do not follow the chase; -
How do we see the badgers hanging up in your courtyards?
O that prince!
He would not eat the bread of idleness!

The wood chopper on the clear river bank engages in unbearably burdensome physical labor, an even heavier pressure is coming from the brutal exploitation of the "prince," who never works, but can reap three hundred bundles of grain, and whose courtyard is always hanging full with hunted creatures. Why is this ultimately? How can they get away with neither ploughing nor hunting themselves and yet cleanly occupying the fruits? This is the laborer's resistance to the products of labor being ruthlessly seized, which is the maintenance of human being's right to survive! When the workers have no way to survive under ruthless oppression, *Odes of Wei* "Shuo Shu" lets out the call yearning for "happy land"!

Large rats! Large rats!
Do not eat our millet.
Three years have we had to do with you,
And you have not been willing to show any regard for us.
We will leave you,
And go to that happy land.
Happy land! Happy land!
There we shall find our place.

The workers were no longer capable of enduring the brutal exploitation waged ruthlessly by the "Large rats," and decisively choose the route of fleeing, to look for their own so-called "happy land." When people choose flight, it proves they most basic right to survive is impossible to safeguard! But where is the "happy land" belonging to the workers? In a society of exploitation, the right to family life and love for working people similarly faces the danger of being taken away at every moment. The many "poems of abandoning wives," "poems of leaving wives" and "poems of leaving people" preserved in *Book of Poetry* profoundly describes for us all the brutal cases of social and ecological balances suffering destruction in the frequency of war campaigns at the time along with the collapse of the ritual and music institution, the intensification of exploitation, the difficulty in finding subsistence and the instability of households. This formidable artistic trend already far exceeded the scope of the Confucian poetic education of "odes to move people, and education to cultivate people," touching on the severely deteriorating state of survival for people in the lowest strata of society at the time, and furthermore touching on the severe disbalances in society and ecology. This is the unique value of the "ode style poetry" in *Book of Poetry*.

13.2 The "On the Pu River Among the Mulberries" Poem Reflecting First People's Authentic Love

The "ode style poetry" in the *Book of Poetry* not only expresses the cry of the broader lower stratum of people fighting for their right to survive; it also expresses people's extremely authentic pursuit of love. This is the "On the Pu River Among the Mulberries" poem, which is the "excessive poetry" so criticized by the feudal literary elite for so long. In actuality, love is the expression of the human being's essence; it is the eternal mother theme of art. Especially in the early period of human history over 3,000 years ago, love was intimately related to primitive early people's reproductive breeding, and even to primitive religious activity, further reflecting some kind of ecological characteristic of the human being. As everyone knows, sexual reproduction is the natural disposition of human beings, and in the early stage of first peoples, reproduction concerned the life and death of clans and tribes, so it found full expression in the mystical worship cultures of humankind. Traditional Chinese "*Book of Changes* Culture" reduced the creation of all creatures in the universe to "the mutual productivity of *yin* and *yang*." In such cultural ideas, *yin* and *yang* affecting and being affected by one another and the transformation of the life of all creatures is intrinsically one with the combination and growth of human beings. At the time, there was a greater degree of freedom in interactions with the other sex, and there were even opportunities provided on holidays for men and women to get to know one another and interact. According to the recording in *Zhou Book of Rites*: "In the mid-spring month, there is a general order for men and women [of marrying age] to wed. At this time, those engaging in the *ben* 奔 ritual will not be forbidden." The ancients thought that dense mulberry forests could cultivate silk worms and bring humans good fortune; they were also tied to reproduction, so mulberry forests were considered somewhat mysterious and sacred places for people to sacrifice and for men and women to meet erotically. Cultural anthropology's "carnival" theory also has an explanation of this cultural phenomenon that integrates worship of sexual reproduction. "Among the Mulberries" (*sang zhong* 桑中) in *Book of Poetry* states:

I am going to gather the dodder,
In the fields of Mei.
But of whom are my thoughts?
Of that beauty, the eldest of the Jiang.
She made an appointment with me among the mulberries;
She will meet me in Shanggong;
She will accompany me to Qishang.

The following two sections repeatedly chant. This poem vividly describes the erotic plot of a young man and women meeting, uniting and departing among the mulberries. The *Mao Commentary* insists that the licentious eloping of this poem is actually a distortion. In actuality, the said poem is the expression of an eloping between man and woman that is related to the ritual of sacrifice; it is a description of authentic human love. In *Research on Bone Inscription Writing* (*jiaguwen yanjiu* 甲骨文研究), Guo Moruo insists that, "among the mulberries is the place where the mulberry forests are. Shanggong is the ancestral hall among the mulberries for sacrifices. Men and women met conjugally there." Some scholars argue, in high antiquity, people made offerings to the god of agriculture and the god of reproduction, "believing that the conjugal union of man and woman could spur the breeding of all creatures, so at many sacrificial services where offerings were made to the god of agriculture, communal marriage meetings of men and women always accompanied," and what "Among the Mulberries" describes is precisely a surviving instance of this type of custom." The *Mozi* "Explaining Ghosts Part II" states: "The state of Yan has the [ceremony at the] ancestral hall, the state of Qi [the ceremony at the alter of] the gods of earth and grain, the state of Song the mulberry forest and the state of Chu the Yunmeng, which is where large assemblies of men and women are observed." The secret erotic meeting of men and women described in "Among the Mulberries," the private communions between people or impregnations in the *Zuo Commentary* were called 'having happiness among the mulberries,' and in *Master Lü's Spring and Autumn Annals,* the episode of Tang of Shang wiping out the Xia and seizing the realm, then the huge drought and 5 years without harvest, at which point 'Tang took himself to pray at the alter in the mulberry forest, and the rains fell in great quantity,' all of this illustrates that the mulberry forest was a sacred sacrificial site, and was also a place for people to elope and fulfill joy. *Record of Rites*, "Record of Music," states: 'The airs near the river Pu, at the mulberry forest, were those of a state going to ruin…,' which is also an allusion to the erotic and merry singing and dancing among men and women at a sacrificial site. Because the place was fixed, as time went on and on, when people brought up this place they would think of those joyful and entertaining things, and would directly use the place name (because sacrifices were often made at the Luo forest, they liked to call the place name by the tree name [hence among the mulberries is a place name 'Sangzhong'-Trans.]) to express that pleasant experience."[5] In *Odes of Chen*, "the Fen Elms at the Eastern Gate," the protagonist more explicitly invites lovers to meet at "the plains of the South" at some specific auspicious occasion. *Book of Poetry* states:

> A good morning having been chosen,
> For the plain in the South,
> She leaves twisting her hemp,
> And dances to it through the market-place.

Here, "good morning" [literally 'grain morning'] "was the sacrifice carnival used to make offerings to the god of reproduction and to pray for robust breeding," "similarly, the place in the poem 'the plain in the South' is not a common place either," "which is also related to the place that the sacrificial ritual demands."[6] Men and women lovers would reach the specific "plain in the South" at this specific day for sacrificial offerings to the god of reproduction, and sing and dance in a carnivalesque meeting. "The Fen Elms at the Eastern Gate" fully expresses the authentic erotic activities conducted by first peoples in mystical and sacred plots of singing, dancing and divining.

13.3 The Artistic Expressive Techniques of "Analogy and Incitement" Established on Ancient Ecological Equality

Rhapsody, Analogy and Incitement are the "three methods" or three expressive techniques of the *Book of Poetry*, among which analogy and incitement have greater significance and which fully reflects that the first Chinese peoples already had rather mature literary expressive techniques early on, and were of continuous influence upon later generations all the way till modern times. The facts suggest that the "earnest thoughts" in "poetry expresses earnest thoughts" were principally expressed by means of artistic "analogy and incitement." "Analogy and incitement" also reflected the idea of ecological equality contained in the ancient Chinese notion of "integral oneness of Nature and humans." The character for analogy (*b* 比) is unfolded in *Explaining Graphs and Analyzing Characters* as two people depending on one another, and is explained as "intimacy. Two people [*er ren* 二人] make up the character *cong* 从 [follow], the reverse of which is the character *bi* 比." Duan Yucai of Qing commented on *Explaining Graphs*, and made the explanation "analogizing (*bi* 比) means drawing close," "its basic meaning is mutual intimacy. Extended meaning: "assisting by side, involving, ordering, comparing, exemplifying, classifying,

[5]Shuangxin [5], p. 178.
[6]Liangfu et al. [6], p. 206.

repeating, choosing the good match and following it, conspiring." He also insists the ancient character *bi* 比 for analogy "should come from two greats (*er da* 二大), which is two people (*ren* 人)." Thus, the basic meaning of *bi* 比 is the intimate and close coexistence of two people. The *bi* 比 used in *Book of Poetry* has the meaning of "making an analogy to something" (*Zhouli*). For example, "Tao Yao" in *Odes of Zhou and the South*:

> The peach tree is young and elegant;
> Brilliant are its flowers.
> This young lady is going to her future home,
> And will order well her chamber and house.

This is a poem describing a young maiden getting married; it uses the gorgeous peach tree flowers that blossom in March to metaphorically refer to the beauty of a newly wed woman, and simultaneously bless her in establishing a happy household. The following two sections separately use substantial fruits and dense leafage to bless the newly wed with many children, more happiness and a flourishing household. The said poem uses the peach tree flower as a metaphor for a beautiful girl, which henceforth became a famous metaphor in the history of Chinese literature, which influenced later generations like the famous poem by Cui Hu of Tang, "Last year on this day through this door, before one the image of peach tree flowers shine red. Before one who knows where to go, the peach tree flowers smile in the spring breeze as of old," and in this way marvelous verses budded forth. More importantly, the poem metaphorically referring to the young maiden as a peach tree flower is an analogy made in the sense of the two being close equals. The "peach tree" in traditional Chinese culture usually has had the meaning of blessing longevity, and up till now, we still gift the elderly with "longevity peaches" to wish them longevity. So, using the peach tree flower as a metaphor not only invokes beauty, but also has the meaning of wishing for one's family and person a long and happy life, which could be called deep implications. This is also the said poem pinning "affective will" on things through the artistic technique of "analogy."

"bi 比" [analogy] was also related to "analogizing virtue" in ancient Chinese aesthetics. Analogizing virtue is the comparing of natural things with the better virtues of man. Confucius said in the "Yong Ye" chapter of Analects: "[t]he wise enjoy waters, the humane enjoy mountains. The wise are active, the humane are still. The wise find delight, the humane find longevity." Xunzi inherits this theory and develops it, saying in "The Proper Model and Proper Conduct" (faxing 法行) that "[a]s for jade, the gentleman compares it with being-virtuous. The way it is gentle, smooth, and lustrous resembles being-humane. The way it has structure and contains patterns resembles being-wise. The way it is firm and unyielding resembles being-righteous. The way it is sharp but does not cut resembles acting-appropriately. The way it can be broken but does not bend resembles acting-courageously. The way that even its flaws are visible resembles being-genuine. In the way that when struck, its sound is pure, rises high, and can be heard far away, but when it stops, it finishes completely, it resembles speaking-properly." (Xunzi: The Complete Text,

Eric Hulton, p. 331 [modified]) Here, Xunzi uses the natural qualities of this natural thing, jade, its warmth, smoothness and lustrousness, its firmness and unyieldingness, its patterns and structure, its being-sharp but not-cutting, its breaking but not bending, its being-flawed but visible, its pure rising sound that when stops finishes completely to metaphorically illustrate the outstanding virtue of someone being-humane, being-wise, being-righteous, acting-appropriately, acting-courageously, being-genuine and speaking-properly. "When I think of my gentleman, he is gentle like jade" cited in this text comes from Odes of Qin "Xiao Rong." The said poem writes of a wife thinking about her husband going out on an expedition. The poem uses the gentle smoothness of jade as a metaphor for her husband's better qualities, and implies deep love and affection through this metaphor. Afterward, Chinese arts would widely make use of such techniques as analogy, incitement and analogizing virtues. For instance, Chinese painting used "pine, bamboo and plum" as metaphors, "the three winter companions," which is another expression of humans and Nature as companions in the realm of art. The influence of the artistic technique of "analogy" that began in Book of Poetry has been far and wide, and through the artistic techniques of analogical incitement and analogizing virtues, the idea of equality and friendly relations between man and nature in Chinese culture based on "the integral oneness of Nature and humans" and the deeper sense of "integrating the virtues of Heaven and Earth" were metaphorically instilled.

Below let us take another look at "incitement" (xing 兴). Zheng Zhong of the Eastern Han stated: "the character xing 興, Explaining Graphs and Analyzing Characters explains as 'raising,' the character is a pictograph of two people holding one thing. Duan Yucai, commenting on Explaining Graphs, states: "incitement is raising. Broad Rhymes states: 'filling, raising, refining.' The 'six poems' of Rites of Zhou, are analogies and incitements. Inciting is entrusting an affair on something." "Incitement" in Book of Poetry was always the use of natural things to incite the person writing to raise a deep implication together through this artistic technique, which is where the implications of the poetic art are found. For instance, Odes of Shao and the North, "Piao You Mei":

Dropping are the fruits from the plum-tree;
There are [but] seven [tenths] of them left!
For the gentlemen who seek me,
This is the fortunate time!

This is a poem of a young girl becoming sexually awakened; it uses the plum-tree's fruit dropping to incite the feeling of a passing phase, the briefness of young girls in puberty, and hence the eagerness to find a mate, informing young boys not to hesitate, so as not to miss the opportune time. The following two sections repeat the chant, adding the lines, "now is the time," "this is the fortunate time," asking young boys not to miss the opportunity or feel ashamed about opening one's mouth. In this way, "dropping are the fruits from the plum-tree" and "for the gentlemen to seek me" collectively incite the feeling of urgency during puberty, expressing the deeper meaning of marrying a mate at the right time and the authentic state of survival in early humanity focusing on breeding and reproduction. "Poems of sexual awakening" began with the poem "Piao You Mei," and became an important "mother theme" of ancient Chinese literature. From the perspective of Chinese paleography, the meaning of the characters bi 比 and xing 兴 as "two people" "mutually close" "being raised together" not only speaks to human being's relation to other humans, but also speaks to human being's relation to other non-human beings. The use of analogical incitement in Book of Poetry mostly uses natural phenomena as analogies for the human heart and human relations, which involves using artistic expressive means to regard Nature as companion and see Nature as a friend who is equal to and neither above nor beneath man. This encapsulates an ancient aesthetic thought of "intersubjectivity," and the richness of Eastern ecological wisdom can be spotted from this.

13.4 "Homesickness" Poems about the Yearning for Home, "Where One is Born and Where One Grows Up"

When Heidegger analyzes human being's state of existence, he defines it by "being-in-the-world." His explanation of this "being-in" goes as follows: "'In' is derived from "innan"—"to reside," habitare," "to dwell" [sich auf halten], "An' signifies "I am accustomed," "I am familiar with," "I look after something." It has the signification of "colo" in the senses of "habito" and "diligo." The entity to which Being-in in this signification belongs is one which we have characterized as that entity which in each case I myself am [bin]. The expression 'bin' is connected with 'bei,' and so 'ich bin' ['I am'] means in its turn "I reside" or "dwell alongside" the world, as that which is familiar to me in such and such a way."[7] Human being's existence originally entails the meaning of being at "home." The term "home" is closely connected to ecology. From the perspective of etymology, the German word "okologie" came from the Greek word "oikos," whose original meaning is the human abode, the family's property, the house and housework. Because of this, from the ecological perspective, the human abode is the "home" fit for symbiosis between humans and other natural creatures and suitable for human survival. Whether it is a material home or a spiritual home, both are that which happy human survival relies. Because of this, the literary theme about "home" has become the mother theme of literature since ancient times. There are many poems about "home" in Book of Poetry, at the time of the writing of which society existed under times of rapid differentiation, and due to the frequency of war campaigns and the burden of forced labor, the broad swath of people had to leave home for long periods of time, and even became homeless. Because of this, poems about "homesickness" in Book of Poetry are particularly many, and became the source of literature about homecoming and homesickness in the history of Chinese literature. Minor Hymns "Si Mu" is an extremely famous "homesickness" poem.

---

[7] Heidegger [7], pp. 63–64.

My four steeds advanced without stopping;
The way from Zhou was winding and tedious.
Oh did I have the wish of return?
But the king's business was not to be slackly performed;
And my heart was wounded with sadness.

The protagonist expressing his emotions in this poem was a person rushing about toiling for the king for long periods of time away from home. Riding a swift steed along a long endless path, he aches in his heart thinking of home. The tiredness of the horse, the length of the road and the endlessness of the king's affairs set off the person's homesick feelings, and so he lets out the inner cry "Oh how could I not have the wish of return." What was the reason why the one away wished to return? Originally it was "no leisure to avoid caring for father" and "no leisure to avoid caring for mother," which is to say, he felt particularly homesick because he needed to care for his elderly parents. Because of this, the one away spots a "crow" piebald hovering as if on wing while speedily walking down the road and becomes even sadder, genuinely sighing that man is inferior to bird. The basic reason why the protagonist feels "homesick" is the most important theme of the said poem: one is homesick in the thought of taking care of one's parents. In the age when Book of Poetry was produced, the social economy was still extremely backward, and the entire society still relied on blood relations to sustain it. For this reason, during that age "fatherly affection and children's filial piety" became the most important moral standards as well as an important cause sustaining the chain between ecology and human society, and the feelings of "desiring return" and "homesickness" connected to this "fatherly affection and children's filial piety" became the common emotions moving countless people's hearts. Let's take a look at the traveler's homesick mood passing among the rain and snow in Minor Hymns "Cai Wei":

At first, when we set out,
The willows were fresh and green;
Now, when we shall be returning,
The snow will be falling in the clouds.
Long and tedious will be our marching;
We shall hunger; we shall thirst.
Our hearts are wounded with grief,
Our sadness unknown to all.

This traveler eager to return home endures the long bitter struggles of the road, suffers unending attacks of hunger and thirst, and bears the grief of remembering parents and wife, but recalls the fresh and green willows upon leaving home and the snow falling in the clouds upon returning home now, the contrast between which only adds grief to sadness. "At first, when we set out, the willows were fresh and green; Now, when we shall be returning, the snow will be falling in the clouds" became a famous verse on homesickness uttered by thousands of mouths over generations; it has hence become deeply emotionally contagious. For sure, no matter how long and far from home anyone happens to be, homesickness is an eternally enduring thought in the hidden recesses of our minds. This is the commonly uttered feeling of "one's native land" [designated in Chinese by "mulberry and catalpa trees"] Minor Hymns "Xiao Bian" goes as follows:

Even the mulberry and catalpa trees
Must be regarded with reverence
But no one is to be looked up to like a father
No one is to be depended on like a mother.

That land where originally grows mulberry and catalpa trees is precisely that land where my parents gave birth to me and raised me and furthermore where they still live, which is an eternal memory and yearning in all of us.

13.5 Poems of "Building Home" Reflecting Early People's Building of an Habitable Environment

Akin to poems about "homesickness" are some poems about "building home" preserved in Book of Poetry. Most of such poems are odes of praise, used to praise the king of Zhou's feat of leading tribes to build a territory. When describing the choice of place to erect the capital, the odes of praise reflect the first people's ancient eco-humanist thought in choosing a place to settle and building a habitable environment guided by the idea of "integrating humanity and Nature into one" at the time. As is widely known, the ancient Chinese focused much on the choice of environment and structure of architecture in building houses, and strove for the harmonious unity of nature and humanity, Qian and Kun and yin and yang. The hexagram judgment of the Tai hexagram in Zhou Book of Changes states, "In Tai, we see the little gone and the great come. (It indicates that) there will be good fortune, with progress and success. The Tuan Zhuan commentary says of Tai, "It is thus Heaven and Earth inter-crossing, and all creatures interlinking; the higher and lower (in society) cross paths, and their wills converge. Yang on the inside and yin on the inside, assertive within and compliant without, superior master within and minor human being without. The authoritative person's path stretches long and the petty man's path disappears quick." Considering the architecture of human inhabited environments, these words inform us that the ancients wanted to achieve "safety" (tai 泰) by handling well relationships between the heavens and the earth, major and minor, yin and yang and inside and outside, and thereby reach the goal a happy survival beneficial to both family and its members. Major Hymns "Mian" describes the story of the Zhou dynasty relocating the capital from Fen to Qi in the Wei River Plains:

The plain of Zhou looked beautiful and rich,
With its violets and sowthistles [sweet] as dumplings.
There he began with consulting [his followers];
There he singed the tortoise-shell, [and divined].

The responses were – there to stay, and then;
And they proceeded there to build their houses.

Here written is the reason for selecting the Wei River Plains, which was because of the fertile soil and rich produce there. So, after consulting divination and receiving the omen from the singed tortoise shell, he decided to "build home here." Minor Hymns "Si Gan" introduces the habitable advantages of the aristocratic mansions:

By the graceful sweep of these banks,
With the southern hill, so calm in the distance,
[Has the palace arisen], firm as the roots of a clump of bamboos,
[With its roof] like the luxuriant head of a pine tree.
May the brothers [here],
Be loving among themselves,
And have no scheming against one another!

Spoken about here is clear flowing waters, calm southern hills, and dense bamboo forest as well as the peaceful and harmonious coexistence among brothers and kin. Such an environment unifying Nature and humanity is the proper habitation of gentlemen, so "it is where the gentleman dwells."

13.6 Poems of "Farming Matters" Reflecting the Laws of Ancient Agricultural Production

China is an ancient civilized state rooted in agriculture, which always felt the greatest importance in agricultural affairs, and all agricultural activities there put great emphasis on farming in accordance with the ecological laws of Nature. Record of Rites "Lunar Ordinances" records, in the first month of spring, "the son of Heaven on the first (xin) day prays to God for a good year; and afterwards, the day of the first conjunction of the sun and moon having been chosen, with the handle and share of the plough in the carriage, placed between the man-at-arms who is its third occupant and the driver, he conducts his three ducal ministers, his nine high ministers, the feudal princes and his Great officers, all with their own hands to plough the field of God. The son of Heaven turns up three furrows, each of the ducal ministers five, and the other ministers and feudal princes nine. When they return, he takes in his hand a cup in the great chamber, all of the others being in attendance on him and the Great officers, and says, 'Drink this cup of comfort after your toil.'" Recordings in Lunar Ordinances of Record of Rites, the twelfth Chronicle of Master Lü's Spring and Autumn Annals, and Seasonal Regulations in Huainanzi all demonstrate that ancient China had the eco-cultural tradition of arranging agricultural affairs according to seasonal timing and hunting in line with natural laws. The poem "Agricultural Affairs" (nong shi 农事) in Book of Poetry was produced precisely under this tradition, reflecting the production activities and ecological ideas of the time. Sacrificial Odes of Zhou "Zai

Shan" describes in detail the such process in agricultural production at the time from reclaiming land, spring ploughing, sowing, inter-plot management and harvesting to sacrifices to Heaven and ancestors. The poem writes:

They clear away the grass and the bushes;
And the ground is laid open by their ploughs.
In thousands of pairs they remove the roots,
Some in the low wet lands, some along the dykes.

This is the majestic scene of two thousand people engaged in clearing away brush and ploughing in the spring, and "[i]t is not now only that there is such a time, from of old it has been thus," namely it had been done in this way since ancient times. Odes of Bin "Qi Yue" is the most typical poem on agricultural affairs, which in extremely intricate detail describes the more or less complete list of agricultural activities such as ploughing the earth, gathering mulberries, spinning, dying cloth, sewing clothes, gathering medicinal herbs, picking fruit, planting vegetables, threshing, renovating, fermenting and sacrificial offering, which all must be conducted following the monthly ordinances and agricultural timing. The pome also describes in these grounds the relations between social classes at the time, expressing the pain of poor peasants having to sew and weave clothes for the noble dukes, while lacking food and clothing themselves to the extent that one's own wife can also be taken by force. The first verse of the poem goes:

In the seventh month, the Fire Star passes the meridian;
In the 9th month, clothes are given out.
In the days of [our] first month, the wind blows cold;
In the days of [our] second month, the air is cold; -
Without the clothes and garments of hair,
How could we get to the end of the year?
In the days of [our] third month, they take their ploughs in hand;
In the days of [our] fourth, they take their way to the fields.
Along with my wife and children,
I carry food to them in those south-lying acres.
The surveyor of the fields comes, and is glad.

In ancient China, the position of the stars determined the divisions of the solar year, the ordinances of the lunar calendar and the timing of agricultural activity, and in the time of the ninth month of the agricultural calendar, Mars already passes the meridian, in November the cold air makes it necessary to wear winter cloths, but how would the toiling peasants survive the winter without such clothes? In the start of spring in March, the people should repair ploughing instruments, and in April arrive in the fields, the wife and kids accompany by serving food, and the field official sees everyone busy working with radiant joy. In what follows is every agricultural activity that needs to be conducted each season, reminding people not to move against the agricultural timing. Because the state had just been established with agriculture at that time, early Chinese people had a special feeling of affection for the land, which is implied in the ode of praise for the meritorious cultivation of the earth as indicated in the hexagram judgment of the Kun hexagram in Zhou Book of Changes, "Kun, the earth's thickness bears all things." Minor Hymns "Xin Nanshan" is an ode filled with deep sentiment regarding the good land of Xin Nan Shan where early people ploughed and cultivated in the Zhou dynasty. The poem goes:

Yes, [all about] that southern hill,
Was made manageable by Yu.
Its plains and marshes being opened up,
It was made into fields by the distant descendant.
We define their boundaries, we form their smaller divisions,
And make the acres lie, here to the south, there to the east.

The heavens overhead are one arch of clouds,
Snowing in multitudinous flakes.
There is superadded the drizzling rain.
When [the land] has received the moistening,
Soaking influence abundantly,
It produces all our kinds of grain.

One could say, this poem fully expresses the deep emotion early people felt about this fertile land at the foot of the southern hill, praising their ancestor the Great Yu for gifting them this fertile soil. This land is wide open and even, the rain water is ample with strong crops and strong sprouts, which made it an excellent home for progeny to dwell, breed, exist and develop in.

13.7 Poems Revering "tian bao" in the Heavens Overhead

The age in which Book of Poetry was produced was a pre-modern agricultural society with low productivity and extremely undeveloped science. People had a deep reverence for natural spirits in thought. The people were pantheists, revered Nature to the extreme, and pinned their own destiny on the blessing of Heaven overhead. Because of this, there were many "Heaven Blessing" poems pleasing for shang di's blessing. For instance, Minor Hymns "Tian Bao" is poem by a minister asking for his rulers blessing; in it we find the important component of asking for blessings from Heaven.

Heaven protects and establishes thee,
It grants thee all excellence,
So that thine every matter is right,
And thou receives every heavenly favor.
It sends down to thee long-during happiness,
Which the days are not sufficient to enjoy.

Heaven protects and establishes thee,
So that in everything thou dost prosper,
Like the high hills, and the mountain masses,
Like the topmost ridges, and the greatest bulks;
That, as the stream ever coming on,
Such is thine increase.

Here, the poet clearly indicates that the state can only safely stabilize and the ruler enjoy fortune and peace by receiving the blessing of the heavens, and moreover, by fervently rejoicing in this fortune that befalls from the heavens through comparing it with the highest mountain peaks and the thickest hills. Vice versa, if we violate the way of Nature, punishment is inevitable. Minor Hymns, "Yu Wu Zheng" is a work "mocking King You," a criticism by a feudal official about King You's violations against natural developmental tendencies; King You "stands not in awe at Heaven" and hence Heaven unleashes disasters that bring about chaos in the state and loss of subsistence for the people. The poem states:

How, Oh great Heaven,
Untrustworthy the monarch's words are?
He is like a man going [astray],
[Who knows] not where he will proceed to,
All ye officers,
Let each of you reverently attend to his duties.
How do ye not stand in awe of one another?
Ye do not stand in awe of Heaven.

Facing the chaos and famines among the people, the downfall of the Zhou palace, the relocation of the officers and the befalling of all kinds of disasters, the poet believes the fundamental cause behind it is "the monarch's words are untrustworthy" and he "does not stand in awe at Heaven." It is perfectly clear that what the poet here expresses is an early human "conception of the mandate of Heaven," which is limited and backward due to the times. Of course, we cannot pin the destiny of humankind all on "the mandate of Heaven," nor can we devotedly stand in awe at Heaven. But if the "mandate of Heaven" could be reinterpreted as natural laws independent of the human will, then this poem tells us that humans should actively take the initiative to follow these laws to live, and moreover maintain due respect and awe for Nature and the earth as the mother of humankind. If we could do this, humans would certainly achieve better survival. This is perhaps precisely what such poems as "The Blessing of Heaven" tell us so inspiringly.

13.8 "Epic Poems" of Grasping Nature and Erecting the State

Many peoples have their own epic poems composed of myth, legend and historical story like the Homeric Hymns of ancient Greece. Book of Poetry also holds some poems of epic Chinese quality like Birth to the People, Duke Liu, Continuous, How Great, King Wen, The Great Illumination, etc. The main theme of all of these poems is praise of the founder of the Chinese people through which runs an idea of "grasping Nature and erecting the state," which is one of the origins of the spirit of the Chinese people. Birth to the People is the people of Zhou praising the ancestral origin of their people, Hou Ji; it is a long poem narrating his legendary historical experience and his contribution in agriculture. The said poem first of all narrates the legendary birth of Hou Ji:

The first birth of [our] people,
Was from Jiang Yuan.
How did she give birth to [our] people?
She had presented a pure offering and sacrificed,
That her childlessness might be taken away.
She then trod on a toe-print made by God, and was moved,
In the large place where she rested.
She became pregnant; she dwelt retired;
She gave birth to, and nourished [a son],
Who was Hou Ji.

What is spoken about here is the legendary birth of Hou Ji, whose mother, Jiang Yuan treaded onto God's footprint and became pregnant with Hou Ji, which is somewhat similar to the birth of Jesus in the Bible. Any sage is a song of Heaven, which is the basis of Hou Ji grasping Heaven or Nature and erecting the state. Many scholars of later generations actively researched the concrete sense of "she trod on a toe-print made by God, and was moved" in the attempt to clarify whether this implies adultery or sexual intercourse with the body of God and became pregnant, but actually there was no massively great need for it. Because what is spoken about here is only the mythological legend of the birth of a people's earliest ancestor. Behind it, the tales of Hou Ji being abandoned thrice and saved thrice is identical to the legendary under-goings of the first ancestors of many other peoples. Moreover, the telling of Hou Ji leading the offspring of Cathay to engage in agricultural planting, which was done with the help of Heaven:

He gave his people the beautiful grains:
The black millet, and the double-kernelled;
The tall red, and the white.
They planted extensively the black and the double-kernelled,
Which were reaped and stacked on the ground.
They Planted extensively the tail red and the white,

Which were carried on their shoulders and backs,
Home for the sacrifices which he founded.

The content of the poem is Heaven gifting good seeds and a bountiful harvest, and so after the harvest the people should offer sacrifices to Heaven and ancestors. The two poems that follow are Duke Liu and Continuous. The former mainly describes how Hou Ji's descendant, Duke Liu, moved the capital from Tai to Bin, and laid the foundation. For instance, duke Liu choosing the address and establishing the capital:

Of generous devotion to the people was duke Liu,
He went there [to the place of] the hundred springs,
And saw [around him[the wide plain.
He ascended the ridge on the south,
And looked at a large [level] height,
A height affording space for multitudes.
Here places could be occupied;
Here shelter for travelers;
Here discussions could be had;
Here debates could be held.

In the poem it states that honest duke Liu established the capital at such a great place with springs, open plain and a ridge as Bin. This is indeed the best choice with advantages to the people's survival, "here places could be occupied; here shelter for travelers," and hence everyone high and low could speak cheerfully, "here discussions could be had; here debates could be held." Major Hymns "Continuous" describes the story of the thirteenth generation patriarch of the Zhou dynasty, King Tai of Zhou, leading his people to settle in the Wei River Plains. What follows is a passage explaining the choice of fertile land good enough for the people's growth:

The ancient duke Dan-fu,
Came in the morning on galloping horses
Along the banks of the Western rivers,
To the foot of [mount] Qi;
And there, he and the lady Jiang,
Came, and together looked out for a site on which to settle.

The poem writes of the ancient duke Dan-fu and newlywed wife riding together at the break of dawn to the banks of the river Wei and the foot of mount Qi to look for and determine the terrain of their people's settlement, illustrating land as the root of development of the people's survival; it was precisely the billowing Wei River waters and the wide-open plains that nurtured the ancestors of the Chinese people.

13.9 "Musical Poems" Expressing the Unity of Ancient Song and Dance

In ancient China, song and dance were one, a unity which was also people's most important way of life at the time. Shamanistic and religious sacrifices were the most important content of people's lives at the time, which, one could argue, ran through every aspect of human life from birth, love and marriage to productive labor to customs and holidays. The first peoples continuously realized their own desires to communicate with the heavens in this song-like, dance-like and poem-like atmosphere of a religious quality. Zhou Book of Changes "Xici I" points out with Confucius's words: "the sage prescribed the signs to fully express the meaning, set up the [64] hexagrams to fully express reality and false appearance, and makes a judgment to fully express its message, changes and connects it to fully examine the advantages, then drums and dances to it to fulfill the spirit." "Drumming and dancing to it" is what actually takes place in the sacrifice; it was the main way in which humans communicated with Heaven and spirits. Book of Poetry preserves a considerable amount of such song-like and dance-like sacrificial poems. Minor Hymns "Chu Ci" describes the song of sacrifice for the ancestors. After it describes in detail the preparations before the sacrifice, it writes of the song and dance during the sacrifice:

The ceremonies having thus been completed,
And the bells and drums having given their warning,
The filial descendants goes to his place,
And the able priest makes his announcement,
'The Spirits have drunk to the full.'
The great representative of the dead then rises,
And the bells and drums escort his withdrawal,
[On which] the Spirits tranquilly return [to their place].

Here, after all the preparation work is completed, the sacrifice begins and the drums and bells ring in unison, completing the sacrificial ritual through the sound of music, then music is used to escort the master of the sacrifice. Odes of Zhou "Zhi Jing" describes the ceremony for the sacrifice to the prior king, which is conducted likewise to the sound of musical poetry:

The bells and drums sound in harmony;
The sounding stones and flutes blend their notes;
Abundant blessing is sent down.
Blessing is sent down in large measure;

Here, bells, drums, stones and flutes, four instruments, sound off in a blended harmony and reach intense solemnity in the activity of sacrifice, fully expressing the scene of the sacrificial ode's "likeness of beauty overflowing with virtue, announcing to the gods with its success." Minor Hymns "Drums and Bells" recounts the playing of the hymn:

His bells ring out qin-qin;
His lutes, large and small, give their notes;
The tones of his organs and sounding stones are in unison.
They sing the Ya and the Nan,
Dancing to their flutes without error.

Here, there are 7 kinds of musical instrument used in the hymn, drums, Pi-pa, Qin, the sheng reed wind instrument, the sounding stones, and organs, all of which ring out accompanying one another in unison and rhythmically pleasing to the ear, a spectacular event. These are all "temple songs" played at sacrifices, but in everyday life there was also banquet music. Odes of Wang "Jun Zi Yang Yang" specifically describes music played at aristocratic banquets:

The masters look around full of satisfaction.
In their left hands holding reed-organs,
With their right calling me to the room.
Oh the joy!

The masters look delighted.
In their left hands holding screens of feathers,
With their right calling me to the stage.
Oh the joy!

Here is described the music at a family banquet, a scene that is equal part dance and song with musicians singing, dancing and playing instruments, some holding reed instruments, some holding streamers while singing and dancing in infinite joy. Common people also have their own musical lives, Odes of Chen "Wan Qiu" describes the ceremonial singing, dancing and meeting for merriment at the beginning of spring to commemorate the god of reproduction at the Pu river among the mulberries with one female dancer elegantly dancing by the mountain slope in the wilderness:

How gay and dissipated you are,
There on the top of Wanqiu!
You are full of kindly affection indeed,
But you have nothing to make you looked up to!

How your blows on the drum resound,
At the foot of Wanqiu!
Be it winter, be it summer,
You are holding your egret's feather!

How you beat your earthen vessel,
On the way to Wanqiu!
Be it winter, be it summer,
You are holding your egret-fan!

Who in the end is this pretty woman dancing and singing in the wilderness? It is generally thought that she is a witch. We could also speculate that she is perhaps also a woman "on the Pu river among the mulberries," who is admired by many young men.

(3) .

In summary, interpreting Book of Poetry from the perspective of eco-ontological aesthetics truly makes us feel discover everything anew and bountifully so. Overall, what the Book of Poetry exhibits is an "affective will" to integrally unify man and nature, which is an ancient form of eco-humanism. We could understand it from the three standpoints of "poetic style," "poetic mood" and "poetic method." From the perspective of "poetic style," Book of Poetry provides us with "ode-style poetry," which is a unique proto-ecological art of poetry whose implications reflect human being's authentic state of existence; it is an ancient art combining shamanistic song and dance, which was the basic way of life of ancient Chinese people. From the perspective of "poetic mood," Book of Poetry describes nearly from all sides the life of first Chinese peoples, reflects their emotional feelings, and expresses the condition of life and passion of desire intimately related to the common people, Nature and the essence of human being in particular. Overall it envelops a variety of aspects from emotion, family and food to labor, witchcraft and music. So-called "emotions" mainly refers to innocent authentic love, namely poems "among the mulberries on the Pu river"; "family" then refers to the feeling of "home," falling under whose scope includes homesickness poems, traveler poems, poems of resentment at one's wife and poems of building homes; "food" refers to "the hungry singings about their food" or those songs of hunger moving people's hearts; "labor" refers to "workers singing about what they have to do," including work songs and songs attacking the exploiters; "witchcraft" principally refers to odes describing sacrifices. In those times, the sacrifice constituted an important content of people's lives, "the great affairs of state are sacrifice and military campaigns" (Zuo Commentary, "13th Year of Duke Cheng"). Sacrifice was even more so the main channel of communication between people and the heavens, so there are many poems describing sacrifices in Book of Poetry. Finally "music" was tightly connected to witchcraft. While witchcraft involved mainly the temple and the imperial activity of the aristocratic class, "music" was the basic way of life of common people in those times; it reflected the authentic condition of life of the common people. From the perspective of "poetic method," Book of Poetry mainly gives us such expressive poetic techniques as "analogy and incitement," which were moreover deployed from the ancient "intersubjective" perspective of equality between Nature and human being, involving the spirit of companionship with Nature; both rare and commendable, it became the fine enduring tradition of Chinese poetry creating poetic charm in the equal communication between humans and nature. The methods of "analogy and incitement" directly influenced the later doctrine of "poetic state of mind," implying poems of profound friendship in the fusional merging of man and object, poetic intention and state of affairs as suggested by the statements "intention is between the lines" and "the emotion beyond the state of affairs."

Reinterpretations of Book of Poetry gives us much inspiration and makes us further recognize that the "practical aesthetics" so widely influential for such a long time and the beauty that it emphasizes involves the theories of "the objectification of the essential forces of man" and that of "subjectivity," which only have partial correctness, and using these theories makes it impossible to properly explain such an ancient literary classic as Book of Poetry. Book of Poetry is not completely composed of poems about labor nor is it some product of the objectification of the essential forces of man; rather, it is mainly composed of proto-ecological songs sung from the natural disposition of human being. It is neither some product of humans reforming and conquering Nature nor entirely the self-praise of human being. It is much rather songs about life and survival arising from human being's natural disposition; it is the expectation of "the integral oneness of nature and humans," and even prayers to the uncertain cosmos and heavenly movements. Its odes to the heavens are far beyond praises of man, and no anthropocentrism can be found in them. Thus, Book of Poetry is composed of songs about life, songs wishing for harmony between nature and man with extremely rich implications of eco-ontological aesthetics. It is precisely this perspective, we believe, that made Heidegger in the middle of the 20th century switch from "anthropocentrism" to ecoholism under the inspiration of Eastern philosophy and aesthetics, and especially Chinese Daoist thought, making the famous doctrine of the fourfold play of sky, earth, gods and mortals so extremely significant. We insist that just as Heidegger made this breakthrough in thought from inspiration gathered from the East, we shall continue to search for inspiration from Eastern arts by further deepening our exposition through the perspective of eco-ontological aesthetics, and our research on Book of Poetry is just one effective attempt among them. The cultural context produced by the Book of Poetry is overall akin to Daoist thought, and the basic implications of its thought are also correlated with the Daoist doctrine of Dao emulates what is naturally so. Because of this, the techniques of "ode-style poetry," "poems among the mulberries," "homesickness poems" and "analogical incitement" exhibited to us in Book of Poetry all envelop concrete art forms and aesthetic experiences with a thick spirit of "integral oneness of Nature and humans." Such experiences give us extremely important inspiration for building contemporary eco-ontological aesthetics.

Of course, Book of Poetry was ultimately a product created over 3,000 years ago, at which time our first peoples still lived in extremely backward, premodern living conditions, and their thought was also in a rather unenlightened state, still stuck to which are many hues of mysticism and superstition. All of this is inevitably reflected still in Book of Poetry and pervades its artistic aesthetic experience, and thus it is inevitably still riddled with many limitations. But, this cannot erase its monumental value nor erase its important role as resources of thought for the construction of contemporary eco-ontological aesthetics.

1. **Looking Back on Home: An Ecological Aesthetics Interpretation of The Right bank of erguna river**

Chi Zijian's long novel The Right Bank of Erguna River (Hereafter abbreviated as Righ Bank) is an outstanding epic novel whose source material is the life of the Ewenki ethnic group and which won the seventh Mao Dun Literary Prize. The achievements of this novel are multifaceted, but what we shockingly discover is that it is an outstanding work of ecological literature rarely noticed in the field of contemporary Chinese literary studies. The author puts to use his plentiful accumulation of life and diverse artistic techniques to exhibit the important theme of contemporary humanity's "reflection on home," revealing contemporary humanity's yearning for "poetic dwelling" in a state of having lost their home. This novel makes a special contribution to our effort to construct contemporary ecological aesthetics and ecological literature with its successful practice in creation.

Bone inscription writing already had the jia 家 character, whose two meanings are: one, "where humans reside," and two "what is connected to ancestors, the ancestral temple of prior kings."[8] This illustrates that the basic meaning of jia 家 is the place where humans reside and the resting place of ancestors, which is the root of humankind. From the microscopic standpoint, "home" is the "place" where everyone of us is born, grows and lives. But from the macroscopic perspective, "home" should be the greater natural world humans rely on to survive. Through the progression of modern industrialization and urbanization, our homes have been scarred and scarred to the point of making us homeless. Because of this, through the horizon of contemporary history, "reflecting on home" has become an extremely important theme in the literary arts and humanities. Heidegger's famous book, Elucidations of Hölderlin's Poetry has a special section elucidating Hölderlin's poem Homecoming, and insisting that by "homecoming" is meant looking for "the best most essential things".[9] Because of this, "looking back" or "looking for" is actually a yearning, and even more so a critique and reflection. Anthropologists of aesthetics believe, "research on past civilization in actuality complexly reflects people's thought, critique and denial of reality."[10] Chi Zijian expresses his own poetic sentiments about the future destiny of humanity and his own deeper reflections on real life through "looking back" on the hundred year history of the rise and fall of the Ewenki people in Righ Bank. The cultural anthropologist Frazer compares his own works metaphorically to the golden bough that the Troy hero, Aeneas, got from the hand of a goddess—not only could help him find his father, but could also help understand his own fate—Chi Zijian also tried to provide humanity's exploration of their own future destiny with a "golden bough" through the description of the Ewenki people's life and destiny.

(1) The Unique Perspective of "Looking Back"—Exploring the Primordial Character of "Home"

"Homecoming poems" are frequent source material in literary arts. Early on in the pre-Qin Book of Poetry we find the verse "At first, when we set out, the willows were fresh and green," after which we find Li Bai's verse "life my head and see the moon, lower my head and miss my home" in Quiet Night Thought, which an even more famous homesickness verse. Righ Bank is a reflective novel in the "postmodern"

8Zhongshu [8], p. 799.

9Heidegger [9], p. 12.

10Jie [10], p. 192.

perspective written at the beginning of the 21st century. The modernization and industrialization that began in the 18th century actually brought good tidings to humanity, but simultaneously brought humankind massive disasters as well. This is undoubtedly an antinomy of the beautiful and not beautiful. In one respect, the condition of human life was largely improved; in another respect, while enjoying modern civilization, natural destruction, spiritual tension and traditional moral decay brought a series of disasters to humanity. The material and spiritual "home" humans have relied upon to survive has already changed beyond recognition, and humankind currently faces the danger of losing "home." As Heidegger put it, "In anxiety one feels 'uncanny'. Here the peculiar indefiniteness of that which Dasein finds itself alongside in anxiety, comes proximally to expression: the 'nothing and nowhere'. But here 'uncanniness' also means 'not-being-at-home' [das Nicht-zuhause-sein]." He then states that uncanniness is a basic way of being-in-the-world.[11] In the case that uncanniness becomes a basic way of humans being-in-the-world, reflective works of "looking back on home" come into being.

Early on in 1962, there were famous ecological writers in America, and in like manner, the female writer Rachel Carson wrote a work of ecological literature with landmark significance reflecting on the disaster of agrochemicals, entitled Silent Spring, which played a huge role in awakening people at the time. Today, Chi Zijian's Right Bank is a reflection on the nomadic hunting Ewenki people losing the home of their survival and being forced to relocate. The writer notes in Postface that the cause that triggered her to write the book was the children of the Greater Kinghan already a sense of the severe degradation and deterioration of forests brought about by the continuous cutting of primeval forests for 30 years. The first to be harmed by this was of course the Ewenki people as nomadic hunters of the mountain forest. She states: "Those suffering the greatest harm are the nomadic hunting people who live in the mountain forest. More specifically, it is that group of Auluguya Ewenki epeople we call the last nomadic group who live off of the reindeer." And its direct fate was the writer receiving a friend's information about the female Ewenki painter Liu Ba leaving the forest, returning to the forest and finally being buried in the river as well as all of the writer's experiences of losing minority groups and the human spirit, which she became familiar with in Australia and Ireland. All of this made the writer sense deeply the common feeling nearly universal among contemporary humanity that the original home is lost. This is what made the author find the resolve to write this important work. When she deeply embedded herself at the fixed settlement of the Ewenki people, Genhe city, the group by group return of the hunting people even more so hardened her resolve. Thus, the writer began her arduous and delicate journey of creation.

The author adopts the technique of the epic poem, and tells the history of the hundred year rise and fall of the Auluguya Ewenki people of the Arguna river through the mouth of an over 90 year old elderly grandmother of the Ewenki people and the wife of the last chief. From beginning to end, the main thread of this account is the tracing back of the primordial Ewenki people's survival, and the keynote of this

[11]Heidegger [7], pp. 218, 318.

account is the great tolerance and gentleness of human nature unique to the sons of the great forest. The entire account splits up into four parts, precisely summarizing the whole process of the entire people from the rise to the fall and hope for the future. Just as the narrator's husband, the last chief Waluojia, sang on that warm night:

> The morning dewdrops moisten the eyes,
> The sun at noon scorches the back,
> The reindeer bells at dusk most refreshing,
> Small birds at night return to the forest.

The ballad implies the entire people being born in the warmth of the morning as well as growing and surviving in the scorching heat at noon and refreshing cool at dusk, and looking forward to journeys in the moonlight at night with each journey implying the foundation of the people's survival. In the early morning account, the elderly Ewenki grandmother recounts the genesis of her people and its natural foundation. According to legend, the Ewenki people emerged in Lamu Lake, which is Lake Baikal:

There are 8 big rivers pouring into the lake, and the lake water is deep blue. Growing in the Lumu Lake are many deep-green aquatic plants, the sun is very close to the lake water, and sunlight floats on the surface year round. There are also pink and white lotus flowers. Surrounding Lamu Lake are towering mountains. Our ancestor, an Ewenki with a long braid resides there.

But 300 years ago, the Russian military invasion made their ancestors move from the Lena River in the Yakutia province to the right bank of the Arguna River, while diminishing from 12 clans to 6. From this point, the Arguna became the dwelling place of living Ewenki peole. She states.

We cannot possibly leave this river. It has always been the center for us, who live beside its so many tributaries. While this river is the palm, its tributaries are the unfolding five fingers, which stretch out in different directions like flash after flash of lightning illuminating our life.

Here, the narrator says the Arguna River is closely related to the breeding life of the Ewenki people; it is the center of the whole people, illuminating their lives generation after generation. The great mountains surrounding the sides of the Arguna River—the Lesser Khingan is also the lands nourishing the Ewenki people. The narrator says:

In my eyes, every mountain on the right bank of the Arguna is a star shining on the earth. These stars are green in the summer season, and yellow in autumn, but by Winter they are silver white. I love them. They are like us, they also have their own character and posture… The trees on the mountain are clump after clump of flesh and blood.

It is this great mountain with "clump after clump of flesh and blood" became the survival and living space of the Ewenki people, who are people of the reindeer, which provide them with deer milk, deer antlers and also excellent transportation and hunting mates. Reindeer are domestic animals unique to the Lesser Khingan, because the forest is dense there, and mosses and lichen grows there, which provide the reindeer with plentiful food. Because of this, the narrator says:

The reindeer were definitely gifted to us by God, without them, we would not exist. Although they once carried away my relatives, I still love them so much. Not seeing their eyes is like not seeing the sun during the day or not seeing the stars at night, which would make the human heart let out a deep sigh.

On the mountains surrounding the Arguna River, rest the remains of Ewenki ancestors. The narrator vividly tells of the extraordinary life journey of her father, mother, husband, uncle and nephew. First her husband Link was struck and killed by lightning in a storm when switching for a stronger reindeer to go down the mountain; his burial remains were set atop a tall pine tree to weather; her mother Damala died dancing painfully after suffering the loss of her husband and, with him, love; her burial remains were set atop a Betula tree to weather away; Of the narrator's two husbands, one froze to death en route to look for the reindeer, the other died in the process of fighting a bear to save another person; Her uncle Nidusaman died putting in the last of his strength to fight off the Japanese; Her nephew, Guogeli died because his mother, the shaman Nihao, had to offer her own child to Heaven in order to heal the sick child of the Han Chinese person He Baolin. These relatives finally returned to Nature to accompany the stars, the moon and the milky way. Just as the shaman Nihao sang in a weathering burial requiem:

Oh ye, whose soul has gone far away,
Do not fear the dark night,
There is a bundle of fire here,
To light your journey.
Oh ye, whose soul has gone far away,
Do not worry of your relatives any longer,
There are stars, the Milky Way, clouds and the moon,
Singing for your arrival.

The "weathering burial" spoken of here is a burial form unique to the Ewenki, which involves choosing four big trees with right angles opposing one another, cutting down some tree trunks and suspending them on the branches to build a raft for the one passing away. Afterward, they raise the one passing away onto that raft with white cloth wrapping, head to the north and feet to the south, then covers the tree branches, and places burial objects on them. Finally the one passing away is sent away by the ceremony held by the shaman. This weathering burial actually illustrates the Ewenki's way of survival in coming from Nature and returning to Nature—They see themselves as the children of Nature.

The Arguna River and the Lesser Khingan have also witnessed the Ewenki people's love and works. The narrator tells of the love, birth and death of her own father's generation and generation after generation of descendants among these beautiful mountain waters. Her father and uncle simultaneously fell in love with the most beautiful and most lovely dancer, the maiden Damala, but ultimately her uncle Niman lost to Link through an arrow shooting competition to decide who becomes the new groom—In actuality he actively gave up his own love. In the second year, when Damala and Link got married, the wedding present that Damala's father gave to her was a bundle of "tinder" so important to nomadic hunting peoples. Later, this "kindle" was given to Damala's own son as a wedding present. When Link got married, Niman scratched his own fingers and became the tribe's shaman. After Link died, Niman's love for Damala awoke again; he used pheasant feathers he gathered for 2 years to meticulously weave the most beautiful skirt, which was the beautiful, eye-catching image of the Arguna River and its surrounding mountains. The narrator describes:

Looking from top to down, this skirt is as if it were composed of three parts; the top part is the grey river, the middle is the green forest, and the bottom is the blue sky.

When Damala received this precious gift, she was extremely happy, filled with astonishment, joy and excitement, telling that this was the prettiest skirt she had seen in the world. But their love ultimately failed due to the common customs of the people (the custom that does not permit widows remarry with the husband's uncles and older brothers), Damala said goodbye to the world grief stricken in the end, and Niman also rashly ended his own life as well. At Damala's funeral ceremony, Niman's requiem sadly wails, the song expresses the Ewenki people's loyalty and selflessness toward love, and also expresses Niman's willingness to cross the fabled "river of blood" and accept any punishment in order to make Damala enter another better world. The song goes:

> Oh torrential river of blood,
> Please erect a bridge,
> Walking before you,
> Is a kind-hearted girl!
> If her feet are soaking with fresh blood,
> What she steps in,
> Is her own fresh blood;
> If at the bottom of her heart there are still tears,
> What she retains,
> Is also her own tears!
> If you would not like a girl
> Blood on feet,
> And tears at the bottom of a heart,
> And if you erect a boulder for her,
> Please also allow her
> To safely jump over it.
> If you want to blame someone,
> Blame me!
> As long as you let her reach the other happy shore beyond,
> No matter how, if you let me dissolve in the river of blood,
> I will not sob!

We can see from this that the Ewenki people are true children of Nature. Nature has witnessed their loves, and the tokens and gifts of their love also come entirely from Nature. The Ewenki people have already fully melded themselves into the surrounding mountains and waters. Their life and flesh and blood have already fused into one with Nature. The Arguna River and the Lesser Khingan have already become an inseparable part of their life and survival. Helena was the first youth to receive higher education among the Ewennki. She is a renown painter, who also has dignified work in the city. But she ultimately quit the job, and returned home to the Arguna River bank. Because, "she was sick of the job, sick of the city, sick of men. She said she has already fully realized that only the reindeer, trees, the moon and breeze do not make one sick." She used the full 2 years of time to paint the history of two hundred years of the Ewenki people's trials and tribulations, then finally eternally rested in the Jiliu River, tributary of her homeland Arguna River.

Having undergone 30 years of development on increasingly grander scale, the environment in which the Ewenki people have survived has already suffered massive damage; there aren't even 200 hunters still living in the mountains, and there are only six or seven hundred reindeer left, so they decided to relocate down the mountain. When mobilizing people to relocate, someone said, hunters and reindeer moving down the mountain is also protecting the forest, when reindeer freely roam, damage to vegetation is unavoidable, resulting in ecological disequilibrium; the person then said, we must now enact protections for animals, we cannot hunt; only a people who has laid down the gun is a civilized people, a people with a future. The narrator responded in her heart:

We and our reindeer have always been wedded to the forest. Compared to countless loggers, we are some dragonflies lightly skipping upon the surface of the water. If the river in the forest is polluted, how could it be caused by some dragonflies brushing along the river?

The narrator also tells us that the reindeer are originally the children of the forest, who treasure the grassland when they eat, who lightly graze upon the grass while walking over it, so the grassland always appears completely undamaged—what should be green is still green. They eat the leaves of Betula trees and Willow trees, but only graze a few bites then leave—the leaves on the branches of those trees are still flush. Why would the reindeer destroy the vegetation? As to the Ewenki people, they are even more so the children of the forest. When they hunt, they do not kill the young, and they protect the small otters; When they light a fire, they only burn the dried brush, the trees that through being struck by lightning and felled by winds have lost all vitality—they use these "wind fallen trees"—unlike the loggers who fell those trees still filled with life. Every time the relocate, they fill-in the earth they dug out to build fire pits and erect chums, then clean up waste and burry it together deeply in the earth so that these places don't let out the stench of trash because they resided there. They maintain respect for Nature. Even if they hunt big wild game, they will eat it after the ritual sacrifice and maintain many taboos. For instance, the Evenk worship the bear, so when they eat bear meat, they call out several crow-like "ya-ya-ya" sounds, to let the spirit of the bear know that it is not humans eating their meat, but crows eating their meat. Cited repeatedly in the book is a bear sacrifice song of the Evenk people:

> Oh, ancestral mother bear,
> You have fallen,
> May you peacefully sleep!
> Those who eat your flesh,
> Are black crows.
> We piously place your eyes
> Among the trees,
> Like arranging a spirit lamp!

The development of the mountain forest has compelled the Evenk people to leave the forest down the mountain in the effort of relocation, but the reindeer cannot live without the moss in the mountain forest and the Evenk people cannot be without the mountain forest either, so they returned to the forest with the reindeer, but what will the future hold? Only the narrator and her nephew live in the vast unpopulated campsite. When the narrator suddenly noticed their little white reindeer Mukulian return, she excitedly said:

And when I saw that reindeer only getting closer and closer to us again, I felt as if it were that pale crescent moon fallen to the earth, I cried, because I already could not distinguish the heavens from the earth.

The small reindeer returned like that crescent moon, but would it tomorrow? The work leaves us space for imagination and think as well. We realized through the narration given by the wife of the last chief of the Evenk that the right bank of the Arguna, the lesser Khingan, all those mountains and waters, already became the flesh and marrow of the Evenk people, the source of their life and survival. Zooming out to the perspective of cultural anthropology, the source of human life and survival is Nature. How should we handle the source and origin of our own life and survival? In today's rapidly worsening environmental destruction and pollution, this is no longer just the problem of the fate of the Evenk people, but has expanded to the scope of all of humanity's fate. This is one of the most important pieces of inspiration that this novel gives us.

(2) The Unique Place "Looked Back" Upon—Exploring the Uniqueness of "Home"

"Home" is the world closely tied to human life and survival, but "place," as the place of concrete human life, has an important characteristic in ecological and environmental literature, which is being one's own special "horizon." The American philosopher of environmental aesthetics, Arnold Berleant, states in Aesthetics of Environment, "Here is familiar ground, personal ground, whose streets and buildings are unified by customary association, recognized with relief, experienced with ease, and remembered with affection. Whatever aids a neighborhood in achieving identity and in generating the warmth of personality helps turn it into a place with which we can associate ourselves and be at ease."[12] The environmental literature theorist Slovic furthermore defines "place" in Going Away to Think as "the local," namely what is nearby, here and now.[13] Right Bank of Arguna River profoundly describes this specific place where the Evenk live and dwell by the Arguna River. According to Heidegger's explanation of places [which he calls "regions"], "This discovery of regions beforehand is co-determined [mitbestimmt] by the totality of involvements for which the ready-to-hand, as something encountered, is freed."[14] "Totality of involvements" and "read-to-hand" are the two basic elements of "places" or regions, which is to say, human being's involvements with the world compose an inseparable totality, and all beings in the world become "things ready-to-hand"; of course, many of these things are things ready-to-hand inseparably tied to the human being in a particular place. Right Bank profoundly describes the inseparable links between the Evenk people and the mountains and waters of the right bank of Arguna River, which thus determines their particular way of life in the sense that each stretch of grass and forest integrates with their flesh, blood, life and survival with a particular irreplaceable quality. This is a deeply meaningful exploration into the uniqueness of a human "home." The Evenk people's clothing, eating and inhabiting is characterized by a specificity linked to the region in which they survive. They use fur for clothing, and moreover, reindeer fur at that; what they eat is mainly meat, because hunting is their basic mode of survival. The "morning" part of the novel specifically describes Link carrying his two children to hunt big game like the moose. They ride in a Betula raft floating down the small river, then wait for a long time in darkness of night. Finally, Link quickly and bravely shoots a moose and takes its life. The capture of a moose brings joy to the entire camp. Everyone takes part in drying the meat, "those dark red strips of meat, like red and white flower petals blown by the wind." Of course, they also consume reindeer milk, grey squirrels, and buy clothing, grains and other food stuffs from Chinese merchants. They have a special food storage warehouse as well—"the dependable old treasure," which is a storage sustained for the needs of the local people or other people at any time. With four pine tree trunks erected as posts, they make the base and four frames, and cover it with Betula bark, and leave an opening at the bottom to place luxury items and items of abundance in. Not only the local people can access it, other peoples can access it as well. The Evenk elders passed down the following two sentences:

You will not carry your own house with you when you go out, and those coming from elsewhere will not carry their own pots along with them on their backs; someone will only enter a hut with a fire inside, only trees with branches invite birds to roost.

This is the Evenk people's specific way of life determined by the specific conditions of mountain forests like massive snow and frigid cold, reflecting the great concerns of this mountain people. When the narrator was young and got lost in the forest, she relied on this "dependable old treasure house" to get food and ran into her husband. The Evenk people's residence is totally special. What they imple-

[12]Berleant [9], p. 66.

[13]Slovic [12], p. 183.

[14]Heidegger [7], p. 121.

ment is a primitive communist institution, organizing into collectives or tribes with neighboring families. Each collective implements a system of primitive communist production and life, dividing labor among the men, women, elderly and infirm, and everyone gets an equal share. They lodge in one house per family—"a chum," also called "an immortal post," which uses 20–30 leafless pine tree trunks, saws them into two human height posts, and directs one end of each to the sky, which converge together with the other ends diverging into the ground equidistantly like countless dancing legs forming a great circle, around which a weather protective covering is stretched. The narrators says:

I enjoy living in the chum, there is a hole in its roof, which naturally becomes a chimney channel for the smoke emitting from the fire pit. I often look at the stars through this little aperture at night. One only sees a few stars from here, but they are extraordinarily bright like oil lamps suspended from the roof of the chum.

The riding tool the Evenk mostly use when going out is the reindeer, whom are generally only ridden by women, children and the infirm. For many survival reasons like the reindeers' need to eat, they have to move the lodging after passing a certain length. The narrators described one case of moving:

When moving, the white Maruwang walks at the forefront, after whom follows the reindeer carrying kindling; then closely following is the reindeer pack carrying our family. The men and stronger women generally move in synchronicity with the reindeer, and really only ride on top of them when too tired. Haxie carries an axe, and makes a notch sign in a big tree after passing a certain stretch.

When the Evenk women give birth, they specifically build a delivery room called a "Yatazhu," which men are absolutely barred from entering during delivery such that women will make their husbands die anxiously when they enter it. Because of this, child bearing is generally always handled by the women themselves in Nature. But they do get cared for by the whole tribe after they get old. By the time most of the tribe go to the settlement point, the narrator already over 90 years old stays behind. So the tribesmen leaves her nephew Ancao'er behind to take care of her, and leave her ample reindeer and food as well; they even fear she will be lonely, and intentionally leave behind two grey cranes, allowing her to see beautiful birds, unlikely to hurt her eyes. When the Evenk get sick they seek treatment through the shaman dancing in a trance, but without taking any medicine. After dying, the weathering funeral is arranged, the body is laid to rest in a tree for the wind to carry it away back to Nature.

Clearly, the Evenks have their own unique manner of clothing, eating, lodging, moving, living, aging and dying. This is their mode of survival; it is their own special living region, in which they suffer pains, but more importantly comfort and adaptation in survival. When the novel describes the scene of the narrator riding with her father together at night that year to hunt moose, it writes:

The Betula raft does not sit deep in the water, it is extremely light like a dragonfly touching down on the water surface nearly without a sound, just a slight sway. When the raft takes its long-drawn-out start, I feel as if a breeze is brushing past my ears, so soothing. When I look at the trees on the bank when moving in the water, each tree is as if it grew legs, one stepping back after another. It's like the river is a warrior, and the trees are the vanquished soldiers. There isn't a trace of clouds surrounding the moon, extremely bright and clean, making people worry that it will fall to the earth suddenly without anything to block it. The river is straight at the outset, then it slightly begins to curve with the river flow hastening and the river widening as the curvature increases in magnitude.

This is truly a beautiful picture unifying Nature and man. Of course, Nature will also bring disasters to the Evenks like "snow disasters," "dry disasters," "pestilences" and "wolf misfortunes." But these are ultimately component parts organically connected to the world in which humans live. Take wolf misfortunes for instance, although they endanger the Evenk people, the wolf is closely tied to them still, they are unavoidable. As the narrator put it:

In our life, the wolf is stream after stream of frigid wind charging at us. But we cannot eliminate them any more than we can stop the winter from coming.

But overall, this incomparably beautiful natural environment of the Arguna bank is the true homeland of the Evenks; it is the home that births and raises them. The mountains and waters here have already fused into the life and blood of every one of them. Here, Nature being "not-ready-to-hand" is only temporary for them, and what is more primordial is that which is so familiar it does not catch your eye, which is the way of life inseparably tied to them. Divorced from this way of life, divorced from the mountains and waters, from the reindeer, tribe and chum, they would be immediately lost without home, unable to adapt, which is even more so for the elderly. When the narrator talks about the event of relocation, she states:

I am unwilling to sleep in rooms where I can't see the stars. At this age, I pass through the night with the company of the stars. If I awake from sleep at midnight and see a pitch-black roof, my eyes will be blinded; my reindeer did not commit any crime, I also do not want to see them kneeling in "prison." If I do not hear that sound of reindeer bells like flowing water, I will definitely go deaf; My legs and feet have gotten used to the uneven mountain passes; if I were made to walk on the flat paths of towns and cities, they would definitely weaken my body to the point of no longer carrying my load, paralyzing me; I always breathe the fresh and clean mountain air; if you made me go smell those "smelly farts" that cars emit, I definitely wouldn't breathe. My body is God-given, I want to be in the mountains, and return my body to the gods.

This is the Evenk people's "home," and the uniqueness of this "place" is even irreplaceable. The unique and important implications of "home" and "place" in ecological aesthetics and ecological literature is figuratively and profoundly expressed in Right Bank.

(3) The Unique Aesthetic Properties of "Looking Back"—Exploring the Beauty of the Ecological Existence Uniquely Possessed by "the Home"

Chi Zijian makes an artistic depiction of Evenk life from the entirely new perspective of ecological aesthetics. In the Evenk people's life that she constructs, humans and Nature are not separate or opposed, and "nature" is not just the object of human cognition, nor is it "humanized nature," "imitated nature" or "picturesque scenery," but is rather proto-ecological Nature ontologically one with human being. It is in this ontological relationship of being-in-the-world between humans and Nature where the

Truth of beings sets itself up in the work,[15] and thereby present the special beauty of ecological being. The "beings" here are the Evenk people, "Truth" here is authentic human nature, and "sets itself up" thus refers to the gradual unfolding of authentic human nature from concealed to cleared. The beauty of this "Truth setting itself up" described in Right Bank is no static, material beauty of symmetrical proportions, nor is it a purely artistic beauty, but is rather the beauty of the ecological being and the beauty of a specifically human nature in the "complete oneness of Nature and humans," in human being's relationship with Nature.

There are two forms of this beauty expressed by Chi Zijian in the work, one of which is the beauty of the supple and sound yin. At this time, humans and nature are found in a state of harmonious blending, whether in the fulfilment of victory after hunting or in the wedding after finding love. Right Bank vividly describes many such happy scenes, for instance, the novel writes of the annual joyful celebration of plentiful reindeer birthing:

This year, we clearly saw by the mountain stream the consecutive birth of 12 reindeer. Generally speaking, one mother reindeer only produces one offspring per embryo, but that year four mothers produced two offspring per embryo, and the young deer were so strong they truly made people beam with joy. We named that nameless mountains stream flowing in the dark-green mountain valley Rolls Gully, to commemorate that Russian Anda so incomparably friendly to us. Its waters are so clean and sweet, not only the reindeer love to drink from it, humans do too.

Because the reindeer produced so plentifully, the Evenk were beaming with joy, the dark-green mountain valley, the clear spring water and sound state of human survival leaps on to the page. This is obviously an auspiciously sound and happy state of survival both in line with the seasons and thriving for humans and their domestic companions; this is the supple beauty of sound yin, reflecting the happy aspect of human life and "the integral oneness of Nature and humans." But most cases are that of the unyielding beauty of strong yang, a special "eco-sublime." Slovic introduces the new implications that contemporary American environmental literature adds to "the eco-sublime" in Going Away to Think. The ecosublime means the unusual combination of joyful awe and fear of death incited by a specific experience of Nature.[16] There are principally two aspects to it, one of which is brave resistance to and fearless sacrifice in a battle with an ugly natural environment. Previously, we already mentioned that mournful scene of Link being struck by lightning when switching to a stronger reindeer in the forest, but even more soul-stirring was the life and death battle between the Evenk Daxi and wolves. Daxi, an outstanding Evenk hunter, once went to look for three lost young reindeer, whom he found surrounded by three wolves on the side of a cliff, trembling, the situation was extremely dangerous.

[15] Heidegger [11], p. 21.

[16] Slovic [12], p. 197.

At the time, Daxi was not carrying his gun, and only had a hunting knife with him; he had to battle with three hungry wolves alone. Although he ultimately killed the older wolves, one of his legs was bitten off by a younger wolf, so he had to bring the three saved young reindeer by crawling back to the camp, from which point on he was disabled. But he made the resolve for revenge, and specially domesticated a falcon to fight wolves to death when attacking the tribe and protect the tribe's interests. One year, due to prolonged pestilence, the number of wild animals dropped along with the number of reindeer as well, and humans lived together with wolves in survival predicament. At this time, the wolf pack always followed the tribe with an eye on the reindeer, and tried to attack. Right when the wolves were preparing to attack, Daxi and his falcon mounted a counterattack, out of which a life-and-death struggle unfolded, ending in the extremely tragic death of both man and wolf. Take a look at this extremely tragic scene of struggle that Right Bank presents to us:

> Many small white birch trees were snapped in half, there were spots of blood on the tree branches; the wormwood between the snowy grounds were trampled as well, one could imagine how tragic the fight was at that time. That battlefield was crossed with four mutilated skeletons, two of them wolf, one of them human, and also another falcon.
>
> I saw Daxi at the weathering burial ground with Yi Fulin, or rather, I saw his pile of bones. The biggest was the skull, the next biggest was a pile of bones differing in width and length still with red flesh attached to them like a pile of dry firewood.... The wolves died, they cannot come back.

This is a typical expression of the natural environment's "unreadiness-to-hand" in relation to humans. At this time, humans fiercely battle the ugly natural environment, and show the human being's obstinate courage and faith in survival. Here, what is specifically shown is Daxi protecting the interests of the tribe, and sacrificing his own human radiance. What the work presents before us is a painting of man and forest characterized by death in battle and bones all around; it presents another sublime beauty of the spirit to survive in the Evenk people. Right Bank also most prominently shows human being's awe at Nature, which is pre-modern clearly. This awe in most clear fashion is expressed in the Evenk people's worship of the shaman religion and in extremely heroic rituals. Shamanism is a primitive religion, an expression of primitive tribes' worship of Nature. The shaman in this kind of religion is basically a witch with the power and magical art of communicating between Nature and humans; the shaman dances trancelike as if drunk, mad and bizarrely mysterious. Right Bank colorfully depicts two generations of mysterious and odd shamanistic rituals, especially the trancelike dance, which exhibits the shaman's sacrificial act of saving humans from danger and composes a sublime beauty of the mysteriously odd. The narrator's uncle, Nidu, is the first generation shaman that the novel describes. The sublime beauty that he exhibits in the religious ritual is concentrated most fully in an extremely unusual trancelike dance he enters in order to counter the Japanese invasion. The novel writes that 2 years after the beginning of the war, the Japanese captured Dongbei province, and one day, Yoshida of the Japanese occupying army brought people up the mountain in the attempt to tame the Evenks. He demanded that the shaman Nidu fix his wounded foot through the trancelike dance or he would otherwise have to burn his own magical instruments and clothing and bend down to beg for his mercy.

At the crucial moment when this implicated the future fate of the tribe, the shaman Nidu without any hesitation accepted the challenge, and said, he would not only fix Yoshida's injured leg with the dance, but would also make the battle horses die with it as well. He said: "I wanted him to know that I would bring a dark night, but that dark night was not mine, but his!" Once the dark night came, the shaman Nidu entered the soul-stirring trance:

The dark night came, the shaman Nidu began beating on the spirit drum, and started to dance… Now and again, he looked up at the sky in great laughter, then lowering his head in a deep chant. When he approached the firepit, we I saw the smoke pocket hanging between his waste, which his mother sewed for him. He does not look as old as you would normally see, his waste straightens up miraculously; he makes the spirit drum emit an intensified drumbeat; his two feet are also so nimble that I find it hard to believe how someone would transform into another posture while dancing. He seems so filled with vitality like the shaman Nidu I saw when I was young… When the dancing stopped, Yoshida moved closer to the firepit, and propped up his leg, at which time we heard him let out a strange yell, because the wound on his leg truly could not be seen! That scar was just like a brightly colored flower, but as of now it withered to nothing in the wind summoned up by Nidu the shaman… That warhorse of Yoshida's already fell to the ground without a breath of life left… Yoshida stroked that dead warhorse without a scar or wound on it, and charged the shaman Nidu screaming out loudly. Wang Lu said, what Yoshida said was, spirit man, spirit man…

The shaman Nidu coughed several times, turned his body and left us. His waist stooped once again. While walking away, he threw things, first the drumstick, then the spirit drum, then the spirit clothing and spirit skirt…..When there wasn't one magical instrument or spirit clothing on his body, he fell to the ground.

This was an Evenk shaman who sacrificed his own life in a trance for the interests of the tribe and the greater right of the people. The great image of his self-sacrifice, that mysterious and bizarre dance of his in the trance along with its unimaginable effect, fabricates a sublime beauty of bizarre oddity. This is what could be called "ecosublime." It is such that I cannot but recall that unspeakable feeling of mysteriousness when young while entering a temple, the feeling of people's insignificance, the horribleness of evil and the necessity of becoming good before such mysterious forces. Although shamanism is a superstitious belief, it is one that controls the spiritual world of the Evenks and often calls up infinite tranquility and sublimity in their hearts. The one following in the shaman Nidu's footsteps, was the wife of his nephew, the shaman Nihao, who, when becoming the new shaman, expressed in front of the whole tribe that should would have to use her own life and god-given abilities to protect her own tribe and make the tribe's population flourish, the fill in the flocks of reindeer and make the hunting bountiful year by year. She actually did this indeed. She offered up the lives of her own three children for the peace of the tribe. The novel writes that the tribe member "Manure Pouch" choked on a bear bone, and was about to die, when the people in the tribe turned their gaze to the shaman Nihao, who alone could save "Manure Pouch." Trembling, Nihao sadly buried her head in her husband's bosom, because she knew that she would have to offer up her own daughter to save "Manure Pouch." But, she still put on her magic cloak and went into a trance:

> After Nihao danced for about two hours, a gust of cool wind suddenly stirred up among the tribe, it called out like a northern wind during the winter, at which time the light casting down from the top of the "post" was no longer white, but was twilight as if the sun had already set. When that strange wind began to gust, it permeated all around, but later it concentrated, yelling in one area, namely on the head of "Manure Pouch." I had the presentiment that that gust of wind would blow out the bear bone. As expected, when Nihao set down the spirit drum and stopped dancing, Manure Pouch suddenly sat up, "ahh" he yelled loudly, spitting out the bear bone……Nihao fell silent for a moment, then began chanting a spell. She was not singing for Manure Pouch coming back to life from death, but for that lily of hers, who withered away too young—Jiaoku Tuokan.

Her lily—her beautiful daughter was forever gone. Before autumn had arrived, there were still a good many happy summer days left, but she made her own flower petals whither to nothing and fall. A life for a life. This is the cruel reality of life, and also the heavy cost that Nihao had to pay to be a shaman. Before the law of the mysterious, man is so insignificant! The strange effect of the shamanistic trance described here could be a coincidence and could also be a psychological effect made by mystical religions and beliefs, but it does show us the special feeling of sublimity produced by awe and powerlessness before nature, which is unique to the hunting tribe. Because this sublimity envelops the fearless spirit of sacrifice shown by the shaman Nihao, it radiates a uniquely human radiance, and hence possesses an aesthetic implication. It is soul-stirring and so deeply moving! The final trance the shaman Nihao went into was in the early spring of 1988 for the sake of extinguishing a forest fire triggered by two logging workers who carelessly threw away their cigarette butts after smoking. The fire was so fierce, smoke surged in clouds, and all birds trying to escape were painted grey by the soot, the Arguna River and Lesser Khingan were taken over in a disaster. Nihao was already old, but she still threw on her magic cape:

When Nihao went into the trance, the thick smoke in the sky was billowing, the flock of reindeer were standing with lowered heads on the bank of the Arguna. The drumbeat roused, but Nihao's two legs were not as nimble as they were before, she danced and danced, but would cough for a fit. Originally she was bent at the waist, and bent even more at each cough. Her magic skirt dragged on the forest floor soaking with a layer of grey soot…After Nihao danced for an hour, a dark cloud appeared in the sky; she then danced for another hour, and the thick cloud covered around densely; after another hour had passed, lightening appeared in a flash. Nihao stopped dancing. Shaking, she walked to the bank of the Arguna, lifted those two damp woodpeckers, and hung them from a strong pine tree. After she had just finished doing all of this, the sound of thunder and flashes of lightening alternated, and the great rain came pouring down. Nihao sang the final spell of her life in the rain. But she never finished singing that spell, having fallen down in the rain—

Oh Arguna,
Why not flow into the Milky Way,
Dry world….

The mountain fire went out, and Nihao left us. Throughout her whole life, she presided over many funerals, but she could not send herself off in the end.

Here, the author sculpts for us a tragic image of the last Evenk shaman offering her own life to the Arguna River and the Evenk people, an image filled with a special beauty of sublimity. With this depiction as the conclusion of the novel, the beauty of sublimity becomes the end of the novel, and rubs onto the work a thickly tragic spectrum of color, which deeply engraves into our hearts the beauty of human survival in the Evenk people of the right bank of the Arguna.

"Looking Back at Home" is the special perspective of Right Bank of Arguna River, which gives us a series of deep impressions that tell us that continuously looking back at home while treading so largely forward in waves of modernization is the attitude humans ought to have. Looking back is a kind of love; it is us forever remembering mother earth's nurturing of humankind; looking back is a kind of reflection urging us to continuously reflect on our own actions. Looking back at Right Shore of Arguna River tells us that there are many forms of civilization in our home the earth, which holds many ways of life, and only in this way does the earth present the even blossoming of so many flowers and diverse colors and postures. Because of this, preserving the diversity of civilizations is also the need of the earth's homes for ecological balance. Can we leave those special mountain paths of the Evenk while erecting high speed freeways? At the same time, Right Bank of Arguna River also tells us that we should never forget that we are ourselves children of Nature. Perhaps, Nature is a belligerent parent at times, but our identity as children is forever unchangeable. The reality that we can only survive relying on such a parent is also unchangeable. Treasuring Nature, loving Nature is to treasure and love our parent, and treasure and love us humans ourselves too!

References

1. Xingpei, Y., Erdong, M., & Fang, D. (1994). *General theory of Chinese poetics*. Anhui Education Press.
2. [Ancient Greece] Aristotle. (1982). *Poetics*, translated by Luo Niansheng. People's Literature Publishing House.
3. [Germany] Ephraim Lessing, G. (1979). *La Ocon*, translated by Zhu Guangqian. People's Literature Publishing House.
4. Heng, G. (1980). *Note on the book of poetry*. Shanghai Ancient Books Publishing House.
5. Shuangxin, C. (2002). *Western Zhou Bronze musical instrument*. Hebei University Press.
6. Liangfu, J. (1998). *Dictionary of appreciation of Pre-Qin poetry*. Shanghai Dictionary Press.
7. [Germany] Heidegger, M. (2006). *Being and time* (revised translation), translated by Chen Jiaying, Wang Qingjie. Sanlian Bookstore.
8. Zhongshu, X. (Ed.). (2003). *Oracle dictionary*. Sichuan Dictionary Press.

9. [U.S.] Berleant, A. (2006). *The aesthetics of environment*, translated by Zhang Min, Zhou Yu. Hunan Science and Technology Press.
10. Jie, W. (2002). *The aesthetic illusion and aesthetic anthropology*. Guangxi Normal University Press.
11. [Germany] Heidegger, M. (2004). *Off the beaten track,* translated by Sun Zhouxing. Shanghai Translation Publishing House.
12. [U. S.] Slovic, S. (2010). *Going away to think*, translated by Wei Qingqi. Peking University Press.

Chapter 14
Interpretation of the Ecological Aesthetic of Foreign Literary Works

14.1 Ecological Aesthetic Interpretation of *Lady Chatterley's Lover*

For Lawrence's ***Lady Chatterley's Lover***, there are many kinds of interpretations by domestic and overseas academic circles, which of course include the perspective of ecological aesthetics. Lin Yutang once said that the book criticizes the British, industrial society, machine civilization, and reason. Lawrence belongs the school of to nature's art and emotional life. Westerner Matthew David pointed out that "D.H. Lawrence is another 'writer' who can be called an early eco-critic because, he is positive to the original nature."

(1) **The conditions of Lawrence's concept of ecological aesthetic**

Lawrence's ecological aesthetic concept has its own historical and personal conditions. Lawrence (1885–1930) was born in Eastwood, a small mining town near Nottinghamshire in the central British city. The first industrial revolution in England was basically completed, but it had become the most developed industrial country in Europe. Of course, the environmental pollution caused by the Industrial Revolution was also the most serious in the developed countries at that time. According to records, the annual output of British coal in 1789 was 10 million tons, compared to 700,000 tons in France. At the same time, with the development of the "Enclosure Movement", the preserved forests in Britain were being depleted, and the environment was seriously polluted. This caused the famous "London Fog" to appear. Moreover, tuberculosis and pneumonia caused by air pollution were unprecedentedly serious and harmful people's health. Lawrence's hometown of Eastwood was a coal mine town. His father was a coal miner, mining underground all year round. His brother died of pneumonia at the age of 23. Lawrence was physically weak and suffering from bronchitis at birth, and was knocked down by pneumonia six weeks before his brother's death. Since then, he did not get rid of lung disease problems throughout his life. Lawrence's life was a long journey with his wife Frida of fleeing urban civilization for his own health and interest, looking for "country shelters" and

F. Zeng, *Introduction to Ecological Aesthetics*,
https://doi.org/10.1007/978-981-13-8984-9_14

"spiritual homes." In a letter to his friends, he said, "I dare not go to London for the sake of life. It's like going into the gas chamber, where the lungs can not stand." This determines that Lawrence early established his initial ecological concept. He once shouted: "Nature, I hope I can write him bigger."

(2) **The Criticism of the Ecological Disease Caused by Industrial Civilization**

Lawrence's critique of capitalism in his works is comprehensive, especially in the critique of the ecological consequences of industrialization. Criticism first began with the destruction of the natural environment, chapter two of which wrote about Wragby, the home of Connie and Clifford. It was a coal mining area and a heavily polluted environment. Written by Connie's observations, "left it at what it was: unbelievable and not to be thought about. From the rather dismal rooms at Wragby she heard the rattle-rattle of the screens at the pit, the puff of the winding-engine, the clink-clink of shunting trucks, and the hoarse little whistle of the colliery locomotives. Tevershall pit-bank was burning, had been burning for years, and it would cost thousands to put it out. So it had to burn. And when the wind was that way, which was often, the house was full of the stench of this sulphurous combustion of the earth's excrement. The air also carries what cacophony under the cellar. even on the Christmas roses the smuts settled persistently, incredible, like black manna from the skies of doom." Lawrence regarded the destruction of the environment and society caused by the chase of capitalism as a "rape" of all good things. In chapter eight, Lawrence described a dialogue between Connie and Clifford. The comprehensive criticism of the capitalist destruction of the natural ecology, social ecology, and spiritual ecology was attributed to, "just kills the vitality in the air", "it's man that poisons the universe", "Fouls his own nest". Lawrence pointed out that the development of capitalism had caused the loss of people's "homeland." Chapter VI describes Connie's inner soliloquy as she returns to her intriguing home from a filthy worker's village: "Connie went slowly home to Wragby. 'Home!' … it was a warm word to use for that great, weary warren. But then it was a word that had had its day. It was somehow cancelled. All the great words, it seemed to Connie, were cancelled for her generation: love, joy, happiness, home, mother, father, husband, all these great, dynamic words were half dead now, and dying from day to day."

(3) **Passionate Praise of Nature**

Lawrence and his family were endangered by the pollution of the environment. Additionally, he suffered in a harsh environment, health hazards, and therefore he had a longing for and love for the beautiful, especially that of original nature. From childhood he regarded the Higgs farm of his girlfriend Jessie's home as his last refuge, wherein he would hide himself! In the novel, "The Woods" was his last refuge for Connie to evade Legoland's stifled old home, "To hide herself there!" In her eyes, this was the most beautiful purification of the blessed land, and she praised it. Lawrence wrote in Chapter XII: "Connie went to the wood directly after lunch. It was really a lovely day, the first dandelions making suns, the first daisies so white. The hazel thicket was a lace-work, of half-open leaves, and the last dusty perpendicular of the catkins. Yellow celandines now were in crowds, flat open, pressed back in urgency, and the yellow glitter of themselves. It was the yellow, the powerful yellow of early summer. And primroses were broad, and full of pale abandon, thick-clustered primroses no longer shy. The lush, dark green of hyacinths was a sea, with buds rising like pale corn, while in the riding the forget-me-nots were fluffing up, and columbines were unfolding their ink-purple ruches, and there were bits of blue bird's eggshell under a bush. Everywhere the bud-knots and the leap of life!"

(4) **Praise of the Love that is in Line with the Ecological Nature of Human Beings**

The most controversial part of the book is a description of sex. Now it seems that this description is profound. It strongly criticizes the ecological features of a deprived human sexuality determined by the mechanized life of capitalism, and offers a warm tribute to the sexual love that is in line with the ecological nature of human beings. The reason why sex is in line with people's ecological nature is because of the necessary adjustments in their philosophy. From the point of view of the subject/object dichotomy between human being and the world, of course, we can conclude that sexual love is negative. However, from the perspective of "human" as a whole, living in the world, then sexuality is in line with human beings' ecological nature. What's more, human reproduction is also a prerequisite for the continuation of ecology. However, in capitalist society, due to the proliferation of instrumental rationality and the chase of money, there has been a spread of corruption, on the one hand, and a negation of normal sexual love, on the other hand. In a society where money and interests are paramount, the natural nature of humankind is strangled, and the human's reproductive capacity is brutally "castrated." Chapter VII of the book reproduces a wonderful discussion of sex in the living room of Cliff's family. Surprisingly, some upper-class people involved in this discussion have completely denied sexual love. Cliff said: "I do think sufficient civilization ought to eliminate a lot of the physical disabilities," He then added: "all the love-business for example, it might just as well go." Another Lady Bennerley said, "So long as you can forget your body you are happy," and added: "So, if civilization is any good, it has to help us to forget our bodies." Winterslow said, "It's quite time man began to improve on his own nature, especially the physical side of it." What he refers to as an improvement is actually a "castration" of sex. It is under the guidance of the theory of dichotomous castration that Cliff proposed a so-called "plan" of reproduction and family division. Given his lack of fertility and the desire to own a so-called property heir, he advised Connie to find a man to have a child to be his own heir! What an incredible idea! Because of this incomprehensible idea, at a young age Connie plunged into extreme anguish and pain. It was "inevitable" for her to have the lingering love affair with the game-keeper Mellors. The work has a passionate tribute to this sexual love, and Lawrence sees sex as "the original place of life." The book focuses on Connie's psychological feelings, Lawrence wrote: "She felt, now, she had come to the real bed-rock of her nature, and was essentially shameless. She was her sensual self, naked and unashamed. She felt a triumph, almost a vainglory. So! That was how it was! That was life!" When Connie describes her view of sex to her sister, she even describes it as "when you feel you live, and are in the very middle of creation". The author also uses the words of the artist Hobbes to describe "sex as a natural and vital thing." In Chapter 18, the author expresses the essence of the creation of the human nature of love through the psychological activities of Mellors: "I stand for the touch of bodily awareness between human beings,' he said to himself, 'and the touch of tenderness. And she is my mate,—Thank God I've got a woman!

Thank God I've got a woman who is with me, and tender and aware of me. Thank God she's a tender, aware woman.' And as his seed sprang in her, his soul sprang towards her too, in the creative act that is far more than procreative.

(5) **The pursuit of ecological aesthetic survival relative to mechanical survival**

The core of the work is a battle between Clifford and Mellors over Connie, which actually reflects the sharp struggle between the two civilizations and the two types of people and the two social ideals. The author's position is thus clearly stated. Specifically, he focuses on the sharp struggle between the increasingly flourishing industrial civilization and the ecological civilization of the future—that is, the ecological state in which people live in harmony with nature with industrial machines, and the ideal of money-worship and the ideal of building a home for human existence. Obviously, the author's position and attitude are inclined to the latter, which shows the author's great forward-looking tendency towards ecological consciousness and shows the extremely important value and significance of the book. The author shows us in the work two absolutely different worlds: one is the so-called "industrial civilization" world with roaring machinery, heavy pollution, and labor suffering, which is Clifford's; and the other is Mellors's woods, where far away from the hubbub and the secular, full of blossoming, and of birds singing: "Everywhere the bud-knots and the leap of life!" There are two kinds of people living in these two worlds. One is Clifford, who is the upper class person of the manor, the coal mine owner. He is a bourgeois person, who pursues money, fame, and fortune for his life. Although he has a well-developed and profit-making mind, he is a "robot" with disabled crib and wheelchair. He is also infertile. The other is the "game-keeper " Mellors—a lower class person employed by Clifford. Although as a lieutenant, he ranks among the upper class, but he banished himself to the woods for his dislike of modern civilization. As a "person," Mellors has a love for ordinary life. He has normal living abilities, including a handsome appearance and a healthy body, and is ecologically integrated with nature. During the fight for Connie between Clifford and Mellors, Connie also went through the psychological journey of "Living in—the shake of thought—shake of action—a complete departure." At the very beginning, she followed Clifford back to the quaint old house of Leggbe, built in the 18th century, and lived a tedious, asexual housewife life. Connie at this time in general is living a "very quiet life," until one day her twenty-year-old young life agitated, "Unjust! Unjust! The sense of deep physical injustice burned to her very soul". Her thoughts began to shake and she wanted to "resist". Chapter ten of the book elaborates on the combination of her with Mellors in the woods, and feels the combination of "relieving her perennial backlog of depression and feeling of peace and tranquility", "that is love" and "resurrection of body." But Connie hesitated in the face of the bourgeois but affluent life of the upper class regarding her return to nature and to her hard life. After the final decision, she chose to go away completely, to divorce Cliff and married Mellors, and become a co-owner of a small farm. This may be Lawrence's so-called return to the "spiritual hometown" of the road. This road has guided the reconstruction of the "homeland" of human material and spirit, and is the pursuit of the ideal of living an aesthetic life. However, after all, the world of industrial civilization is extremely powerful. The conclusion of the work has not given people any conclusion, but only gives people an expectation.

14.2 The Ecological Aesthetic Interpretation of *Moby Dick*

Moby Dick is a masterpiece of the famous American novelist Herman Melville (1819–1891). Published in October 1851, the book did not receive enough attention until it was published on the 100th anniversary of the birth of the author with his first biography in 1921, when it was then widely known and highly regarded by the community, and finally hailed as a monument to English literature, to American epics, and so on.[1]

Into the late 20th century, there have been some changes in the academic evaluation of *Moby Dick.* In his book *Environmental Imagination,* Lawrence Buyier, a famous contemporary American Eco-literary critic, pointed out that "the novel *Moby Dick* is more prominent than any other piece of contemporary work … and shows human beings' atrocities against the animal world."[2] *The New York Times* once said that one of the ironic things in our literature is that the greatest American novel, Moby Dick, is a career and way of life that has been completely eradicated in the United States. From the vigorous development of the mid-19th-century industrial revolution written of in this book to today's new era of "ecological civilization" of the post-industrial revolution, there has been indeed a corresponding change in the evaluation of this book. Because based on today's literature evaluation position, in any case it should include the ecological dimension. So, for *Moby Dick* we should also consider it from the ecological dimension and give it a new evaluation.

Any successful work provided to us is a living art image. From the perspective of the image being bigger than the thought, *Moby Dick,* with its unique aesthetic image, not only provides us with a horrifying picture of the cruel killing of human beings, but also provides a powerful warning to humankind regarding how to treat nature properly. In the cultural atmosphere of the emerging nation of the United States, Melville broke the routine and used unique methods of combining realism, scientism and symbolism in Moby Dick not only to vividly record the cruel process of the human plunder of whales, but also to utilize with its unique, complex feelings ideas about the complex relationship between man and nature, and to give us a powerful warning with its rich atmosphere of tragedy and symbolic strokes.

[1]Fuheng et al. [1], pp. 170–193.

[2]Nuo [2], p. 164.

(1) To Truly Reproduce the Cruel Predation of Whales and Other Natural Resources Caused by the Deformity of the Whaling Industry in the Early Stage of Capitalism

Moby Dick is based on the American whaling industry. The description of the novel is based on the landmarks of Nantucket and New Bedford in the United States, and the whale-hunting Pequod is its most important maritime base. The land-sea bases of these whaling industries are all very vivid illustrations of the tremendous impetus promoted by the vigorous whaling industry in the early capitalist phase of the United States over the natural resources such as plundering whales. Nantucket, an island 48 km south of Cape Cod, Massachusetts, has a whaling industry that peaked on the eve of the eighteenth century American independence. It was home to 125 whaling ships. *Moby Dick,* wrote: "What wonder, then, that these Nantucketers, born on a beach, should take to the sea for a livelihood!— and in all seasons and all oceans declared everlasting war with the mightiest animated mass that has survived the flood; most monstrous and most mountainous", "parcelling out among them the Atlantic, Pacific, and Indian oceans, as the three pirate powers did Poland." This also shows that prior to the advent of gas and electricity, and in an era where whale oil was the main source of lighting energy, the United States had been involved in the process of carving out whales and dividing the world with other capitalist countries. Nantucket was its earliest whaling base. Subsequently, the New Bedford replaced and quickly became a huge new whaling base. *Moby Dick* states that the city became a luxurious new town drawn from a rubbish-like stone: "It is a land of oil","nowhere in all America will you find more patrician-like houses; parks and gardens more opulent, than in New Bedford. Whence came they? how planted upon this once scraggy scoria of a country? Go and gaze upon the iron emblematical harpoons round yonder lofty mansion, and your question will be answered. Yes; all these brave houses and flowery gardens came from the Atlantic, Pacific, and Indian oceans. One and all, they were harpooned and dragged up hither from the bottom of the sea." Added: "In New Bedford, fathers, they say, give whales for dowers to their daughters, and portion off their nieces with a few porpoises a-piece. You must go to New Bedford to see a brilliant wedding; for, they say, they have reservoirs of oil in every house, and every night recklessly burn their lengths in spermaceti candles." This is evidenced by the spectacular New Bedford development and prosperity through the whaling industry. The book also describes in detail that the people at the time not only had an amazing demand for whale oil, but also for the whale bone, whale brain, whale horns, and even whale-specific ambergris, which were also in high demand. The huge profits of whaling were a powerful driving force for its development. Whales were even needed for urban construction.

History tells us that prior to the massive extraction of oil from humankind in 1861, lighting energy relied on whale oil. It therefore became a matter of course that the whaling industry became the pillar industry of that time. The United States, as a newly emerging maritime nation, was rapidly emerging in the whaling industry. At the time, the whaling industry also became the engine for the U.S. economy to take off, and supported the rapid transformation of a young, once-crushed colonial country into a global hegemon. The United States had become the world's leading whaling nation: in 1846, there were 900 whaling ships in the world, 755 in the United States. In 1853, the United States caught 8,000 whales in the first year, earning a profit of 11 million U.S. dollars. *Moby Dick* compares the size of the United States with Britain in terms of whaling. "the Yankees in one day, collectively, kill more whales than all the English, collectively, in 10 years." The huge scale of the U.S. whaling industry is inseparable from the state's support when the whaling industry in the United States became a national act with direct national intervention and mapping of whales. The author states specifically in the note to the book that he was fortunate to have received an official circular signed by the Washington Meteorological Service on April 16, 1856, by Lieutenant Colonel Moreau indicating that the National Weather Service is completing the charting of a whale reef tour, "This chart divides the ocean into districts of five degrees of latitude by five degrees of longitude; perpendicularly through each of which districts are twelve columns for the twelve months; and horizontally through each of which districts are three lines; one to show the number of days that have been spent in each month in every district, and the two others to show the number of days in which whales, sperm or right, have been seen." The fact that the country was directly involved in whaling is clearly recorded here. It is this large-scale national whaling that greatly promoted the rapid development of the whaling industry and caused brutal assault on whales. *Moby Dick* tells us that a large number of capture and killings have made the whales dare not to act alone in favor of acting as a group.

(2) **Realistically Reproduce the Specific Process by Which Whaling Ships Plunder Whales**

Moby Dick is a literary work with unique characteristics of the United States. It is not only a realistic literary work, but also contains many elements of scientific evidence. For example, the book gives readers a detailed introduction to the classification of whales and the use of various whaling tools in a particularly real way, reproducing the complete process of whaling, whaling watching, whale-killing, and oil refining. This positivist style of writing is rare in the creation of novels. Some critics have argued that this method of writing had some impact on the literary character of the work, and criticized its "vulgarity and lengthiness." Some critics even criticized the book as, "a pot of porridge with romance, philosophy, natural history, beauty, graceful feelings and vulgar language".[3] But it is undeniable that this method does reproduce the detailed process of whaling, and has become a valuable record of whaling history.

A detailed description of the whaling process by *Moby Dick* provides us with a true record of the hunt for whales. At the author's discretion, the Pequod whaling ship is a whale-hunting base. It was a slaughterhouse for killing whales and a refining whale oil factory. Its cruelty is shocking. In fact, whales are a kind of very tame animal in the ocean. In addition to foraging, whales generally do not attack other creatures, and their nature is not to try to fight back when hit. Usually when whales are cruising in a strong group and one is captured, the other whales do not leave. When a

[3]Melville [3], "Foreword", p. 3.

young whale is caught, the female whales do not leave. Whales are not aggressive to humans, but humans hunt whales madly and cruelly for utilitarian needs. Chapter 61 of *Moby Dick* "Stubb Kills a Whale" depicts a complete and detailed whale hunting process. Pequod cruised over the ocean near Java and suddenly found a sperm whale. Immediately the boat chased, the second mate Stubb was close to the whales, darted dart after dart into the flying fish;

The boat alternately sterning out of the way of the whale's horrible wallow, and then ranging up for another fling. For wounded whales being darted, the book reads: "The red tide now poured from all sides of the monster like brooks down a hill. His tormented body rolled not in brine but in blood, which bubbled and seethed for furlongs behind in their wake" In the end, the whale came to death, "spasmodically dilating and contracting his spout-hole, with sharp, cracking, agonized respirations. At last, gush after gush of clotted red gore, as if it had been the purple lees of red wine, shot into the frightened air; and falling back again, ran dripping down his motionless flanks into the sea. His heart had burst!" This is a cruel picture of bloody spewing and painful struggle depicts a living creature strangled in the pursuit of the interests of humankind! Not only that, whaling ships do not leave behind baby whales. *Moby Dick* also portrays the situation in which the Chief Mate Starbuck catches a whale with a young whale during a whaling: "so now, Starbuck saw long coils of the umbilical cord of Madame Leviathan, by which the young cub seemed still tethered to its dam. Not seldom in the rapid vicissitudes of the chase, this natural line, with the maternal end loose, becomes entangled with the hempen one, so that the cub is thereby trapped." Mother whales and young whales being captured together is in itself, is contrary to ecological laws. *The Analects of Confucius* contained: Confucius only used the fishing rod (with a hook) to fish, instead of fishing (with many hooks), only shot flying birds, do not shoot birds nesting in the nest." China has a tradition of protecting young animals since ancient times. In *Moby Dick*, the author documents the process of Pequod whaling: When a whale that has been speared was exhausted, people also wanted to sweep around it, ravaged. The novel reads: "and the lances were darted into him, they were followed by steady jets from the new made wound, which kept continually playing, while the natural spout-hole in his head was only at intervals, however rapid, sending its affrighted moisture into the air". "His eyes, or rather the places where his eyes had been, were beheld. As strange misgrown masses gather in the knot-holes of the noblest oaks when prostrate, so from the points which the whale's eyes had once occupied, now protruded blind bulbs, horribly pitiable to see. But pity there was none. For all his old age, and his one arm, and his blind eyes, he must die the death and be murdered, in order to light the gay bridals and other merry-makings of men, and also to illuminate the solemn churches that preach unconditional inoffensiveness by all to all. Still rolling in his blood, at last he partially disclosed a strangely discolored bunch or protuberance, the size of a bushel, low down on the flank." This whale had been swept on when it was on the verge of death, making it "rolling in his blood". This is no longer whaling, but an outspoken whale capturing, completely deviated from the "humanitarianism" that contains compassion, which is really scary. This further proves that in the economic chase of the "profit-first" capitalism, norms of ethics and morality such as "fraternity" advocated by it must inevitably go toward nothingness.

Moby Dick clearly shows that in order to maximize economic efficiency, the Pequod has actually become a processing plant for marine whale slaughterhouses and whale oil refining. The first is a whale slaughterhouse. Chapter sixty-seven writes about the whale reprocessing after it was captured, "it was a Saturday night, and such a Sabbath as followed! Ex officio professors of Sabbath breaking are all whalemen. The ivory Pequod was turned into what seemed a shamble; every sailor a butcher. You would have thought we were offering up ten thousand red oxen to the sea gods." This is not to the sea God; this is entirely brutal killings in pursuit of high profits. Next, the book details the entire process of slaughtering whales in great detail, including the division of dead whales, cuttings, cuttlefish brains, digging ambergris, and finally discarding the entire remaining whale. This is followed by a refinement of whale oil, which is written in Chapter 96 of *Moby Dick*: "The Try-Works": Besides her hoisted boats, an American whaler is outwardly distinguished by her try-works. She presents the curious anomaly of the most solid masonry joining with oak and hemp in constituting the completed ship. It is as if from the open field a brick-kiln were transported to her planks." "By midnight the works were in full operation. We were clear from the carcass; sail had been made; the wind was freshening; the wild ocean darkness was intense. But that darkness was licked up by the fierce flames, which at intervals forked forth from the sooty flues," After refineries load their barrels, the whaling vessel is full of whale barrels. When fully loaded, it is also time to win high profits. As the book said: "He goes and hunts for his oil, so as to be sure of its freshness and genuineness, even as the traveler on the prairie hunts up his own supper of game." The whaling base, slaughterhouse, and refinery constitute the description of the whaling ship in *Moby Dick.* The whaling ships undoubtedly became an outpost hostile to the oceans and sea creatures.

(3) Actually Reproduced that the Action and Psychology of Whalers in a Particular Era Are Crazy about Whales

The center thread of *Moby Dick* is the vicious fight between the whalers, represented by Ahab, and Moby Dick. The novel tells us that the main purpose of whaling people to choose whaling, a dangerous job, was, in addition to curiosity, exploration, and other reasons, most importantly the need for economic profit. At that time, the whaling industry in the United States adopted a quota system that linked the interests of whaling people with the efficiency of whaling. In this regard, Chapter 16 of *Moby Dick* gave a more detailed description. In the mouth of Ishmael, the narrator, it says, "I was already aware that in the whaling business they paid no wages; but all hands, including the captain, received certain shares of the profits called lays, and that these lays were proportioned to the degree of importance pertaining to the respective duties of the ship's company. I was also aware that being a green hand at whaling, my own lay would not be very large; –that is, the 275th part of the clear net proceeds of the voyage." Such a distribution system will inevitably lead to the result that the more whales make more profits, and the whalers desperately whale their whales to make more profits. However, *Moby Dick* does not stop at the simple economic aspect of the relationship between the whaler and the whale. Instead, it goes deep into the relationship between human beings and nature. In that particular era of industrial revolution, whalers participated in whaling not only for profit, but also strive to overcome nature, even adopting the psychology of antagonism with nature as a manifestation of human antagonism to nature and self-esteem. This is mainly reflected in Captain Ahab of the Pequod whaling ship. According to the book, the name of Ahab is taken from the King of Israel who was punished by God in *Bible • Old Testament • 1 Kings*. Ahab was an orphan whose widowed mother died a year after his birth. He lives in a stormy sea all year round. For more than 40 years of whaling, he lived on shore for less than 3 years until he married a young girl in his fifties. He is a well-known, courageously determined whaling man who lost one leg as a result of battling Moby Dick, but still fought on the whaling front. Such a captain made the Pequod whaler more mysterious. Captain Ahab does not appear until the middle of the novel, and the novel uses the wording of "hearing his/her voice before seeing the person." When the narrator, Ishmael arrived on board, he heard Peleg, one of the ship owners introduced Ahab: "Ahab's above the common; Ahab's been in colleges, as well as 'mong the cannibals; been used to deeper wonders than the waves; fixed his fiery lance in mightier, stranger foes than whales. His lance! aye, the keenest and the surest that out of all our isle! Oh! he ain't Captain Bildad; no, and he ain't Captain Peleg; he's Ahab, boy; and Ahab of old, thou knowest, was a crowned king!" "And a very vile one." Here, the author makes a basic description of the status and character of Ahab as an extraordinary and ruthless whaler leader. In chapter 28, a few days after the Pequod had left Nantucket, Ahab finally made his first appearance. The author gave a lot of ground to his appearance, first making people feel his existence through arbitrary orders, and the novel states that the first, second and third officer "only they sometimes issued from the cabin with orders so sudden and peremptory, that after all it was plain they but commanded vicariously. Yes, their supreme lord and dictator was there, though hitherto unseen by any eyes not permitted to penetrate into the now sacred retreat of the cabin". Then came his formal appearance, and on a gray morning Ishmael the narrator saw Ahab standing tall and steady like a mountain on the deck. "He looked like a man cut away from the stake, when the fire has overrunningly wasted all the limbs without consuming them, or taking away one particle from their compacted aged robustness. His whole high, broad form, seemed made of solid bronze, and shaped in an unalterable mould, like Cellini's cast Perseus. Threading its way out from among his grey hairs, and continuing right down one side of his tawny scorched face and neck, till it disappeared in his clothing". This appearance is extraordinary. First, the author described Ahab as a martyr, but an uncompromising martyr in truth. At the same time, the author described him as the hero of Parsus in Greek mythology. Perseus is a hero in Greek mythology, the son of Argos Princess Danae and Zeus, who killed the sea monster Medusa, became king of Tilins, and finally ascended the throne, representing bravery, justice and perseverance. In the eyes of the author, Ahab is a Perseus hero who killed the monster Moby Dick, a martyr without any fear of any difficulties. This established the determination and tone of the main character Ahab's pledge to kill Moby Dick. As the whalers entered whale concentration, Ahab called a full-fledged crew of men in a pro-warlike manner. This is actually a swearing-in meeting with Moby Dick. The thirty-sixth chapter wrote, Ahab stopped suddenly shouted:

> "What do ye do when ye see a whale, men?"
>
> "Sing out for him!"
>
> was the impulsive rejoinder from a score of clubbed voices.
>
> ""Good!" cried Ahab, with a wild approval in his tones; observing the hearty animation into which his unexpected question had so magnetically thrown them."
>
> "And what do ye next, men?"
>
> "Lower away, and after him!"
>
> "And what tune is it ye pull to, men?"
>
> "A dead whale or a stove boat!"

Here is the oath of "A dead whale or a stove boat!", but Ahab did not stop. To further target his real rival, Moby Dick, He said:

> Aye, Starbuck; aye, my hearties all round; it was Moby Dick that dismasted me; Moby Dick that brought me to this dead stump I stand on now. Aye, aye," he shouted with a terrific, loud, animal sob, like that of a heart-stricken moose; "Aye, aye! it was that accursed white whale that razeed me; made a poor pegging lubber of me for ever and a day!" Then tossing both arms, with measureless imprecations he shouted out: "Aye, aye! and I'll chase him round Good Hope, and round the Horn, and round the Norway Maelstrom, and round perdition's flames before I give him up. And this is what ye have shipped for, men! to chase that white whale on both sides of land, and over all sides of earth, till he spouts black blood and rolls fin out. What say ye, men, will ye splice hands on it, now? I think ye do look brave." "Yes, yes!" "Aye, aye!" shouted the harpooneers and seamen, running closer to the excited old man: "A sharp eye for the White Whale; a sharp lance for Moby Dick!"

At this point, the purpose of this sea-going Pequod, the whaling vessel, represented by Ahab, has focused very clearly on one goal: dead fight with the whale Moby Dick! In addition to its economic benefits, the underlying causes include the reason why humans and nature are hostile and seek vengeance upon one another. With the further clarification of the theme of *Moby Dick*, we cannot but ponder: what exactly is the most fundamental relationship between humankind and nature? Because the entire ship is to kill Moby Dick as the ultimate goal, though Pequod has captured a few whales, refined enough whale oil, and can return home, Ahab was still unwilling to give up, and still insisted on cruising upon the ocean, waiting to catch Moby Dick. The turning point in the plot occurred when the Pequod sailed to the Sea of Japan during a typhoon and a raging night of thunderstorms. At one point, the east wind blew up. If it sailed down, Pequod would return to Nantucket without any problems and the windward voyage will have to die. "we can turn it into a fair wind that will drive us towards home. Yonder, to windward, all is blackness of doom;" At this time, Starbuck, who has been the Chief mate and natural reconciliation, told Ahab by grasping Ahab by the arm "–"God, God is against thee, old man; forbear! 't is an ill voyage! ill begun, ill continued; let me square the yards, while we may, old man, and make a fair wind of it homewards, to go on a better voyage than this." However, Ahab categorically rejected the views of the chief mate and practically resorted to violent measures, to force vessels, to face upwind and eventually embarked on a path of destruction. *Moby Dick* finally wrote in Chapter 119:

> But dashing the rattling lightning links to the deck, and snatching the burning harpoon, Ahab waved it like a torch among them; swearing to transfix with it the first sailor that but cast loose a rope's end. Petrified by his aspect, and still more shrinking from the fiery dart that he held, the men fell back in dismay, and Ahab again spoke:–"All your oaths to hunt the White Whale are as binding as mine; and heart, soul, and body, lungs and life, old Ahab is bound. And that ye may know to what tune this heart beats: look ye here; thus I blow out the last fear!" And with one blast of his breath he extinguished the flame.

Since then, Pequod continued its voyage to the Sea of Japan, often cruising to sperm whales, in search of Moby Dick. After finally encountering it, Pequod continued fighting with Moby Dick for 3 days. The humans and whale were injured at the same time on the first 2 days. On the third, the day of the decisive battle, both sides tried their very best and the situation was very dangerous. With all the others on the boat, Ahab led the fight. Despite the fatal blow to Moby Dick, at the last moment, the big whale also overturned the Pequod, and the ship was covered in surging sea. Ahab was also dragged into the sea and buried in the sea, at last, both man and fish died, only the narrator Ishmael survived by turning a coffin into a lifeboat.

In this tragedy of both man and whale death, although the author's tendency has some wandering, the general tendency is still obvious, that is the tribute to the whalers represented by Ahab and the negation of the great whale of Moby Dick. *Moby Dick* is a great tribute to Ahab. We can see from chapter 44, which discusses when Ahab was so disturbed that he was unable to rest and sleep on behalf of Moby Dick:

> God help thee, old man, thy thoughts have created a creature in thee; and he whose intense thinking thus makes him a Prometheus; a vulture feeds upon that heart for ever; that vulture the very creature he creates.

Instead, at the same time that the author mystified Moby Dick, he further demonized it. The book not only describes the whalers' perception of Moby Dick as a "monster, immortal," but also describes it as "a paranoid incarnation of vicious power." Chapter 41 reads:

> Nor was it his unwonted magnitude, nor his remarkable hue, nor yet his deformed lower jaw, that so much invested the whale with natural terror, as that unexampled, intelligent malignity which, according to specific accounts, he had over and over again evinced in his assaults. More than all, his treacherous retreats struck more of dismay than perhaps aught else. For, when swimming before his exulting pursuers, with every apparent symptom of alarm, he had several times been known to turn around suddenly, and, bearing down upon them, either stave their boats to splinters, or drive them back in consternation to their ship.

Here, we can see the author's sharp attitude: On the one hand, the whale hunter, Ahab, was deliberately hostile to the whale, searching everywhere, the act of hunting and killing it being seen as the act of stealing fire from the gods like the hero Prometheus. On the other hand, the tame, passive enemy of the whale is regarded as the embodiment of the "evil spirit" of "malicious power". Obviously, the theme and ideological inclination of the novel *Moby Dick* is a tribute to the behavior of whalers represented by Ahab and scourge the so-called "sinful" natural forces represented by Moby Dick.

(4) A True Representation of the Whaling Culture of the West, and Especially the United States, that Led to the Wild Development of the Whaling Industry

The development of the whaling industry, the formation of special whalers, and the frenzied killing of marine life such as whales all have certain cultural support. This is the special whaling culture in the United States at the time of *Moby Dick*. Not only is *Moby Dic*k valuable in being able to tell us a thrilling whaling story, but it also, more importantly, allows us to recognize the special era of special whaling culture. Despite the vivid twists and turns of the battle between human beings and whales, but after all, what is superficial has become the dust of history. Culture is a deeper content. It not only reveals the motivations of the incident to us, but also profoundly enlightens us today for further thinking about the relationship between human beings and nature.

Moby Dick tells us that the unprecedented development of the whaling industry of the 18th and 19th centuries was the result of a special pioneering culture in the United States. As we all know, the United States is an immigrant country and a colonial country. After European immigrants arrived in the Americas, they would not only drive off indigenous peoples and colonize them, but also opened up territory and industrial development. Herman Melville's era is precisely the era when the United States rapidly advanced from the Ohio Valley to the Pacific Ocean, and was an era that continued to expand into the oceans. It is an era of fighting for colonies with established imperialists. This pioneering culture has created a spirit of human beings in war with nature. From its merits, this is a spirit of hard work and courage. However, from the problems left behind, we think this is a despicable act of looking only at results, recklessly, and endlessly, to nature. Whalers, represented by Ahab, possess this pioneering spirit. This American-style pioneering spirit is addressed in "the Debate on Whaling" in *Moby Dick*. The chapter criticizes some of the critics of the whaling industry as "unnecessary", "dirty", and "horrifying". In the author's opinion, whaling is not only economically significant, "for almost all the tapers, lamps, and candles that burn round the globe, burn, as before so many shrines, to our glory!", and, more importantly, It is the whaling industry that opens up the boundaries so that capitalism can gain more colonies. The novel writes:

> For many years past the whale-ship has been the pioneer in ferreting out the remotest and least known parts of the earth. She has explored seas and archipelagoes which had no chart, where no Cooke or Vancouver had ever sailed. If American and European men-of-war now peacefully ride in once savage harbors, let them fire salutes to the honor and glory of the whale-ship, which originally showed them the way, and first interpreted between them and the savages.
>
> That great America on the other side of the sphere, Australia, was given to the enlightened world by the whaleman. After its first blunder-born discovery by a Dutchman, all other ships, long shunned those shores as pestiferously barbarous; but the whale-ship touched there. The whale-ship is the true mother of that now mighty colony.

Chapter 89, "Fast-Fish and Loose-Fish," discusses mainly the ownership of whales in the open sea and the resulting colonial development issues. The book reads:

> What was America in 1492 but a Loose-Fish, in which Columbus struck the Spanish standard by way of wailing it for his royal master and mistress? What was Poland to the Czar? What Greece to the Turk? What India to England? What at last will Mexico be to the United States? All Loose-Fish.

The author even said, "What is the great globe itself but a Loose-Fish?" Such a pioneering, possessing, colonizing, and governing, "pioneering culture" is exactly important motivation that Ahab and other whalers should do to capture Moby Dick, surrender them and make them the" Fast-Fish ". The famous saying of a pioneering culture is, "I would rather die and not live." Chapter 23 offers very important narrative language for this:

> But as in landlessness alone resides the highest truth, shoreless, indefinite as God–so better is it to perish in that howling infinite, than be ingloriously dashed upon the lee, even if that were safety! For worm-like, then, oh! who would craven crawl to land!

Another whale culture that Moby Dick offers us is the American-style Christian culture of the time. Chapters 7, 8, and 9 introduce the American-style Christian culture closely related to whaling. In order to comfort the whaler's mind, New Bedford built a whaler's church. The peculiar thing about this church is that its teaching lecture room resembles a whaler's mast, and the church's staircase is similar to the one on the ship's ladder. The mural behind the lecture hall is also a big ship that is breaking the waves. The church carries all the duties associated with the whaler's spiritual comfort, and the walls of the church are covered with inscriptions of whaling victims, bearing the grief of their loved ones. The pastor's sermons also deal with whaling. While using chants to soothe the whalers, the priest sings:

> The ribs and terrors in the whale,
> Arched over me a dismal gloom,
> While all God's sun-lit waves rolled by,
> And lift me deepening down to doom.
>
> I saw the opening maw of hell,
> With endless pains and sorrows there;
> Which none but they that feel can tell–
> Oh, I was plunging to despair.
> In black distress, I called my God,
> When I could scarce believe him mine,
> He bowed his ear to my complaints–
> No more the whale did me confine.
> …………

Obviously, the priest here is attempting to use hymn to soothe the whalers, to convince them that they will receive God's blessing. Although the sea was choppy, the whales opened their mouths and the whalers were in dire catastrophe. The almighty God could hear the wailing beetles and the whales could eventually be hunted. Priest would combine whaling practice to further their spiritual comfort in the process of lectures, first by *Old Testament·Jonah,* the story of Jonah for preaching and revelation. The Prophet Jonah went against God, and tried to escape and God let the big whale swallow him, but he succumbed to regret in the whale's belly and finally was forgiven. More importantly, the priest preached in his lectures a strong philosophy, directed at the connotation of the "pioneering spirit" of the United States. Pastor said:

> Enthusiasm,–"But oh! shipmates! on the starboard hand of every woe, there is a sure delight; and higher the top of that delight, than the bottom of the woe is deep. Is not the main-truck higher than the kelson is low? Delight is to him–a far, far upward, and inward delight– who against the proud gods and commodores of this earth, ever stands forth his own inexorable self. Delight is to him whose strong arms yet support him, when the ship of this base treacherous world has gone down beneath him.

The priest said in sermons "higher the top of that delight, than the bottom of the woe is deep", which is to encourage the whaling people not to be afraid of disaster in the pursuit of happiness, a kind of American-style Christian motto.

At the time of the whaling culture, the most powerful should be "anthropocentrism", which is the most popular theoretical concept during the industrial revolution. It is also the basic foundation for mankind's attitude towards nature. The most fundamental connotation of this "anthropocentrism" is that "mankind is great, omnipotent, and will win everything." Such thoughts are embodied in *Moby Dick*: people are superior to whales and all things, and people will win whales, and will win in spirit. Chapter 99 "The Doubloon" writes that Ahab once said that whoever found the Moby Dick first got the gold coins. Just at the crucial moment of the life and death struggle with the whale, Ahab stared at the scene in the pattern of gold coins. The relevant passage reads:

> There's something ever egotistical in mountain-tops and towers, and all other grand and lofty things; look here,– three peaks as proud as Lucifer. The firm tower, that is Ahab; the volcano, that is Ahab; the courageous, the undaunted, and victorious fowl, that, too, is Ahab; all are Ahab; and this round gold is but the image of the rounder globe, which, like a magician's glass, to each and every man in turn but mirrors back his own mysterious self.

This section of interpol explicitly states an anthropocentric approach to nature. Whether it is a solid mountain, victorious fowl, or a round earth, all reflect the mysterious self. This self is Ahab, who becomes the "self" and "center" of all things. In fact, the "pioneering culture" mentioned earlier and the American-style whaling Christian culture are largely "anthropocentric", and provides theoretical support for the human plunder of whales. Comparing Melville's *Moby Dick* with his contemporary Thoreau in *Walden*, which proposed "human beings and nature as friends," Melville still maintained pace with the era of the industrial revolution. As history has evolved, his "anthropocentrism" and its related whaling culture have been shown to be wrong.

(5) **The Tragic Atmosphere Created by the Artistry of a Combination of Realism and Symbolism Gives a Strong Warning to Humankind**

Through the combination of realism and symbolism, Melville portrayed the tragic scenes of the death of both human beings and whales in *Moby Dick*, giving a strong warning to human beings. First of all, he skillfully used the realistic writing technique to truly record the tragic scenes of the mermaid battle and the death of the last 3 days, which is basically a truthful, day by day account. On the first day, whales were found and wrestled with, with the result that Ahab's dinghy was snapped by Moby Dick and that Ahab was rescued. On the second day, a rare whale jumped: the whale from the depths of the sea jumped out of the sea as quickly as possible, resulting in waves transfiguring and resulting in the collision and destruction of the two ships. Ahab's boat was broken, his false leg is also broken off. The third day was the last decisive battle, resulting in a melancholy but meritorious ending. The novel reads:

"The ship? Great God, where is the ship?" Soon they through dim, bewildering mediums saw her sidelong fading phantom, as in the gaseous Fata Morgana; only the uppermost masts out of water; while fixed by infatuation, or fidelity, or fate, to their once lofty perches, the pagan harpooneers still maintained their sinking look-outs on the sea. And now, concentric circles seized the lone boat itself, and all its crew, and each floating oar, and every lancepole, and spinning, animate and inanimate, all round and round in one vortex, carried the smallest chip of the Pequod out of sight.

In addition to using realistic artistry, Melville also makes better use of symbolic artistry, especially metaphorical approach. *Moby Dick* uses metaphors in names: Ishmael, the narrator, borrowed from Ishmael, who was expelled from the house in *Old Testament·Genesis,* suffered from rough ups and downs and unfair social treatment. The hero Ahab, the king of Israel, was brought to disaster with the help of the *Old Testament·1Kings*. As for the ship's name, "Pequod", it was borrowed from an already extinct name of the Indian tribes in Massachusetts. In addition, the place names written in the book also contain many meanings. Examples include the Nantucket whalers dining at the "Oil Cooper Inn", the gallows-mounted mid-mast standing in front of the inn, the nearby shopkeeper named "Coffin" (coffin), etc.. All of these contain the moral of a tragedy. Other aspects of metaphor in *Moby Dick* are almost everywhere. For example, the sudden drop of water in chapter 126 by the former mast watcher, the loss of lifeguards, and the replacement of lifeguards with coffins are seen as "fulfillment of a previously foreseen bad thing", thus with a metaphorical nature. Other encounters such as "La Xie" and "Double Happiness" on the sea are metaphorical. Because the so-called "La Xie" is the use of *Old Testament • Jeremiah,* Yago's wife La Xie crying for the lost children. The "Double Happiness" is the sea burial of the five sailors who died in battle with Moby Dick. These are the ominous signs.

The work also uses some of predictions, whose prophecies are also somewhat mysterious and serve to create some sort of tragic atmosphere. In chapter 99, the black sailor Pip after the madness in the sea says something to Ahab of prophecy:

> Ha! ha! old Ahab! the White Whale; he'll nail ye! This is a pine tree. My father, in old Tolland county, cut down a pine tree once, and found a silver ring grown over in it; some old darkey's wedding ring. How did it get there? And so they'll say in the resurrection, when they come to fish up this old mast, and find a doubloon lodged in it, with bedded oysters for the shaggy bark. Oh, the gold! the precious, precious gold!–

This is obviously a prediction of the tragic ending of Pequod. Chapter 105, does the Whale's Magnitude Diminish?—Will He Perish? predicts that will whales go extinct because of overfishing:

> whether Leviathan can long endure so wide a chase, and so remorseless a havoc; whether he must not at last be exterminated from the waters, and the last whale, like the last man, smoke his last pipe, and then himself evaporate in the final puff.
>
> This talk of extinction that resulted from overfishing is indeed an important prophecy of warning value. *Moby Dick* also uses a symbolic approach. In chapter 119, Thunderbolts Hitting the masts and lighting up three huge fires is a symbol of the sacrifice of the deceased, creating a deep tragic atmosphere. The novel writes:
>
> All the yard-arms were tipped with a pallid fire; and touched at each tri-pointed lightning-rod-end with three tapering white flames, each of the three tall masts was silently burning in that sulphurous air, like three gigantic wax tapers before an altar.
>
> At the end of the description of the sunken Pequod rolling into the sea, the use of "vulture" pennants also adds the mythical final color to the end of the story.

More importantly, these realistic and symbolic artistic techniques used by *Moby Dick* are all aimed at creating a tragic tone for the entire work. The greatest contribution of *Moby Dick* is that through the tragic clashes between the whaler Ahab and Moby Dick, and the tragedy of mermaid's death.I It shows to humankind that the assault on nature will inevitably lead to dual destruction, which is the tragedy of fate. It should be said that *Moby Dick* still uses the traditional concept of tragedy. As Hegel said, "that is to say, through such conflicts, eternal justice makes use of the tragedy characters and their purposes to show that their individual particularity (one-sidedness) which undermines the ethical entity and the peaceful state of unity; With the destruction of this particularity, eternal justice has restored the entity and unity of the ethics."[4] *Moby Dick* created two special images of Ahab and Moby Dick. While Ahab represented humankind's predatory vengeance on nature, Moby Dick represented natural punishment of humanity through anthropomorphic depictions by the author. These two kinds of images represent rational and one-sided ethical concepts. Such desperate conflicts and struggles cannot solve the problem; they can only come to an end by the death of both. This reveals to humankind a more everlasting concept of reconciliation between human beings and nature that only through reconciliation can coexist. This may not be clearly understood by the author. However, the rich, tragic atmosphere created by the artistic image and the works it created have given us such inspiration. Moreover, the only conscious clerk, the Starbucks, the only lucid person of the Pequod, also told us of this concept. At the last moment of the third day of the white whale hunting, he told Ahab:

> "not too late is it, even now, the third day, to desist. See! Moby Dick seeks thee not. It is thou, thou, that madly seekest him!"

However, Ahab rejected the request for reconciliation and finally ended up dying with the whale. The novel last wrote:

[4]Hegel [4], (part 2 Vol. 3), p. 287.

> Now small fowls flew screaming over the yet yawning gulf; a sullen white surf beat against its steep sides; then all collapsed, and the great shroud of the sea rolled on as it rolled five thousand years ago.

A fierce battle between mermaids has finally turned into smoke and dust, with only the boundless sea still billowing forward. By its unique power, the novel tells us that only through reconciliation between man and nature can we truly conform to the laws of the universe. This is the enlightenment given by the artistic image of *Moby Dick* and the greatest contribution of the work.

References

1. Fuheng, W. et al. (1999). *On American writers*. Shandong Education Press.
2. Nuo, W. (2003). *European and American ecological literature*. Peking University Press (王诺:《欧美生态文学》,北京大学出版社2003年版).
3. [U. S.] Melville, H. (2001). *Moby Dick*. People's Literature Publishing House.
4. [Germany] Hegel. (1981). *Aesthetics*, translated by Zhu Guangqian. Commercial Press.

Part VII
Reflections on the Construction of Ecological Aesthetics

Chapter 15
Construction and Development of Ecological Aesthetics

15.1 Reflections on the Discipline Construction of Ecological Aesthetics

If we begin from the mid-20th century, the proposition of Ecological Aesthetics has been 50 years so far. All along, there have been different views on the formulation and development of ecological aesthetics. This is because ecological aesthetics is a new aesthetic concept, and its understanding and acceptance must proceed over time. At the same time, more importantly, the academic colleagues engaged in the study of ecological aesthetics have not done enough in the construction of ecological aesthetics; they have not yet provided more explanatory results for the rationality of ecological aesthetics. Therefore, we need to redouble our efforts. We believe that for the construction of ecological aesthetics, there is need for reflection on two aspects:

(1) **Ecological aesthetics is an emerging discipline in the post-modern context**

There is a lot of controversy in the academic field as to whether ecological aesthetics can become a discipline. The crucial reason is that it has not been fully understood in the postmodern context regarding the nature of the emerging discipline. Regarding its historical location, ecological aesthetics is the product of post-industrial society in the late 20th century, especially since the 21st century, and is an emerging discipline different from the traditional disciplines in the postmodern context. As a new discipline in the postmodern context, it has several characteristics: one is its transcendence of reflection. In other words, ecological aesthetics is the result of transcending reflection on traditional aesthetics. It is both a reflection and the transcendence of epistemology of "traditional anthropology" and "anthropocentrism," as well as a transcendence of traditional aesthetics that completely disregards the ecological dimension and is confined to artistic aesthetics. Therefore, ecological aesthetics is an unprecedented aesthetic form that contains ecological dimensions. The second is open pluralism. As a new discipline in the postmodern context, ecological aesthetics, unlike traditional disciplines, has some kind of super-stability, but as an open and pluralistic system. As the author of deep ecology, Alan Naisi said, his ecological

F. Zeng, *Introduction to Ecological Aesthetics*,
https://doi.org/10.1007/978-981-13-8984-9_15

philosophy is only ecological philosophy A. There will be other theorists to join them, as representatives of ecological philosophy B, C, D, E, F, G… and many more. This book expresses our view of ecological aesthetics, which was put forward for your reference, not and it is not possible to be determined at once. The development of ecological aesthetics needs to have more people join them. The third is interdisciplinarity. As a new discipline in the postmodern context, ecological aesthetics has obvious interdisciplinarity, including aesthetics, ecological philosophy, ecological ethics and so on. The fourth is constructive. As a new discipline in the postmodern context, ecological aesthetics has the characteristic of construction. It has developed with the times and has obvious character of advancing with the times. Therefore, the development of ecological aesthetics welcomes more scholars from other related disciplines involved in the construction.

(2) **The importance of the birth of ecological aesthetics**

Ecological aesthetics appeared in the second half of the 20th century. It is undoubtedly a revolution in the field of aesthetics, and is of great significance.

1. **Have a new role in the construction of the world view**

We believe that contemporary ecological aesthetics should become the most basic view of the human world in the new century and become our basic cultural standpoint and attitude towards life. Facts have shown that since the 1960s, the relationship between humankind and nature has undergone significant changes. That is, since the industrial revolution, a "subject and object" mode of thinking is no longer able to adapt to the new situation requirements. The opposition between human beings and nature has been extremely threatening to the survival of humankind. First of all, the opposition between human beings and nature has led to the frequent occurrence of various ecological crises. From the well-known "London fog" to the amazing Japanese "Minamata disease," as well as the recent "SARS virus," "Indonesian tsunami," "H1N1 influenza" and so on—all of these are natural punishments for human destruction. And in recent years the raging "dust storm," "Huaihe pollution," "Taihu cyanobacteria" and so on in China, have all been recognized as serious environmental crises, which are direct threats to human health and safety, and are even related to the success or failure of China's modernization. To continue to oppose nature or to maintain harmony with nature has become a crucial choice for human beings at the crossroads of success or failure. This was Rachel Carson's warning 45 years ago, one of the pioneers of contemporary ecological theory. The antagonism of human beings among themselves, especially the war crisis caused by the capitalist system, and the great development of science and technology, makes any war enough to cause the destruction of humankind. Human-made nuclear weapons have been enough to destroy the whole of human civilization. The expansion of contemporary capitalism and the severe opposition between the north and the south and the rich and the poor caused by exploitation make hundreds of millions of people live in hunger and pain. In this case, ontology will replace epistemology, and "symbiosis" will replace anthropocentrism as the core of the contemporary values and attitudes towards life.

Regarding "symbiosis," the first is the symbiosis of human beings and nature, as the relationship between human beings and nature is the basis and premise of the relationship between human and the world. The most basic difference between human beings and animals is the unity of animals and nature; the "relationship" between them does not matter. People are distinguished from nature, resulting in the "relationship" between human beings and nature. The pursuit of harmony between humankind and nature is the eternal goal of humankind. Different historical times have ushered in different relationships between human beings and nature, resulting in a different outlook on life and aesthetics. In ancient times, humankind had just diverged from nature, and the power of nature was far superior to that of humankind. Human beings and nature were antagonistic. Human beings had an inexplicable reverence for and fear of nature, which contains the world view that "all things are spiritual." To achieve harmony between human beings and nature with simple and bold artistic creation, the pursuit of poetic dwelling constitutes the period of the "symbolic" aesthetic. In the period of agricultural society, despite the great development of human society, the power of nature was still far better than that of human beings. The desire of humankind to live in harmony with nature and in the hope of religion and its resurrection is to develop a religious worldview that hope for the afterlife. In the aesthetic domain, it is manifested in the combination of suffering and salvation. After the industrial revolution, with the development of science and technology, public rationality, social progress, and life improvement, people obtained a dominant position in the relationship with nature. However, the "anthropocentric" outlook on life also appeared, which takes the form of excessive superstition regarding science and technology, as well as the dwelling of science and technology instead of poetic dwelling. This distorted aesthetics shelters people with things. Since the mid-20th century, especially in the new century, through the negative effects of the reflection on the rationality of "anthropocentrism," human beings put forward the concept of nature as an "ecological whole" and "symbiosis" of the world view. In aesthetics, it is put forward through the "The Fourfold Game of Sky, Earth, Man and Gods" to the ecological aesthetics of people's "poetic dwelling." Since then, the ecological concept of a "symbiotic" relationship between human beings and nature has attracted more and more attention from humankind. In 1972, The United Nations tried to make the harmony between human beings and nature become the consensus of all humankind through the Declaration of the Environment. China has also developed a sustainable development strategy in the 1990s, and recently put forward the strategic concept of "scientific development concept" and "environment-friendly society."

But there is always a difficult contradiction entangled within the relationship between human beings and nature, between development and environmental protection, and between contemporary and future generations. On the one hand, there is the strong advocacy of environmental protection. On the other hand, driven by economic interests, ecological damage and environmental pollution is increasing. The seriousness of the problem has gone far beyond the people's good wishes and ecological affordability, Environmental and ecological disasters are an increasing trend, and the seriousness of the situation may have gone far beyond people's estimates. It is time to face it and change it directly. But the key to the problem lies in the cultural attitude of people and the lifestyles they choose. The famous Roman club sponsor Beyerce has long foreseen and discussed this. It is imperative to change people's outlook on life, their cultural attitudes, and their lifestyles as soon as possible. In contemporary times, we should advocate a kind of "symbiosis" as the main connotation of life, cultural attitudes, and lifestyles. "Poetic dwelling" should be its goal of ecological aesthetic. This is a new aesthetic concept which is a kind of "being in the world" with the existence of existence as its philosophical basis. It treats the relationship between human beings and nature with an aesthetic attitude as its main content, and it promotes combining the current moment of human beings with the future good life as its purpose. It is in line with the reality of the development of society today. The scientific concept of development is the important content of contemporary advanced ecological culture. This kind of ecological aesthetics is a kind of "participatory" aesthetic, which is different from the traditional view of Kant's. It is not only the pleasure of the body as a basis for happiness. Rather, it treats aesthetics as a guide to transform reality, which requires people to follow the laws of ecological aesthetics to transform reality and achieve harmony between human beings and nature. This is exactly what Marx had foreseen, which is the combination of thorough naturalism and thorough humanitarianism in future society. This kind of effort to have ecological aesthetics as the new century's basic view of life is of great significance. However, it is also a very difficult historical task. As an integral part of a grand project, it should be included in the learning and practice of the contemporary, scientific concept of development. Only by treating nature with an aesthetic attitude, as well as the corresponding aesthetic attitude towards others, society, and their own, can contemporary society really move along the road of harmonious development. We believe that it will change people's natural, ecological, aesthetic, and lifestyle views, and achieve the great rejuvenation of the Chinese nation, as well as the aspirations of our people to have a good habitat through this ambitious study.

2. **The Role of Contemporary Aesthetics Discipline Construction**

The formulation and development of ecological aesthetics plays an important role in the construction of contemporary aesthetics. First of all, it renews the traditional concept of natural beauty and constructs a new ecology aesthetics. There is always a debate on this issue regarding whether ecological aesthetics can become an independent discipline. We feel that in the new transition period, ecological aesthetics is at least a new discipline growing point. It may not be mature enough, but it generally already possessed a prototype as an independent discipline. In general, a relatively independent subject should have a relatively independent theoretical category, relatively independent research methods, and a relatively independent scholar group. This criterion is, of course, the product of the intellectual system of modern enlightenment, but it still has its reference value under the current educational and disciplinary system. If the basic requirements are set up from the above disciplines, ecological aesthetics can be said to have been initially equipped with the above three elements. From the relatively independent theoretical category, it first renews the traditional concept of natural beauty, and does not admit that nature is opposed to people. Therefore it does not think that there is so-called "natural beauty," which is a separate from people. Rather, there is only the beauty of natural ecology, which is the harmonious coexistence between human beings and nature. Although natural ecology is closely linked with people and even inseparable, it constitutes a relatively independent area compared with the arts and daily life. The International Society of Aesthetics President Petzwald recently said that there is aesthetics of the ecological environment, the aesthetics of artistic philosophy, and the aesthetics of everyday life. The object of ecological aesthetics is the ecological system that includes people, rather than from the outside of the so-called "natural." Its most basic aesthetic category is the ecological existence of aesthetics, as well as the "poetic dwelling," "the Fourfold Game," "home consciousness," "place consciousness," "participation in aesthetics," "ecological aesthetic education," "ecological literature," "two forms of ecological aesthetics," "ecological aesthetic education," and so on. These aesthetic categories represent a clear breakthrough and innovation compared to the three-point method of traditional aesthetics of "beauty, the feeling of beauty, and art." Regarding its research method, contemporary ecological aesthetics is different from the traditional epistemological research methods. However, it has adopted contemporary ontology and ecological phenomenology as a method. Because only in the sense of the ontology, as well as from the ecological phenomenology approach, can ecology, humanism, and aesthetics can be truly unified. Ecological aesthetic research scholars, it now appears, are also in the gradual growth stage.

In short, from a complete discipline, ecological aesthetics as a new discipline should be said to be in construction and development, and also needs to be further improved and strengthened. On the other hand, the development of ecological aesthetics is of great significance to the whole aesthetic discipline. It broke through the traditional epistemological model that "aesthetics is the philosophy of art," by claiming that natural ecology is a very important research content. It also breaks the tendency of "anthropocentrism" in the long-standing aesthetic research, and promotes "ecological holism" and "ecological humanism" in the study of aesthetics. These basic theoretical concepts will greatly affect the art and the aesthetic of daily life, and will bring about important changes for these areas.

3. **The role of value reconstruction in contemporary aesthetics**

At the end of the 19th century and the beginning of the 20th century, the German philosopher Nietzsche put forward a major idea of "value reconstruction." This means that the rapid transformation of society requires the adaptation of philosophy and aesthetics to the necessary value reconstruction. The proposal and development of ecological aesthetics has an important role in the reconstruction of value in the field of aesthetics. Because the relationship between human beings and nature is the most basic relationship in human life, if the grasp of this relationship is to achieve a major shift from anthropocentrism to ecological holism, as well as from epistemology to ontology, then the associated field of aesthetics is bound to follow the change. Earlier, we have already said that the proposal and development of ecological aesthetics occurred during the development from art to nature and life in the aesthetic research project. From the perspective of research, there has been a change from epistemology to the existential theory of being, etc. Some of the relevant theoretical concepts will also change accordingly, such as the definition of traditional literature and aesthetics based on the established epistemology and the notion of the "mirror." Now it seems to be a problem because they are based on the aesthetic and literary theory of the "subject and object dichotomy." Additionally, the "typical theory" that literature and art are higher than life also needs to be re-examined. From the concept of ecological aesthetics, natural ecology and people are symbiotic, with a relationship of common, mutual construction. Each has its richness and irreplaceability. There is no question of who is higher than anyone else. At the forefront of contemporary ecological aesthetics, Plato's "theory of mind," Hegel's "beauty is the sensibility of the concept of appearance," and Kant's "quiet" aesthetics, all have obvious historical limitations. For Chernyshevsky's theory of "beauty is life," we have generally criticized its "mechanical materialism" tendency for a long time, basically denying it. However, the proposal and development of ecological aesthetics has led us to see that the theory of "beauty is life," also contains valuable features that emphasize ecology and life. For example, Chernyshevsky's emphasis on "exuberant health and balanced physique," and the "beauty of the characteristics," has a special significance and value. Additionally, the establishment of the split between body and spirit is only by virtue of the audio-visual senses or "quiet" aesthetics. Ecological aesthetics emphasizes that aesthetics is based on spiritual unity, which does not rule out the other body senses in the aesthetic role, thus advocating "participation in aesthetics." This is also a new breakthrough. The "symbiotic" concept, "home consciousness" and "place awareness" and so on, under the guidance of ecological aesthetics theory, have also had an important impact on contemporary architectural aesthetics, landscape aesthetics, tourism aesthetics, and so on. It is possible to update many traditional ideas. The proposition and development of ecological aesthetics is also a reversal of the tendency of the west to basically deny the value of oriental aesthetics. No doubt, ecological aesthetics has important significance on the re-evaluation of ancient Chinese aesthetics, and breaking the "Euro-centrism" as well as the so-called "Oriental aesthetics of non-logical" conclusion, and rediscovering the Oriental aesthetics, especially the meaning of life aesthetics of the ancient Chinese classical ecological existence, which is that "life is easy."

4. **The special role of contemporary literary criticism**

The formulation and development of ecological aesthetics has added a new dimension to contemporary literary criticism, and has provided it with a new theoretical weapon. Ecological aesthetics advocates the "aesthetics of ecological existence theory," follow the "green principle," and "home consciousness," which plays a special role in literary criticism. According to these principles, we not only rediscover the special contributions of Rousseau, Thoreau, Leopold, and so on in the development of ecological aesthetics and ecological literature, but, more importantly, we discover the ecological and aesthetic wisdom of Confucianism and Buddhism of ancient China since the pre-Qin Dynasty. To understand and evaluate a series of ancient and modern important literary works by the use of ecological aesthetics and methods is a significant event. For example, we have re-interpretated and evaluated the famous western *Chatterley Lady's Lover*, *Moby Dick* and so on, which are familiar to all of us from the perspective of ecological aesthetics. For the literary works of other Renaissance and Enlightenment periods, we also affirmed the historical status of "humanism" related to "anthropocentrism," while revealing the limitations of its time. For example, we all know Shakespeare's famous drama *Hamlet* for the great praise of the people: "noble reason! The essence of the universe! The spirit of all things!" This obvious "anthropocentrism" should now be re-evaluated according to ecological aesthetics. For ecological aesthetic wisdom in our country with *The Book of Songs* as the representative of ancient art works, we should also be based on the principle of ecological aesthetics to be re-excavated and explained.

5. **The role of practical guidance**

As an organic part of contemporary ecological theory, ecological aesthetics, like other contemporary ecological theories, not only has a very important theoretical character, but also has a very important practical character. Ecological aesthetics is not only a theoretical construction, but also is a guide to our practice. First of all, it should guide our life practice so that each of us can have an ecologically aesthetic attitude in our treatment of natural ecology and real life so that we may lead lives of loving, natural ecology, saving, and simplicity. In this way, a "loving and nursing ecology" may become our standard of living. This kind of ecological aesthetic principle should be further combined with environmental aesthetics, urban aesthetics, and architectural aesthetics, and can be carried out in real life. Furthermore, it is becoming more and more important for people to realize "poetic dwelling" and "good living."

15.2 We Must Lay the Philosophical Foundation of Ecological Aesthetics Firmly on the View of Ecological Ontology

At present, the academic field's doubts about ecological aesthetics mainly focus on how to unify people and nature, or how to unify the ecological view, the humanist view, and the aesthetic view. This is actually related to the principle of whether ecological aesthetics can be established. To solve this problem, we must achieve a major transformation of philosophy and adjustment. Facts have shown that the reason why ecological aesthetics can be established, and that people, nature, ecology, humanities, and aesthetics can be unified, is that the philosophical basis of modern ecological aesthetics is the concept of ecological ontology. This is, first of all, an era and academic progress. It is a breakthrough in traditional epistemology and "anthropocentrism." The history of philosophy tells us that the relationship between human beings and the world has two models. One is the traditional, epistemological model, in which the subject and object are diagonally opposed. It led to the emergence of "anthropocentrism" during the era of the industrial revolution, which believes that human beings can dominate everything, and that the source of later "ecocentrism," which believes that non-human creatures can have greater value than human beings. But this subject and object dichotomy and epistemological model, as well as the resulting "anthropocentrism" and "ecocentrism" have been historically proven to be impracticable. In this way, at the end of the 19th century and the beginning of the 20th century, there was a philosophical view of existentialism different from epistemological philosophy. This view of existential philosophy is a "being in the world" mode of life, which advocates that people and all things are symbiotic, and together constitute the world. People as "this" have the ability to understand and explain the meaning and value of all things. Heidegger pointed out: "Dasein is an entity which, in its very Being, comports itself understanding towards that Being. In saying this, we are calling attention to the formal concept of existence. Dasein exists. Furthermore, Dasein is an entity which in each case I myself am."[1] also said, "The concept of 'facticity'. Implies that an entity 'within-the-world' has Being-in-the-world in such a way that it can understand itself as bound up its 'destiny' with the Being of those entities which it encounters within its own world."[2] Here, Heidegger summed up the "Dasein and the world" mode of life with "within-the-world." Moreover, this "within-the- world" also means that the "Dasein" of that person also has the "being" of all who are in existence, and have the "destiny" that has been within the world, which is bound up by all the people who encounter it. In other words, people and all things on earth are closely linked. They form a symbiosis and common prosperity. They cannot be away even for a few moments, and together they constitute the "world." In fact, this is the concept of ecological ontology: it is a kind of existential philosophy with an ecological dimension. It is the philosophical foundation of human beings and nature and the unity of ecology and humanism. The theory of ecological existence is a new concept of aesthetics. It abandons the substantive concept of beauty, and places beauty in certain world relations, seeing beauty as the process of the gradual emergence of the truth, which is put forward in the famous proposition that "beauty is the self-placed truth." Here, the existence of the existence of "*Dasein*" in the understanding and in interpretation, gradually moves from shelter to clarity. Its truth is to self-display, and the beauty is to be presented. This is the unity of existence, the unity of truth, the appearance of beauty, and the unity of ecology, humanism, and aesthetics. It should be noted that the philosophical basis of ecological aesthetics and of the ecological existence is our basic view of ecological aesthetics. This does not exclude other understandings of ecological aesthetics. The openness of ecological aesthetics makes it completely willing to accept the argument and analysis of various views. For example, there is a "weak anthropocentrism" view in ecological philosophy, ecological literature, and ecological aesthetics, which advocates we should not give up the guidance of epistemology on ecological theory. We believe that this is also a valuable consideration. As long as both theory and practice can work together, we may wish to co-exist with it. The famous American contemporary ecological writer Silovik clearly expressed his theory of epistemology, which does not prevent him from becoming an important ecological writer. Of course, the inherent theory of self-consistency is worthy of scrutiny. However, these are all matters within the scope of academic discussion.

[1]Heidegger [1], pp. 61–62.

[2]Ibid., pp. 65–66.

15.3 The Relationship between Ecological Aesthetics and Environmental Aesthetics

The relationship between ecological aesthetics and environmental aesthetics has always been a common concern of academia both at home and abroad. Contemporary Western aesthetic fields, mainly in Anglo-American academia, have been vigorously advocating environmental aesthetics, and Chinese contemporary aesthetics scholars strongly advocate ecological aesthetics. In 2006, an international aesthetics seminar was held in Chengdu. After an author made a speech on ecological aesthetics, the question that foreign scholars focus on is the relationship between ecological aesthetics and environmental aesthetics. Originally, from the perspective of the natural ecological dimension of aesthetics, ecological aesthetics and environmental aesthetics all belong to the range of natural ecological aesthetics, which is a breakthrough in the traditional concept of "aesthetics is the philosophy of art." They should belong to the coalition of forces that need to be united. There is no need to divide its frontiers too clearly. However, from the perspective of academic research, it is necessary to find out the relationship between the two. This is also the response of Chinese academia to the relevant international academic circles' questions.

Environmental aesthetics is an important reference and resource for the development and construction of contemporary Chinese ecological aesthetics. From a cultural standpoint, ecological aesthetics and environmental aesthetics have two common positions, which is the standpoint of jointly advocating the protection of the ecological environment in the face of serious ecological damage in contemporary China. China's contemporary ecological aesthetics is actually a product of the gradual deepening of China's modernization and its entry into the era of ecological civilization. It regards the construction of ecological civilization as its own goal, and environmental aesthetics is also produced after western environmental problems have been highlighted, taking environmental protection as its own firm stance. As the Finnish environmental esthetician Joseph Seppama said, "We can see more and more clearly that modern environmental aesthetics started only in the 1960s and was a product of the environmental movement and its way of thinking. The emphasis on ecology distinguishes today's environmental aesthetics from the German version of the earlier 100-year-old." He also explicitly regards "ecological principles" as one of the important principles of environmental aesthetics, pointing out that "in nature, when a natural cycle process is continuous and self-contained, the system is a healthy system."[3] Another common position is that they represent a breakthrough for traditional aesthetics that ignore natural aesthetics. China's ecological aesthetics researchers have made it clear that the most basic feature of ecological aesthetics is that it is a kind of aesthetic that contains an ecological dimension.[4] an important

[3]Sepanmaa [2], pp. 221,180.

[4]Fanren [3], p. 303.

position of western contemporary environmental aesthetics is a breakthrough in traditional aesthetic neglect of the natural ecological environmental aesthetics. Allen Carlson, a prominent Canadian environmental aesthetist, points out in his book *Aesthetics and the Environmental*:"Many of the issues in contemporary work on the aesthetics of nature are foreshadowed in one article: Ronald W. Hepburn's seminal "Contemporary Aesthetics and the Neglect of Natural Beauty." (see Footnote 5) After noting that by essentially reducing aesthetics to the philosophy of art, analytic aesthetics virtually ignores the natural world, Hepburn sets the agenda for the discussion of the late twentieth century. ……He furthermore suggests that with nature such serious appreciation may require different approaches that can accommodate not only nature's indeterminate and varying character, but also both our multi-sensory experience and our diverse understanding of it.[5] The development of western environmental aesthetics is earlier. The ecological aesthetics produced in the mid-1990s in our country has obviously accepted the resources of environmental aesthetics. In the process of the construction of ecological aesthetics, environmental aesthetics has given great nourishment. Their discourse on "livable" ideas gives us great inspiration. In particular, Berleant's *Art and Engagement, Ecology Phenomenology* as well as the theoretical point of view that "there is nothing outside of nature" affect us even more. It is from this sense that we believe that environmental aesthetics is an important resource and reference for the construction of contemporary Chinese ecological aesthetics.

Of course, China's ecological aesthetics and western environmental aesthetics still have some obvious differences. First, ecological aesthetics and environmental aesthetics originate in different times and regions. Environmental aesthetics originated in the Western developed countries of the 1960s. At that time, these countries basically completed industrialization, and most of them have relatively rich natural resources. It was in the mid-1990s, especially into the early 21st century that China began to gradually form a certain scale of ecological aesthetics research. Its historical background is a gradual deepening of industrialization, and scholars found that the simple dimension of economic development can't achieve modernization on its own, but must be accompanied by a cultural and aesthetic dimension. This is the reason for the scientific outlook on development and the construction of a harmonious society. The study of ecological aesthetics in our country has been gradually developed. Therefore, the construction of eco-aesthetics in our country is confronted with the current situation of a society in urgent need of economic development. Additionally, the unprecedented shortage of environmental resources, economic development, and environmental protection have all become needs. This is in contrast with the historical and cultural background proposed by western environmental aesthetics. Moreover, in the 21st century, humankind's understanding of ecological theory has also made great developments and changes. It is hard to sustain "anthropocentrism," and simple "ecocentrism" can hardly become a reality. Rather, only the "symbiosis" between

[5]Carlson [4], p. 17.

human beings and nature is the way to go. It is under such circumstances that eco-humanism and eco-holism have emerged that are more in line with the law of social development, and have become the theoretical fulcrum of contemporary ecological aesthetics.

Secondly, from the point of view of word meaning, "ecology" and "environment" also have different meanings. The word "environment," with the meaning of "encirclement, revolving around, and surrounding things," clearly refers to the external being of people and the opposite of human beings. The environmental aesthetician Seppama himself thinks that "even the term 'environment' implies the human view that human beings are at the center and everything else surrounds him."[6] In contrast, "Ecological" has the meaning of "ecology and ecologically protected," while the prefix "eco" means ecologically, family, and economical. Heidegger said in the interpretation of "being," that "'Being' does not mean that ready-made thing is 'one in the middle of another' in space; as far as origin is concerned, the "in" does not mean the spatial relationship of the above methods at all. 'In' is originated from innan-, living, habitare, stay. 'An' means that: I have lived, I am familiar, I am used to, I care; it has the meaning of colo: habito [I live] and diligo [I care]."[7] Here, "colo" already has the "living" and "stay" connotation. The term "ecology" was first combined by the German biologist Haikel in 1866 with the two Greek words okios ['homeland' or 'home'] and logos [research]. It can be seen that the meaning of "ecology" does include the meaning of "home, residence, and stay" and so on. It is more in line with the integration of man and nature than "environment." From the perspective of ecological aesthetics as an ecological ontological aesthetics, the meaning of "living and staying" included in "ecology" more conforms to the connotation of the aesthetics of ecological existence.

Again, from the perspective of aesthetic connotation, "ecology" has a more positive meaning than "environment." Ecological aesthetics originated in the late 20th century and the early 21st century. It has integrated the achievements of long-term exploration on the protection of the ecological environment for more than 100 years. It is well-known that over 100 years ago, humankind has been trying hard to explore the scientific relationship between humankind and the natural ecology, and experienced painful lessons of "anthropocentrism" and "ecocentrism." "Anthropocentrism" has been shown to be an unreasonable path by more than 200 years of industrial revolution. Serious environmental pollution is a painful lesson left to humankind, and "ecocentrism" is also an impossible way. Facts have shown that as a member of the ecological chain, all species, including humans, have only relative equality, and there can be no absolute equality. The absolute equality view of "ecocentrism" is impossible, and can only be a thoroughly "utopian" idea. The only viable way is the path of "ecological holism" and "ecological humanism." As Marxism advocates "the unity of naturalism and humanitarianism," Marx pointed out: "This communism, as completed naturalism, is equal to humanitarianism, and as completed humanitarianism,

[6]Sepanmaa [2], p. 136.

[7]Heidegger [1], p. 63.

equal to naturalism."[8] So called "Ecological holism" and "ecological humanism" are the synthesis and reconciliation of "anthropocentrism" and "ecocentrism," which are the absorption of favorable factors and the abandonment of unfavorable factors. The ecological aesthetics is guided by the theory of "ecological holism" and "ecological humanism" as their own theories.

The history of "environmental aesthetics" is earlier, and is obviously limited by "anthropocentrism" and "ecocentrism." *The Beauty of Environment* by Seppama has a more obvious "anthropocentric" tendency. He not only defines the "environment" as something external to human beings, but also shows the tendency of "anthropocentrism" in the connotation of "environmental aesthetics." In his opinion, "the core area of environmental aesthetics is the issue of aesthetic objects," and "making the environment an aesthetic object is usually based on the audience's choice." He chose to examine the way and object of investigation, and define its scope of time and space. He believes that, "aesthetic objects seem to mean the fact that this thing is at least to a certain extent suitable for aesthetic appreciation."[9] It is clear that Seppama did not completely jump out of traditional aesthetics. He not only examines the aesthetic completely from the main body, but also considers the formation of aesthetic object of environmental aesthetics from the formal aesthetic of the traditional art, such as the proportion of form, symmetry and harmony, etc.. He treats this as the so-called "picturesque landscape," without considering ecological aesthetics' "poetic dwelling" and "homeland consciousness" and so on, which it should have. In contrast, Carlson is a relatively complete environmentalist, and is more inclined to the "ecocentric" point of view. His "Natural All-Beauty Theory" proposed in his book *Aesthetics and the Environment* is a reflection of this "ecocentrism." He said: "all the natural world is beautiful. According to this view, the natural environment, insofar as it is untouched by man, has mainly positive aesthetic qualities; it is, for example, graceful, delicate, intense, unified, and orderly, rather than bland, dull, insipid, incoherent, and chaotic. All virgin nature, in short, is essentially aesthetically good."[10] He calls natural aesthetics "affirmative aesthetics," with the help of many western ecological theorists' points of view. For example, the famous ecological theorist, Marsh, believes that nature is harmonious, and that humankind is an important disturbing body of harmony and nature. Geographer Rodin Tarter believes that humankind is terrible, nature is sublime and so on. Obviously, Carlson's "Natural All-Beauty Theory" is based on the theoretical stand of "ecocentrism," which thus leads to the total negation of human activities, including human artistic activities, which should be said to be very incomplete. Of course, environmental aesthetics also contains many profound and valuable aesthetic connotations. For example, Berleant's theory of "environmental aesthetics" is more scientific and reasonable. He actually advocated a brand new aesthetic concept of natural ecology. He said:"Environment, in the large sense, is not a domain separate and distinct from ourselves as human inhabitants. We are rather continuous with environment, an integral part of its processes. The usual

[8]Marx and Engels [5], (Vol. 42), p. 120.
[9]Sepanmaa [2], pp. 36, 41, 44.
[10]Carlson [4], p. 109.

tradition in aesthetics has difficulty with this, for it claims that appreciation requires a receptive, contemplative attitude. Such an attitude befits an observer, but nature admits of no such observer, for nothing can remain apart and uninvolved."[11] Here, he put forward the famous "Environment, in the large sense," wherein "nothing can remain apart." The "nature" here is not external to people or against people, but rather includes people. In fact, it is what we usually call a "natural system." What he calls "Aesthetics and the Environment" opposes traditional aesthetics. It is an aesthetic of engagement applied to natural aesthetics, which is different from the "static non-utilitarian aesthetics" which emphasizes artistic appreciation represented by Kant. We are more willing to translate it into the "Aesthetics of Engagement," which refers to the active participation of the eyes, ears, and nose in the natural aesthetic. The theory of "nothing can remain apart" and the "Aesthetics of Engagement" have become important resources and references for building ecological aesthetics.

Finally, the reason why ecological aesthetics has its origin in the Chinese cultural atmosphere has a very profound relationship with the traditional Chinese cultural resources. In ancient Chinese philosophy of culture, there was no "environment" external to human beings, only one "heaven" with people. Heaven and humankind are always closely linked. Exploring ancient Chinese cultural history is to explore the history of the relationship between heaven and human beings and ancient and modern, or the so-called "Study [of] the occasion of heaven and earth, [and] know the ancient and modern change." The so-called "occasion of Heaven and Man" and "harmony between man and nature" and "harmony between Heaven and Earth" all refer to the more complex relationship between human beings and nature in the "ecosystem," whether it is Confucianism's "neutralization," of "People and things, Daoism's "Daoist nature." and "Everything in One," or Buddhism's "beings equality," "无尽缘起" and so on. All are talking about the "ecosystem." These oriental ecology theories influenced many Western ecophilosophers and estheticians in modern times, such as Thoreau, Heidegger, and so on. Especially Heidegger's "ecological existence" in philosophy and aesthetics, "the Fourfold Game," and "homeland consciousness" and so on can be said to be "a foreign land interpretation of Lao Zi Daoism." In the soil of such a rich and abundant oriental ecology theory, we believe we must be able to develop ecological aesthetics with Chinese characteristics that combines both modern consciousness and conventions and rich ancient cultural connotations.

We have discussed the relationship between ecological aesthetics and environmental aesthetics above. We have also emphasized that "aesthetics and the environment" has a strong practicality, and has "landscape aesthetics" and "livable environment" as its core connotation. To explore the urban and rural living environment and the construction of a large number of issues, there are many both professional and operational thinking. Due to the current lack of ecological aesthetics, the development of ecological aesthetics urgently needs to be absorbed, drawing on the resources of this aspect of environmental aesthetics, because the fundamental nature of ecological theory is its strong practicality. Of course, how to play the guiding role of ecolog-

[11] Berleant [6], p. 12.

ical aesthetics in urban and rural environment construction in the end can still be discussed.

In short, the two aesthetic forms of ecological aesthetics and environmental aesthetics actually have very close relations, if we learn more from each other in theoretical explanation, we can totally explain the aesthetic relationship between human beings and the natural ecology from different perspectives. As Professor Chen Wangheng, the author of the first "environmental aesthetics" in our country said, "both of these aesthetics have existent value, and they cooperate with each other to jointly promote the development of aesthetics".[12]

15.4 The Future Development of Ecological Aesthetics

The future development of ecological aesthetics is mainly focused on construction and gradually improving in its construction. To this end, we should achieve the following five aspects of convergence:

(1) **Connect with the Construction of Contemporary Ecological Civilization**

Ecological aesthetics is the product of contemporary ecological civilization, which is an organic part of contemporary ecological civilization. Therefore, its development depends on the construction of contemporary ecological civilization to provide realistic and theoretical support. As we all know, ecological aesthetics is the product of ecological civilization since the mid-20th century. Ecological civilization is the arrival of a new era after industrial civilization. Generally speaking, it began in the 1960s. However, its real birth was in the 1970s, marked by the United Nations Conference on Environment and the Environment Statement in 1972. In theory, it is the end of anthropocentrism, and the beginning of new eco-humanism. In practice, it is the end of treating nature as an enemy and the beginning of symbiosis between human beings and nature. Economic development is the end of a simple model of economic growth, though now we have reached the beginning of a win-win situation between economic development and natural ecological protection. This is actually the beginning of a new era. As a post-development country in China, the construction of ecological civilization really started in October 2007, after the national ideology formally proposed the goal of ecological civilization construction. Since then, the construction of ecological civilization has become an advanced culture in line with the direction of social and economic development. Ecological aesthetics is an integral part of this advanced culture. Only by linking up with the construction of ecological civilization in contemporary society can we gain the impetus from the construction of realistic ecological civilization, obtain the resources, obtain abundant information, and find the correct direction so as to make the construction of ecological aesthetics not only obtain some legitimacy and rationality—thus making it firmly based on the reality—but also can at the same time avoid many inconsistencies between the times

[12]Wangheng [7], p. 45.

and the reality in China. In this way, it will not become a kind of utopianism without a realistic interpretation and vitality. Of course, the development of ecological aesthetics is in great need of international dialogue. In a sense, our country currently does not have advanced status in the construction of ecological theory, including ecological aesthetics. The task of learning from foreign advanced theories is still heavy. However, the practicality of ecological theory determines that it must base itself on the land of its own country and conform to its own reality, and to obtain nutrition mainly from the country's ecological civilization.

(2) **Connect with the Development of Contemporary Ecological Theory**

Ecological aesthetics is ecological theory, especially the organic part of ecological philosophy. In this way, the development of ecological aesthetics must be linked with the development of contemporary ecological theory. It has been shown that the development of contemporary ecological aesthetics is synchronized with the development of ecological theory, and the breakthrough of ecological philosophy and ecological ethics leads to the breakthrough of ecological aesthetics. Ecological theory is a very cutting-edge and sensitive theoretical form, involving many very complex theoretical and practical problems, such as natural intrinsic value, natural rights, natural aesthetic attributes, the relationship between ecology and humanism, the relationship between ecology and science and technology, and so on. All of the progress of these theoretical issues will be important for the development of ecological aesthetics. To this end, we must strive to absorb the achievements of ecological theory both home and abroad in this regard. Of course, it must also be closely integrated with the our own situation of ecological aesthetics, especially the special experience of the aesthetic, sensual characteristics, digestion, absorption, and innovation in the construction of ecological aesthetics.

(3) **Link with the Ecological Wisdom of Chinese Traditional Culture**

China is an ancient cultural country with a history of 5,000 years of civilization. With the establishment of agriculture in our country and the development of agricultural civilization, we have a wealth of ecological wisdom. However, at present, we have not done enough to organize and interpret this. There is sufficient evidence to prove that the development of ecological theory in contemporary Western countries, including ecological aesthetics, is obviously related to the influence of ancient Chinese ecological wisdom. Our theoretical circles do not know enough about this. At present, the sorting out and interpretation of China's traditional ecological wisdom has just begun. There has been some progress in document collation and theoretical explanation, but the gap is still large. Facts have shown that the ancient Chinese aesthetics has a morphology different from that of western aesthetics, which is an aesthetics of the epistemology of static object's rather than one of ecological life. Therefore, the development of China's contemporary ecological aesthetics inevitably accompanies the recognition and collation of the life aesthetics of ancient Chinese ecology. On this basis, the theory of ecological aesthetics with Chinese characteristics and Chinese style will be gradually formed to make its own contribution to the development of world aesthetics.

(4) Link with the Development of Contemporary Ecological Literature and Art

The practicality of ecological aesthetics is also manifested in that it is closely related to the practice of literature and art. Its development must be based on summing up the development of ecological literature and art, and play a guiding role in the further development of ecological literature and art creation. At present, the development of ecological literature in the world is gradually showing a flourishing trend. Eco-criticism has gradually become significant school. However, in our country, both the development of ecological literature and art as well as the criticism of ecological literature and art have not developed in a satisfactory way. The gap with the international versions is obvious. We need to better play a guiding role of ecological aesthetics, to promote the development of eco-literary and ecological criticism, and to promote a benign interaction between the two. Only on this basis can the development of ecological aesthetics can have a more solid foundation.

(5) Linking up with the Road of Building a Culture with Chinese Characteristics in the Dialogue between China and the West

In today's era, cultural construction has broken through "Eurocentrism" and has moved toward the era of exchanges, dialogues, and pluralistic coexistence. Under such circumstances, the cultural construction of all countries should absorb and supplement each other on the basis of exchange and dialogue, and at the same time closely combine with national conditions. We should build a cultural pattern that combines both international and national characteristics. The same is true of the construction of ecological aesthetics. We may say that in the initial stage of ecological aesthetics construction, due to the lack of resources, we are more absorbed Western resources. In this way, in our country, the construction of a variety of ecological theories including ecological aesthetics has made remarkable achievements. At the national level, there have been proposed guidelines for the construction of ecological civilization. We should strive to take the road of Chinese characteristics in the construction of ecological aesthetics.

These Chinese characteristics are mainly shown in two aspects: one aspect is that it closely follows the reality of China and is in line with China's reality. China is a post-development country. Modernization and industrialization are still in the mid-term development stage. Economic and social development is the need of the prosperity of nations and of national rejuvenation. In addition, our country is a resource-scarce country. The population is huge, and the resources are relatively poor. Under such circumstances, we can only insist on the win-win development and environmental protection as well as the unification of ecology and humanity as we develop the guiding ideology of ecological aesthetics. Neither can we go on the road of anthropocentrism, nor can we go on the road of ecocentrism. On the other hand, as China is an ancient civilization with relatively rich ecological and cultural resources. Therefore, in the construction of modern ecological aesthetics, we must strive to excavate and absorb the ancient ecological aesthetic wisdom and construct ecological aesthetic discourse with its own characteristics. Facts have shown that although the modern theory of ecology in the developed western countries has come to the forefront of the world, we really should pay attention to absorbing and drawing lessons from western modern ecological theory. However, China's cultural tradition has accumulated a lot more resources of ecological wisdom than the west. Confucianism and Daoism release each of the ecological wisdom showing a unique glory, they will become a valuable asset for the construction of modern ecological civilization of humankind, and have also become valuable assets for the construction of contemporary ecological aesthetics. We should properly summarize and transform and integrate it into the discourse system of modern ecological aesthetics construction, thereby making our ecological aesthetics have an obvious Chinese style that can contribute to the world.

References

1. [Germany] Heidegger, M. (2006). *Being and time* (revised translation), translated by Chen Jiaying, Wang Qingjie. Sanlian Bookstore.
2. [Finland] Sepanmaa, Y. (2006). *The beauty of environment*, translated by Wu Xiaoxi, Zhang Yi. Hunan Science and Technology Press.
3. Fanren, Z. (2007). *Chinese aesthetics in transition*. The Commercial Press (曾繁仁:《转型期的中国美学》,商务印书馆2007年版).
4. [Canada] Carlson, A. (2005). *Aesthetics and the environment—The appreciation of nature, art and architecture*, translated by Yang Ping. Sichuan People's Publishing House.
5. Marx and Engels. (1979). *Collected works*. People's Publishing House.
6. [U.S.] Berleant, A. (2006). *The aesthetics of environment*, translated by Zhang Min, Zhou Yu. Hunan Science and Technology Press.
7. Wangheng, C. (2007). *Ecological aesthetics*. Wuhan University Press (陈望衡:《生态美学》,武汉大学出版社2007年版).

Postscript I

This book is the result of 8 years of hard work since the fall of 2001. Eight years should not seem short for a person over 60 years of age, but an exploration of this length indeed appears too short for one in an emerging discipline. Although I have done my best to make the subject of ecological aesthetics more circumspect, complete and persuasive, ecological aesthetics, as a new and still developing subject, has its own drawbacks. In addition to my own academic qualifications, therefore, the book's exploration is preliminary and immature. But this kind of exploration has also enabled me to understand that ecological aesthetics is full of vitality. I hope that my explorations will play a role in paving the way for the work of future generations.

The style of this book is characterized a teacher's perspective, which has in many ways benefited from other teachers as well as from students. First of all, I have talked about the content of the book to a number of university students, and have listened to their views. Additionally, I have explained the central idea of the book as the main content of the "Ecological Aesthetics Research" course in the doctoral and post-doctoral studies of the Shandong University Literary and Aesthetic Research Center in 2007 and 2008. I have also organized relevant discussions on the topic that have included about 30 doctoral students and post-doctoral researchers as participants in the study and discussion. I learned many good ideas from their discussions, and some of my students helped me sort out the lectures. Therefore, this book embodies the wisdom and labor of these students as well. In the process of writing the book, Dr. Liu Yuedi and Dr. Wang Nuo were asked to review the manuscript, and it has absorbed their valuable opinions. Dr. Li Xiaoming provided me with the relevant foreign literature, and Dr. Qi Haiwen reviewed the relevant ancient literature for proofreading. In the process of organizing the manuscript, I was also assisted by postdoctoral staff from Shandong University's Literary and Artistic Research Center, Dr. Fu Songxue and Dr. Li Yanyan.

F. Zeng, *Introduction to Ecological Aesthetics*,
https://doi.org/10.1007/978-981-13-8984-9

This book is one result of the “Interim Results of the 985 Project” of Shandong University’s Research Center for Literary Theory and Aesthetics. Its publication has received strong support from Shandong University and the Literary and Artistic Research Center of Shandong University. Ms. Cong, of the Commercial Press, has given her consent and support for the publication of the book, as well as Dr. Wang Decheng, who cares deeply about the book's publication. To all of the students and friends, I would like to express my heartfelt thanks. Finally, I would like to thank my wife, Ji Wenyu, for her care and love for me, and for her silent dedication to me! Of course, my greatest expectation is the criticism from a vast readership and peers.

September 5, 2009 Fanren Zeng

Postscript II

Introduction to Ecological Aesthetics was published by commercial press in July, 2010, second edition was in 2019. We successfully applied for the Classic China International Publishing Project in 2018. During the translation process, we enjoyed the cooperation of many people from both within and outside of our Project Team. I would like to take this opportunity to express my heartfelt thanks to all the teachers and friends who offered us help in accomplishing the translation of this book.

My deepest gratitude goes first and foremost to commercial press and the author Prof. Zeng Fanren for their trust and support in my translation, and also for the Prof. Qi Haiwen's revision of the whole book for the English version.

Second, My sincere gratitude should be given to American scholar Chad Austin Meyers, Doctor of Philosophy, who translated some chapters of this book, which includes Chaps. 6–10. (about 65000 Chinese words). I myself, as the main translator, translated the rest part, which includes contents, preface, introduction part, Chaps. 1–5, 11–14, part 7, bibliography, indexes, epilogue (about 260000 Chinese words).

Likewise, Thanks should also be given to Frank Saunders Jr., Doctor of Philosophy, whose native skills in English and excellent grasp of Chinese put the finishing touches on all of my translation. He put in tireless work to ensure that the text stands as an accurate and smooth rendering of the original.

I also owe my gratitude to some friends who have participated in finding all the reference materials. They are Chen Pan and Prof. Qi Haiwen, who truly did indispensible work in helping me accomplish this project.

I also cannot forget my friend Zhu Xiaoer who assisted me to use CAT, Trados, Tmxmall which were most helpful and useful, without his generous help I would not have finished the project in any reasonable timeframe.

F. Zeng, *Introduction to Ecological Aesthetics*,
https://doi.org/10.1007/978-981-13-8984-9

It would be impossible to adequately acknowledge the generosity of so many friends in offering encouragement and support, and in a certain sense this book is as much theirs as it is mine. What I have asked of them far exceeds what I am able to repay.

May 12, 2019 Wu Lihuan

Bibliography

1. [Germany] Heidegger, M. (2000). *Interpretation of Holderlin's poetry*, translated by Sun Zhouxing. The Commercial Press.
2. [Germany] Heidegger, M. (2004). *On the way to language*, translated by Sun Zhouxing. The Commercial Press.
3. [Germany] Moltmann, J. (1999). *God in creation*, translated by Kui Renlian et al. Chinese Christian Culture Institute.
4. [U. S.] Meadows, D. etc. (2006). *The limits to growth*, translated by Li Tao, Wang Zhiyong. Mechanical Industry Press.
5. (2003). *Dictionary of Environmental Sciences*. Science Press (《环境学词典》,科学出版社2003年版).
6. Guying, C. (2001). *Translations of Zhuangzi's present notes*. Zhong Hua Book Company (陈鼓应:《庄子今注今译》,中华书局2001年版).
7. Wangheng, C. (2007). *Environmental aesthetics*. Wuhan University Press (陈望衡:《环境美学》,武汉大学出版社2007年版).
8. Zhihong, H. (2006). *Research on aesthetic ecocriticism*. China Social Sciences Press (胡志红:《美学生态批评研究》,中国社会科学出版社2006年版).
9. Yi, L. (2000). *Ecological ethics*. Shaanxi People's Education Press. (雷毅:《生态伦理学》,陕西人民教育出版社2000年版).
10. Peichao, L. (2004). *Subversion of ethical expansionism: A study of western environmental ethics*. Hunan Normal University (李培超:《伦理拓展主义的颠覆:西方环境伦理思潮研究》,湖南师范大学出版社2004年版).
11. Gangji, L. (1992). *Aesthetics of the book of changes*. Hunan Education Press (刘纲纪:《周易美学》,湖南教育出版社1992年版).
12. Shuyuan, L. (2000). *Ecological literature and art*. Shaanxi People's Education Publishing House (鲁枢元:《生态文艺学》,陕西人民教育出版社2000年版).
13. Chongwen, X. (1986). *Global problems and human dilemma*. Liaoning Publishing House (徐崇温:《全球问题与人类困境》,辽宁出版社1986年版).
14. Heng Chun, X. (2000). *Ecological aesthetics*. Shaanxi People's Education Publishing House (徐恒醇:《生态美学》,陕西人民教育出版社2000年版).
15. Editor-in-chief, Dingsheng, Y. (2002). *Eco-aesthetics*. China Culture and History Press (袁鼎生等主编:《生态审美学》,中国文史出版社2002年版).

F. Zeng, *Introduction to Ecological Aesthetics*,
https://doi.org/10.1007/978-981-13-8984-9

16. Yongcheng, Z. (2000). *Green thought of literature and art: An introduction to literary ecology*. People's Literature Publishing House (曾永成:《文艺的绿色之思——文艺生态学引论》,人民文学出版社2000年版).
17. Shiying, Z. (2002). *Introduction to philosophy*. Peking University Press (张世英:《哲学导论》,北京大学出版社2002年版).

MIX
Papier aus verantwortungsvollen Quellen
Paper from responsible sources
FSC® C105338

If you have any concerns about our products, you can contact us on
ProductSafety@springernature.com

In case Publisher is established outside the EU, the EU authorized representative is:
Springer Nature Customer Service Center GmbH
Europaplatz 3, 69115 Heidelberg, Germany

Printed by Libri Plureos GmbH
in Hamburg, Germany